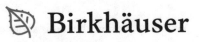

Frontiers in Mathematics

More information about this series at http://www.springer.com/series/5388

Sorin G. Gal • Irene Sabadini

Quaternionic Approximation

With Application to Slice Regular Functions

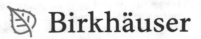 Birkhäuser

Sorin G. Gal
Department of Mathematics
and Computer Science
University of Oradea
Oradea, Romania

Irene Sabadini
Dipartimento di Matematica
Politecnico di Milano
Milano, Italy

ISSN 1660-8046   ISSN 1660-8054 (electronic)
Frontiers in Mathematics
ISBN 978-3-030-10664-5   ISBN 978-3-030-10666-9 (eBook)
https://doi.org/10.1007/978-3-030-10666-9

Mathematics Subject Classification (2010): 30G35, 30E05, 30E10, 41A10, 41A17, 41A25

This book is published under the imprint Birkhäuser, www.birkhauser-science.com by the registered company Springer Nature Switzerland AG.
The registered company address is: Gewerbestrasse 11, 6330 Cham, Switzerland

Preface

Complex analysis is very rich in approximation theory results. In this Preface we briefly list some of them which are nowadays considered classical, with no pretense of completeness, the aim being to show what type of problems we shall discuss in this monograph.

One important result, obtained by Stone [181], is the extension of the Weierstrass theorem to approximate complex-valued continuous functions defined on a compact Hausdorff space. Continuous complex-valued functions can be approximated by generalized polynomials as in the work by Szász [185], who obtained Müntz-type results, and when these functions depend on a real variable they can also be approximated by entire functions, as in the work of Carleman [30] and Kaplan [126]. The particular case of complex-valued analytic functions was studied by several authors. Two classical results in this framework are the celebrated Runge approximation theorem obtained in [172], using rational functions, and the approximation result obtained by Mergelyan [145], using polynomials, for functions holomorphic in the interior of a compact set K and continuous on K with $\mathbb{C} \setminus K$ connected. Approximations of analytic functions on closed (possible unbounded) sets in $\mathbb{C}$ by entire functions were obtained by Arakelian [10], and approximations of analytic functions on compact sets in $\mathbb{C}$ by Faber polynomials were obtained by, e.g., Suetin [183].

Other results on the approximation of analytic functions by polynomials in weighted Bergman spaces on the unit disk can be found in, e.g., Duren and Schuster [66]. More recently, Gal showed that analytic functions can be approximated by polynomial and nonpolynomial convolution type operators and by Bernstein-type operators in compact disks (see the monograph by Gal [77] and the references therein).

Other important phenomena are the overconvergence of polynomial Chebyshev and Legendre expansions attached to analytic functions in an ellipsoid in $\mathbb{C}$, see, e.g., Davis [57], and the equiconvergence for the difference between the Taylor polynomials of degree n and the Lagrange interpolation polynomials of degree n associated to the roots of unity and to analytic functions in disks in $\mathbb{C}$ obtained by Walsh [193].

Universality of complex power series and entire functions, that is uniform approximation of any analytic functions in the interior of a compact set in $\mathbb{C}$ and continuous on K, by subsequences of partial sums of a chosen power series, were obtained by Chui and Parnes [37] and by Luh [137].

Crucial tools, of independent interest, strongly involved in converse results in approximation by complex polynomials are the classical inequalities for complex polynomials of Bernstein [14], [15], Erdös and Lax [133], and Turan [188].

Most of the aforementioned results and many others can be found collected in several research monographs like those of Andrievskii, Belyi and Dzjadyk [8], Gaier [74], Gal [77], [80], Aral, Gupta and Agarwal [11], Gupta and Agarwal [113], Gupta and Tachev [114], Jakimovski, Sharma and Szabados [125], Lorentz [134], Suetin [183], Walsh [193] and others.

Provided such a richness of the complex setting, a natural question is to ask what kind of approximation results in the quaternionic setting can be obtained. This is the goal of this research monograph, where we present extensions to the quaternionic setting of the approximation results in complex analysis discussed above. We have also included the main inequalities regarding the behavior of the derivatives of polynomials with quaternionic coefficients.

Almost all the material in this book comes from recent investigations of the authors concerning the approximation of slice regular (also called slice hyperholomorphic) functions of a quaternionic variable. Exceptions are the Stone–Weierstrass results in Section 2.1, which refer to continuous functions of quaternionic variable, and the results in Chapter 7 concerning approximation for quaternionic- or Clifford algebra-valued functions which belong to the kernel of generalized Cauchy–Riemann operators.

Slice hyperholomorphic functions have been introduced and intensively studied in the past decade. They have proven to be a fertile topic, and their rapid development has been largely driven by the applications to operator theory. In fact, it was soon realized that there is a sound functional calculus associated with such functions which allows one to naturally define powers of operators and so an evolution operator for quaternionic linear operators, and also for n-tuples of linear operators. The works in this direction are numerous and a relevant collection can be found in the book [51]. Further works which illustrate applications to quaternionic operator theory, also in connection with realizations of functions and with Schur analysis, can be found in [7]. It is important to point out that the aforementioned functional calculus is based on a new notion of spectrum, the so-called S-spectrum, see [51]. The introduction of this spectrum was a crucial step in quaternionic spectral theory, in fact several works treated the spectral theorem (see the book by Adler [2] and references therein) but, surprisingly, the notion of spectrum in use was not specified. The introduction and the study of the S-spectrum opened the way to establishing a spectral theorem for normal operators, see [5], and to a systematic study of quaternionic spectral theory (see [40] and references therein to various works in mathematical physics which have originated within this framework). In recent times, applications to fractional powers of linear operators have been studied, see [38] and the book [39].

However, quaternionic-valued functions that are "hyperholomorphic" are studied since the thirties of the past century. More generally, Clifford algebra-valued "hyperholomorphic" functions, are studied since the late sixties. In both cases, by "hyperholomorphic" functions we mean functions in the kernel of a suitable differential operator generalizing the Cauchy–Riemann operator. Applications in mathematical physics and engineering of these more established theories of functions are widely known in the literature, and we refer the interested reader to the books [20, 62, 117] and the references therein. In Chapter 7 we discuss approximation results that can be obtained in this important field of studies.

In short, the contents of the book can be described as follows.

In Chapter 1 we present the main concepts and results in the theory of slice regular functions of a quaternionic variable used in the next chapters. We will discuss the possible definitions which coincide or not according to the choice of the open sets on which we consider the functions. We also recall a few basic important facts on quaternionic polynomials and their zeros. We then introduce slice hyperholomorphic functions (called in this setting slice monogenic) with values in a Clifford algebra. The concept of slice regularity is relatively new, but in the literature quaternionic-valued or Clifford algebra-valued functions are widely studied, so we present other notions extending holomorphy to this setting. This provides the ground for presenting other approximation results in Chapter 7. The section which concludes the chapter shows a connection between slice monogenic functions and monogenic functions via the Fueter–Sce–Qian mapping theorem. These two function theories are not in alternative, in fact they allow to solve different types of problems.

Chapter 2 deals with approximation of continuous functions of a quaternionic variable. We present here two main results: a Stone–Weierstrass type result and a Carleman type result, analogous to the classical results in complex approximation. Also, the Müntz–Szász type result is discussed and presented as an open question in this framework.

In Chapters 3, 4 and 5, which can be considered as the core of the book, the most important results in complex approximation are extended to the quaternionic setting: approximation by q-Bernstein polynomials in compact balls, by convolution operators in compact balls and in Cassini cells, by nonpolynomial quaternion convolutions, Runge type results, Mergelyan type results, Riemann mappings for axially symmetric sets, Arakelian type results, approximation by Faber type polynomials, by polynomials in Bergman spaces, overconvergence of Chebyshev and Legendre polynomials, Walsh equiconvergence type results, universality properties of power series and entire functions. Some of these results are discussed also in the framework of slice monogenic functions.

In Chapter 6 we treat the classical inequalities of Bernstein, Erdős–Lax and Turán in the quaternionic setting. As we have pointed out, the Bernstein's inequality is strongly involved in inverse results in approximation theory.

Chapter 7 contains some approximation results in the setting of nullsolutions of generalized Cauchy–Riemann operators in the quaternionic and in the Clifford algebra setting. This latter case is addressed only marginally in this work, but in this chapter we consider also this more general setting when it is the framework of the original sources. It is obvious that one can always specialize the results to quaternions. As we shall see, the approximation results are different in spirit from the results we proved in the case of slice regular functions. We illustrate some Runge-type results, then we discuss the important topic of Appell systems, which are of crucial importance for approximating monogenic functions. In another section, we consider results in L^2 and Sobolev spaces obtained using monogenic polynomials and also we present an overview of studies on complete orthonormal systems of monogenic polynomials. The next section discusses the Padé approximation for axially monogenic functions. In the case of axially and biaxially monogenic functions one can also obtain some toroidal series expansion, as shown in the fifth section. Finally, in the last section we discuss the polynomial approximation of quasi-conformal monogenic maps.

Except for Chapter 7, this book contains results obtained by the two authors in the past few years. The original sources are explicitly mentioned in each chapter.

The book is addressed to researchers in various areas of mathematical analysis, in particular hypercomplex analysis, and approximation theory, but it is accessible to graduate students and suitable for graduate courses.

Acknowledgments. The authors would like to warmly thank Dorothy Mazlum and Thomas Hempfling for their assistance during the preparation of this manuscript.

Sorin G. Gal
Department of Mathematics and Computer Science, University of Oradea
Romania

Irene Sabadini
Dipartimento di Matematica, Politecnico di Milano
Italy

Contents

Chapter 1

Preliminaries on Hypercomplex Analysis

In this chapter we present some preliminary results which provide the background for the next chapters. Our purpose is to make the book self-contained, but since the material is presented in a concise way, the reader is referred to [7, 51, 54, 99] for a deeper treatment of slice hyperholomorphic functions and to [20, 48, 62, 107, 117] for a treatment of regular functions in the sense of Cauchy–Moisil–Fueter and of monogenic functions with values in a Clifford algebra. Since we regard the contents of the various sections as foundational, we do not quote the sources of all the results. Instead, we refer to monographs where the interested reader can find the original sources.

With no pretense of completeness, we will discuss slice regular functions in the quaternionic and Clifford algebra setting (here the functions are called slice monogenic), provide some useful properties of quaternionic polynomials, and then move to the more classical hyperholomorphic functions, namely those in the kernel of the Cauchy–Moisil–Fueter or Dirac operator. In the last section we show that slice monogenic functions can be related with the more classical hyperholomorphic functions via the Fueter–Scc–Qian mapping theorem. These two function theories are not in alternative, in fact they allow to solve different types of problems and they generate different results and applications.

1.1 An introduction to slice regular functions

This section contains some preliminaries on quaternionic analysis of slice regular functions which will be useful in the sequel. As we shall see, quaternionic polynomials with coefficients on one side are a special case of these functions. For the reader's convenience, the preliminaries on polynomials are provided in a separate section.

© Springer Nature Switzerland AG 2019
S. G. Gal, I. Sabadini, *Quaternionic Approximation*, Frontiers in Mathematics,
https://doi.org/10.1007/978-3-030-10666-9_1

The noncommutative field $\mathbb{H}$ of quaternions consists of elements of the form $q = x_0 + x_1 i + x_2 j + x_3 k$, $x_i \in \mathbb{R}$, $i = 0, 1, 2, 3$, where the imaginary units i, j, k satisfy

$$i^2 = j^2 = k^2 = -1, \ ij = -ji = k, \ jk = -kj = i, \ ki = -ik = j.$$

The real number x_0 is called the real part of q while $x_1 i + x_2 j + x_3 k$ is called the imaginary part, or vector part, of q. Given q as above, the quaternion $\bar{q} = x_0 - x_1 i - x_2 j - x_3 k$ is called the conjugate of q. We define the modulus (or norm) of a quaternion q as

$$|q| = \sqrt{q\bar{q}} = \sqrt{x_0^2 + x_1^2 + x_2^2 + x_3^3}.$$

By $\mathbb{B}$ we denote the open unit ball in $\mathbb{H}$ centered at the origin, i.e.,

$$\mathbb{B} = \{q = x_0 + ix_1 + jx_2 + kx_3, \text{ such that } x_0^2 + x_1^2 + x_2^2 + x_3^3 < 1\},$$

and by $\mathbb{S}$ the unit sphere of purely imaginary quaternion, i.e.,

$$\mathbb{S} = \{q = ix_1 + jx_2 + kx_3, \text{ such that } x_1^2 + x_2^2 + x_3^3 = 1\}.$$

The sphere $\mathbb{S}$ belongs to the hyperplane $x_0 = 0$ and thus is a 2-dimensional sphere in $\mathbb{H}$ identified with $\mathbb{R}^4$. It is an easy calculation to show that if $I \in \mathbb{S}$, then $I^2 = -1$. Thus, for any fixed $I \in \mathbb{S}$, we define

$$\mathbb{C}_I := \{x + Iy, \text{ such that } x, y \in \mathbb{R}\},$$

which can be identified with a complex plane. The real axis belongs to $\mathbb{C}_I$ for every $I \in \mathbb{S}$ and so a real quaternion $q = x_0$ belongs to $\mathbb{C}_I$ for any $I \in \mathbb{S}$. Each non-real quaternion q is uniquely associated to the element $I_q \in \mathbb{S}$ defined by $I_q := (ix_1 + jx_2 + kx_3)/|ix_1 + jx_2 + kx_3|$ and so q belongs to the complex plane $\mathbb{C}_{I_q}$.

The functions we will consider in this section are the so-called slice regular functions of a quaternion variable. The study of this class of functions began in a systematic way about 10 years ago, and the literature around it is continuously growing. We refer the reader to [7, 51, 54, 99] and the references therein, for a deeper treatment of these functions and their applications.

In [100], Gentili and Struppa, inspired by a work of Cullen, introduced the following:

Definition 1.1.1. Let U be an open set in $\mathbb{H}$. A real differentiable function $f : U \to \mathbb{H}$ is said to be *left slice regular* or *left slice hyperholomorphic*, in short, *slice regular* if, for every $I \in \mathbb{S}$, its restriction f_I of f to the complex plane $\mathbb{C}_I$ satisfies

$$\bar{\partial}_I f(x + Iy) = \frac{1}{2}\left(\frac{\partial}{\partial x} + I\frac{\partial}{\partial y}\right) f_I(x + Iy) = 0.$$

The set of slice regular functions on U will be denoted by $\mathcal{R}(U)$.

A real differentiable function $f : U \to \mathbb{H}$ is said to be *right slice regular* if, for every $I \in \mathbb{S}$, its restriction f_I of f to the complex plane $\mathbb{C}_I$ satisfies

$$(f\bar{\partial}_I)(x + Iy) = \frac{1}{2}\left(\frac{\partial}{\partial x} f_I(x + Iy) + \frac{\partial}{\partial y} f_I(x + Iy)I \right) = 0.$$

The following result has a straightforward proof:

Proposition 1.1.2. *Let U be an open set in $\mathbb{H}$. The set $\mathcal{R}(U)$ is a right linear space over $\mathbb{H}$.*

Another simple yet useful property of slice regular functions comes from this observation: fix $I, J \in \mathbb{S}$ with $I \perp J$ and let us write the values of a function as

$$\begin{aligned}
f(q) &= f_0(q) + f_1(q)I + f_2(q)J + f_3(q)IJ \\
&= (f_0(q) + f_1(q)I) + (f_2(q) + f_3(q)I)J \\
&= F(q) + G(q)J,
\end{aligned}$$

where f_ℓ are real valued for $\ell = 0, \ldots, 3$, while F, G are $\mathbb{C}_I$ valued. Let us consider the restriction f_I of f to the complex plane $\mathbb{C}_I$. Then the condition of slice regularity gives:

Lemma 1.1.3 (Splitting Lemma). *Let $I, J \in \mathbb{S}$ with $I \perp J$ and let $f \in \mathcal{R}(U)$. Then there exist two holomorphic functions $F, G : U \cap \mathbb{C}_I \to \mathbb{C}_I$ such that*

$$f_I(z) = F(z) + G(z)J,$$

where $z = x + Iy$.

It is immediate that the class of slice regular functions contains polynomials of the variable q with quaternionic coefficients written on the right, namely polynomials of the form $P(q) = \sum_{n=0}^{N} q^n a_n$, and convergent power series defined on balls centered at the origin. In fact, also the converse statement holds (see Theorem 1.1.5 below) and the proof mimics the one in the complex case, once a suitable notion of derivative is introduced:

Definition 1.1.4. Let U be an open set in $\mathbb{H}$, and let $f : U \to \mathbb{H}$ be a slice regular function. The *slice derivative* $\partial_s f$ of f, is defined by:

$$\partial_s(f)(q) = \begin{cases} \partial_I(f)(q) = \frac{1}{2}\left(\frac{\partial}{\partial x} - I \frac{\partial}{\partial y} \right) f_I(x + Iy), & \text{if } q = x + Iy, \ y \neq 0, \\[2mm] \dfrac{\partial f}{\partial x}(x), & \text{if } q = x \in \mathbb{R}. \end{cases}$$

The definition of slice derivative is well-posed because it is applied only to slice regular functions. Moreover

$$\frac{\partial}{\partial x} f(x + Iy) = -I \frac{\partial}{\partial y} f(x + Iy) \quad \forall I \in \mathbb{S},$$

and therefore, analogously to what happens in the complex case,

$$\partial_s(f)(x + Iy) = \partial_I(f)(x + Iy) = \frac{\partial}{\partial x}(f)(x + Iy).$$

For simplicity, we will write $f'(q)$ instead of $\partial_s(f)(q) = \frac{\partial}{\partial x}(f)(x + Iy)$ and

$$f^{(n)}(x + Iy) := \frac{\partial^n}{\partial x^n} f(x + Iy).$$

Note also that for any $f \in \mathcal{R}(U)$ its derivative f' belongs to $\mathcal{R}(U)$.

Theorem 1.1.5. *Let* $B(0;R) = \{q \in \mathbb{H} \; ; \; |q| < R\}$. *A function* $f : B(0;R) \to \mathbb{H}$ *is left slice regular on* $B(0;R)$ *if and only if it admits a series representation of the form* $f(q) = \sum_{n=0}^{\infty} q^n a_n$, *uniformly convergent on* $B(0;R)$.

We will not give the proof of this result, but we observe that $a_n = \frac{1}{n!} f^{(n)}(0)$. Thus all the functions slice regular in a ball with center at the origin (or, more in general, with center at a real point q_0, respectively) admit power series expansion in the variable q (or $q - q_0$, respectively). If q_0 is a non-real quaternion and the function f is slice regular in the ball $B(q_0;R)$, it is no longer true that f can be written as a power series in $q - q_0$. In fact, it is immediate to check that $(q - q_0)^n$ is not slice regular, if $n > 1$ and so the class of slice regular functions does not coincide with the class of convergent power series. Thus, in the sequel, it will be sometimes convenient to use the following notion.

Definition 1.1.6. The function $f : B(0;R) \to \mathbb{H}$ is *left analytic in the Weierstrass sense in* $B(0;R)$ in short, *left W-analytic*, if $f(q) = \sum_{k=0}^{\infty} c_k q^k$, for all $q \in B(0;R)$, where $c_k \in \mathbb{H}$ for all $k = 0, 1, 2, \ldots$. The function f is called *right W-analytic* in $B(0;R)$, if $f(q) = \sum_{k=0}^{\infty} q^k c_k$, for all $q \in B(0;R)$.

In the definition, the convergence of the partial sums $\sum_{k=0}^{n} c_k q^k$ and $\sum_{k=0}^{n} q^k c_k$ to f is required to be uniform in any closed ball $\bar{B}(0;r) = \{q \in \mathbb{H}; |q| \leq r\}$, $0 < r < R$.

Remark 1.1.7. It is well known that, in the case of complex variable, the two concepts in Definition 1.1.6 coincide (because of the commutative setting) and they coincide with the analyticity in the Weierstrass sense. This is, in turn, equivalent with the concept of holomorphy (differentiability) introduced by Cauchy, based on the usual definition of the derivative. As we already observed, in the present setting the notion of hyperholomorphy we are using, namely the slice regularity, does not necessarily coincide with the concept of Weierstrass analyticity, unless we consider functions slice regular on balls centered at real points.

Note that f is left W-analytic in $B(0;R)$ if and only if it is right slice regular (analytic) in $B(0;R)$. Similarly, f is right W-analytic in $B(0;R)$ if and only if it is left slice regular (analytic) in $B(0;R)$.

In the sequel, for simplicity, we will always write slice regular, W-analytic instead of left slice regular, right W-analytic, respectively.

Definition 1.1.8. By *entire function* we mean a slice regular function defined on the whole $\mathbb{H}$.

To fully understand the nature of slice regular functions, it is necessary to introduce some particular open sets in $\mathbb{H}$. These will be the domains on which slice regular functions possess nice properties. To this end it is useful to introduce the following notation: if $q = x + Iy$ we put

$$[q] = \{q' = x + Jy; \; J \in \mathbb{S}\}. \tag{1.1}$$

As we shall see in Section 1.4, the set $[q]$, which is a 2-dimensional sphere, can be interpreted as the equivalence class of q with respect to a suitable equivalence relation.

Definition 1.1.9. Let $U \subseteq \mathbb{H}$. We say that U is *axially symmetric* if, for any $q \in U$, all the elements in $[q]$ are contained in U. We say that U is a *slice domain* or s-domain for short, if it is a connected set whose intersection with every complex plane $\mathbb{C}_I$ is connected.

We note that balls centered at a real point are axially symmetric slice domains.

The topological property of being an s-domain is crucial for proving the following result.

Theorem 1.1.10 (Identity Principle). *Let f, g be slice regular functions on a slice domain $U \subseteq \mathbb{H}$. If, for some $I \in \mathbb{S}$, f and g coincide on a subset of $U \cap \mathbb{C}_I$ having an accumulation point in $U \cap \mathbb{C}_I$, then f and g coincide in U.*

The values of a slice regular function defined on an axially symmetric s-domain can be computed if the values of a restriction to a complex plane are known:

Theorem 1.1.11 (Representation Formula). *Let f be a slice regular function defined in an axially symmetric s-domain $U \subseteq \mathbb{H}$. Let $J \in \mathbb{S}$ and let $x \pm Jy \in U \cap \mathbb{C}_J$. Then the following equality holds for all $q = x + Iy \in U$:*

$$f(x + Iy) = \frac{1}{2}\left[f(x + Jy) + f(x - Jy)\right] + I\frac{1}{2}\left[J[f(x - Jy) - f(x + Jy)]\right]$$
$$= \frac{1}{2}(1 - IJ)f(x + Jy) + \frac{1}{2}(1 + IJ)f(x - Jy). \tag{1.2}$$

Remark 1.1.12. The representation formula implies that if $U \subseteq \mathbb{H}$ is an axially symmetric s-domain and $f \in \mathcal{R}(U)$, then

$$f(q) = f(x + Iy) = \alpha(x, y) + I\beta(x, y), \tag{1.3}$$

where

$$\alpha(x, y) := \frac{1}{2}[f(x + Jy) + f(x - Jy)]$$

and

$$\beta(x,y) := \frac{1}{2}J[f(x-Jy) - f(x-Jy)]$$

do not depend on $J \in \mathbb{S}$. So α, β are $\mathbb{H}$-valued differentiable functions, $\alpha(x,y) = \alpha(x,-y)$, $\beta(x,y) = -\beta(x,-y)$ for all $x+Iy \in U$, and moreover satisfy the Cauchy–Riemann system.

A meaningful theory of slice regular functions in the sense of Definition 1.1.1 can be developed only on axially symmetric s-domains. However, a function of the form (1.3) is defined on axially symmetric open set, non necessarily an s-domain. Thus these functions deserve a separate study, which is carried out in the next section.

1.2   Slice regular functions, an alternative definition

On axially symmetric s-domains, a slice regular function is of the form (1.3), thus we now study this class of functions. It is interesting to note that these functions are related with the Fueter construction that will be discussed in Section 7.

We begin by introducing some terminology following Ghiloni and Perotti, see [104].

Definition 1.2.1. Let D be an open set in $\mathbb{C}$, symmetric with respect to the real axis. A function $F : D \subseteq \mathbb{C} \to \mathbb{H} \otimes \mathbb{C}$ of the form $F(z) = \alpha(z) + \iota\beta(z)$ where $z = x + \iota y \in D$, $\alpha, \beta : D \to \mathbb{H}$, $\alpha(\bar{z}) = \alpha(z)$ and $\beta(\bar{z}) = -\beta(z)$, is called a *stem function*. Given a stem function F, the function $f = \mathcal{I}(F)$ defined by

$$f(q) = f(x + Iy) := \alpha(x,y) + I\beta(x,y) \qquad (1.4)$$

for any $q \in \Omega_D = \{x + Iy; \ x + \iota y \in D, \ I \in \mathbb{S}\}$, is called the left *slice function* induced by F.

It is worthwhile noting that the term *stem function* appears already in the work of Rinehart [170], though in a more general framework, while it has been used by Qian in relation to Fueter's construction, see [163].

Remark 1.2.2. The function $\mathcal{I}(F)$ is well defined since α and β are, respectively, even and odd functions in the second variable y. The function $\mathcal{I}(F)$ is said to be *real* when the components α and β are real valued.

Definition 1.2.3. Let D be an open set in $\mathbb{C}$, symmetric with respect to the real axis, and let $F(z) = \alpha(z) + \iota\beta(z)$, $z = x + \iota y$, be a (left) stem function. Assume $\alpha, \beta : D \to \mathbb{H}$ are $\mathcal{C}^1$ functions satisfying the Cauchy–Riemann system:

$$\begin{cases} \dfrac{\partial \alpha}{\partial x} - \dfrac{\partial \beta}{\partial y} = 0, \\[2mm] \dfrac{\partial \alpha}{\partial y} + \dfrac{\partial \beta}{\partial x} = 0. \end{cases}$$

The function $f = \mathcal{I}(F) : \Omega_D \to \mathbb{H}$ is called a left *slice regular* function.

The definition of slice regular functions immediately implies the following result, see [104]:

Proposition 1.2.4. *Let* $f : \Omega_D \to \mathbb{H}$ *be a slice regular function in the sense of Definition 1.2.3. Then for every* $I \in \mathbb{S}$ *the function* f *satisfies*

$$\frac{1}{2}\left(\frac{\partial}{\partial x} + I\frac{\partial}{\partial y}\right)f(x + Iy) = 0,$$

on $\Omega_D \cap \mathbb{C}_I$ *and so* f *is slice regular in the sense of Definition 1.1.1.*

Conversely, functions that are slice regular in the sense of Definition 1.1.1 are slice regular in the sense of Definition 1.2.3 only on axially symmetric s-domains. Thus in order to have a larger class of functions, and to avoid the use of stem functions (which, most of the times, is not needed for our purposes) a more suitable definition of slice regular functions is the following:

Definition 1.2.5. Let $U \subseteq \mathbb{H}$ be an axially symmetric open set and let $f : U \to \mathbb{H}$ be a $\mathcal{C}^1$ function of the form

$$f(q) = f(x + Iy) = \alpha(x, y) + I\beta(x, y)$$

where $\alpha(x, y) = \alpha(x, -y)$, $\beta(x, y) = -\beta(x, -y)$ for all $x + Iy \in U$ and moreover satisfy the Cauchy–Riemann system:

$$\begin{cases} \dfrac{\partial}{\partial x}\alpha(x, y) - \dfrac{\partial}{\partial y}\beta(x, y) = 0, \\[2mm] \dfrac{\partial}{\partial y}\alpha(x, y) + \dfrac{\partial}{\partial y}\beta(x, y) = 0. \end{cases}$$

Then f is said to be left *slice regular* on U.

The definitions given so far can be repeated by writing the imaginary unit I on the right, namely for functions of the form

$$f(q) = f(x + Iy) = \alpha(x, y) + \beta(x, y)I$$

in order to define right slice regular functions.

Obviously, a function that is slice regular in the sense of Definition 1.1.1 is slice regular in the sense of Definition 1.2.5 when it is defined in an axially symmetric s-domains, by virtue of the representation formula. Moreover, a function that is slice regular in the sense of Definition 1.2.5 is in the kernel of the Cauchy–Riemann operator $\overline{\partial}_I$ for every $I \in \mathbb{S}$ and thus it satisfies Definition 1.1.1. Note also that f is harmonic in the variables x, y.

The class of functions that are slice regular in the sense of this second definition has been considered in the literature in relation with the Fueter mapping theorem, not only in the quaternionic case, but also in the Clifford algebra case. For a further generalization to the case of functions with values in a real alternative algebra we refer the reader to [104].

In this book we will use both Definition 1.1.1 and 1.2.5, according to the level of generality that we need.

It is worthwhile mentioning that in [104] the authors show that slice regular functions can be seen as a subclass of the following set of functions:

Definition 1.2.6. Let $U \subseteq \mathbb{H}$ be an axially symmetric open set. Functions of the form $f(q) = f(x + Iy) = \alpha(x, y) + I\beta(x, y)$, where α β are continuous $\mathbb{H}$-valued functions such that $\alpha(x, y) = \alpha(x, -y)$, $\beta(x, y) = -\beta(x, -y)$ for all $x + Iy \in U$ are called *(continuous) slice functions.*

It is easy to see that continuous slice functions satisfy the representation formula (1.2) which is, in fact, a property of slice functions.

Functions slice regular in a neighborhood of a point $q_0 \in \mathbb{H} \setminus \mathbb{R}$ admit, according to Definition 1.2.5, a power series expansion. Also in this framework we cannot use an expansion in the powers of $q - q_0$ since $(q - q_0)^n$ is not slice regular for $n > 1$. To this end, we introduce the following definition.

Definition 1.2.7. Let $q_0 = x_0 + Iy_0 \in \mathbb{H}$, with $x_0, y_0 \in \mathbb{R}$, $y_0 > 0$, $I \in \mathbb{S}$ and $R > 0$, and let us consider

$$B(x_0 + \mathbb{S}y_0; R) := \{q \in \mathbb{H}; \ |(q - x_0)^2 + y_0^2| < R^2\}.$$

A set of the form $B(x_0 + \mathbb{S}y_0; R)$ will be called a *Cassini cell*. If $y_0 = 0$, then clearly $B(x_0 + \mathbb{S}y_0; R) = B(x_0; R)$.

It is immediate that a Cassini cell is axially symmetric and it is an s-domain if $R > y_0$. Note that the terminology of Cassini cell comes from the fact that such sets are defined by means of the so-called Cassini pseudo-metric, see [104]. In view of Theorem 5.4 in [104], we can state the following :

Theorem 1.2.8. *Let U be an axially symmetric set in $\mathbb{H}$ and let $f : U \to \mathbb{H}$ be slice regular. Let $q_0 = x_0 + Iy_0 \in U$ and $B(x_0 + \mathbb{S}y_0; R)$ be such that its closure is contained in U. Then f admits the representation*

$$f(q) = \sum_{k=1}^{\infty} [(q - x_0)^2 + y_0^2]^k [c_{2k} + qc_{2k+1}], \ \text{for all } q \in B(x_0 + \mathbb{S}y_0; R), \quad (1.5)$$

where $c_{2k}, c_{2k+1} \in \mathbb{H}$, for all $k \in \mathbb{N}$.

This theorem was originally proved in [180] for functions slice regular defined an axially symmetric s-domains.

1.3 Further properties of slice regular functions

In this section we present additional useful results on slice regular functions. The reader interested in more details may consult [51, 54, 99].

The representation formula yields a method to extend a holomorphic map, namely a quaternionic-valued function in the kernel of the Cauchy–Riemann operator, to a slice regular function. Before to state the extension result we recall the following:

Definition 1.3.1. Let U_J be an open set in $\mathbb{C}_J$ and let

$$U = \bigcup_{x+Jy\in U_J,\ I\in\mathbb{S}} \{x + Iy\}. \tag{1.6}$$

We call U the *axially symmetric completion* of U_J in $\mathbb{H}$.

Proposition 1.3.2. *Let U_J be a domain in $\mathbb{C}_J$ intersecting properly the real axis. Let U be its axially symmetric completion. If $f: U_J \to \mathbb{H}$ satisfies $\overline{\partial}_J f = 0$, then the function*

$$\text{ext}(f)(x + Iy) = \frac{1}{2}\left[f(x + Jy) + f(x - Jy)\right] + I\frac{1}{2}\left[J[f(x - Jy) - f(x + Jy)]\right]$$

is the unique slice regular extension of f to U.

This extension theorem allows us to define a notion of product between slice regular functions.

Let $f, g: U \subseteq \mathbb{H}$ be slice regular functions such that their restrictions to the complex plane $\mathbb{C}_I$ can be written, by virtue of the splitting lemma, as $f_I(z) = F(z) + G(z)J$, $g_I(z) = H(z) + L(z)J$, where $J \in \mathbb{S}$, $J \perp I$. Then F, G, H, L are holomorphic functions of the variable $z \in U \cap \mathbb{C}_I$ and they exist by the splitting lemma. The $*$-product of f and g is defined as the unique slice regular function whose restriction to the complex plane $\mathbb{C}_I$ is given by

$$(F(z) + G(z)J) * (H(z) + L(z)J)$$
$$:= (F(z)H(z) - G(z)\overline{L(\overline{z})}) + (G(z)\overline{H(\overline{z})} + F(z)L(z))J. \tag{1.7}$$

Similarly, given two right slice regular functions it is possible to define their $*_r$-product.

Pointwise multiplication and slice multiplication are different, but they are related as in the following result:

Proposition 1.3.3. *Let $U \subseteq \mathbb{H}$ be an axially symmetric s-domain, and let $f, g: U \to \mathbb{H}$ be slice regular functions on U such that $f(q) \neq 0$. Then*

$$(f * g)(q) = f(q)g(f(q)^{-1}qf(q)), \tag{1.8}$$

for all $q \in U$.

Note that the map $q \mapsto f(q)^{-1}qf(q)$ is a rotation in $\mathbb{H}$, and $|q| = |f(q)^{-1}qf(q)|$. Also, if $(f * g)(q) = 0$, then either $f(q) = 0$ or $g(f(q)^{-1}qf(q)) = 0$. Formula (1.8) shows that if $\alpha \in \mathbb{H}$, then $(f * g)(\alpha)$ is not the product of $f(\alpha)$ and $g(\alpha)$ unless α is a real quaternion and $f(\alpha) \neq 0$. In fact, in this case

$$f(\alpha)g(f(\alpha)^{-1}\alpha f(\alpha)) = f(\alpha)g(\alpha f(\alpha)^{-1}f(\alpha)) = f(\alpha)g(\alpha).$$

Remark 1.3.4. When dealing with convergent power series, the $*$-product reduces to the product where the result has coefficients which are obtained via the Cauchy product of the coefficients of the factors: given $f(q) = \sum_{n=0}^{\infty} q^n a_n$ and $g(q) = \sum_{n=0}^{\infty} q^n b_n$, we have

$$(f * g)(q) = \sum_{n=0}^{\infty} q^n c_n, \quad c_n = \sum_{r=0}^{n} a_r b_{n-r},$$

where the series converges.

Remark 1.3.5. When dealing with slice regular functions according to Definition 1.2.5, the $*$-multiplication of $f(x + Iy) = \alpha(x,y) + I\beta(x,y)$ and $g(x + Iy) = \gamma(x,y) + I\delta(x,y)$ is defined by

$$(f * g)(x + Iy) = (\alpha\gamma - \beta\delta)(x,y) + I(\alpha\delta + \beta\gamma)(x,y).$$

Definition 1.3.6. The subclass $\mathcal{N}$ of slice regular functions on an open set U is defined as

$$\mathcal{N}(U) = \{f \in \mathcal{R}(U) : f(U \cap \mathbb{C}_I) \subseteq \mathbb{C}_I, \ \forall I \in \mathbb{S}\}.$$

The class $\mathcal{N}(U)$ includes all elementary transcendental functions.

The following result is immediate:

Proposition 1.3.7. *A function slice regular on the ball $B(0, R)$ with center at the origin belongs to $\mathcal{N}(B(0, R))$ if and only if its power series expansion has real coefficients.*

Functions as in this proposition are said to be *real*. More generally, if U is an axially symmetric s-domain, then $f \in \mathcal{N}(U)$ if and only if $f(q) = f(x + Iy) = \alpha(x,y) + I\beta(x,y)$ with α, β real valued (see [44], Proposition 2.1 and [103], Lemma 6.8).

Considering the conjugate $\bar{q}$ of a quaternion q, it can be shown that a function f belongs to $\mathcal{N}(U)$ if and only is it satisfies $f(q) = \overline{f(\bar{q})}$. Note that complex functions defined on open sets $U \subset \mathbb{C}$ that are symmetric with respect to the real axis and satisfy $\overline{f(\bar{z})} = f(z)$ are called in the literature intrinsic, see [170]. In analogy with the complex case, we say that the functions in the class $\mathcal{N}$ are *quaternionic intrinsic*. It is interesting to note that this class of functions was already discussed in the literature since they appear in Fueter's construction (see (1.29) in Section 1.7) and also, more generally, in Sce's construction for Clifford algebra-valued functions [176]. They are also called radially holomorphic functions, see, e.g., [117], or holomorphic functions of a paravector variable, see [179].

Remark 1.3.8. As we shall see, quaternionic intrinsic functions play an important role in this book. In fact, as expected in view of the noncommutative setting, the composition of slice regular functions, when defined, does not necessarily give a slice regular function. However, if $f \in \mathcal{N}(U)$, $g \in \mathcal{R}(V)$ and $f(U) \subseteq V$, then the composition $g(f(q))$ is slice regular. Moreover, $(f*g)(q) = (fg)(q)$ when $f \in \mathcal{N}(U)$, $g \in \mathcal{R}(U)$.

Given any slice regular function f, one can define its inverse f^{-*} with respect to the $*$-multiplication. The construction requires some auxiliary functions denoted by f^c and f^s.

Definition 1.3.9. Let $f \in \mathcal{R}(U)$ and let $f_I(z) = F(z) + G(z)J$. We define the function $f_I^c : U \cap \mathbb{C}_I \to \mathbb{H}$ as

$$f_I^c(z) = \overline{F(\bar{z})} - G(z)J. \tag{1.9}$$

Then $f_I^c(z)$ is a holomorphic function and we define the so-called *conjugate function* f^c of f as

$$f^c(q) = \text{ext}(f_I^c)(q).$$

Let $U \subseteq \mathbb{H}$ be an axially symmetric s-domain and let $f : U \to \mathbb{H}$ be a slice regular function. Using the notation in (1.9), we define the function $f_I^s : U \cap \mathbb{C}_I \to \mathbb{C}_I$ by

$$
\begin{aligned}
f_I^s(z) = (f_I \star f_I^c)(z) &= (F(z) + G(z)J) \star (\overline{F(\bar{z})} - G(z)J) \\
&= [F(z)\overline{F(\bar{z})} + G(z)\overline{G(\bar{z})}] + [-F(z)G(z) + G(z)F(z)]J \\
&= F(z)\overline{F(\bar{z})} + G(z)\overline{G(\bar{z})} \\
&= (f_I^c \star f_I)(z). \tag{1.10}
\end{aligned}
$$

It is readily seen that f_I^s is holomorphic.

Definition 1.3.10. Let $U \subseteq \mathbb{H}$ be an axially symmetric s-domain and let $f : U \to \mathbb{H}$ be slice regular. The function

$$f^s(q) := \text{ext}(f_I^s)(q)$$

defined by the extension of (1.10) is called the *symmetrization* or *normal form* of f.

Remark 1.3.11. Note that (1.10) shows that, for all $I \in \mathbb{S}$, $f^s(U \cap \mathbb{C}_I) \subseteq \mathbb{C}_I$. It is also interesting to note that all the zeros of f^s are spheres of type $[x + Iy]$ (real points, in particular) and that, if $x + Iy$ is a zero of f (isolated or not), then $[x + Iy]$ is a zero of f^s.

We finally give:

Definition 1.3.12. Let $U \subseteq \mathbb{H}$ be an axially symmetric s-domain and let $f \in \mathcal{R}(U)$. We define the function f^{-*} as

$$f^{-*}(q) := (f^s(q))^{-1} f^c(q).$$

We note that when the function f is right slice regular, its $*$-inverse is defined by $f^{-*r}(q) := f^c(q)(f^s(q))^{-1}$.

Let us now consider the function $f(q) = s - q$. First of all, we note that for $|q| < |s|$ the function $f^{-*}(q)$ corresponds to the sum of the series $\sum_{n=0}^{\infty} q^n s^{-n-1}$, in fact

$$(s - q) * \sum_{n=0}^{\infty} q^n s^{-n-1} = \sum_{n=0}^{\infty} q^n s^{-n} - \sum_{n=0}^{\infty} q^{n+1} s^{-n-1} = 1.$$

Thus,

$$(s - q)^{-*} = \sum_{n=0}^{\infty} q^n s^{-n-1}, \tag{1.11}$$

and using the formula in Definition 1.3.12 we deduce that

$$(s - q)^{-*} = (q^2 - 2\mathrm{Re}[s]q + |s|^2)^{-1}(\bar{s} - q). \tag{1.12}$$

This expression corresponds to the slice regular Cauchy kernel and it is also denoted by $S^{-1}(s, q)$. It is a function slice regular on the left in q and on the right in s. Since one may define the inverse of $f(q) = s - q$ on the left or on the right, there are two Cauchy kernels. We refer the interested reader to [51].
Formula (1.12) is valid outside the singularities of $(s - q)^{-*}$, namely for $q \notin [s]$, and not only in the ball $|q| < |s|$ as (1.11) may suggest.

The Cauchy kernel is an example of function with singularities. It is then useful to give the following:

Definition 1.3.13. We call *Laurent series* centered at 0 a series of the form

$$\sum_{n \in \mathbb{Z}} q^n a_n, \quad a_n \in \mathbb{H}.$$

A Laurent series is absolutely convergent in an annulus

$$A(0, R_1, R_2) = \{q \in \mathbb{H}; \ 0 \leq R_1 < |q| < R_2\},$$

and the convergence is uniform on compact sets. The sum of the series is slice regular. If we take a real point q_0, then we may consider the series $\sum_{n \in \mathbb{Z}}(q - q_0)^n a_n$ in the annulus $A(q_0, R_1, R_2)$ with obvious meaning of the symbol. We point out that when q_0 is not real, it is possible to give a Laurent expansion of the form (1.5).

In the sequel, we will make use of the Cauchy integral formula:

Theorem 1.3.14. *Let $U \subseteq \mathbb{H}$ be an axially symmetric s-domain such that $\partial(U \cap \mathbb{C}_I)$ is union of a finite number of continuously differentiable Jordan curves, for every $I \in \mathbb{S}$. Let f be a slice regular function on an open set containing $\bar{U}$ and, for any $I \in \mathbb{S}$, set $ds_I = -I ds$. Then for every $q \in U$ we have*

$$f(q) = \frac{1}{2\pi} \int_{\partial(U \cap \mathbb{C}_I)} S^{-1}(s, q) \, ds_I f(s) \tag{1.13}$$

where

$$S^{-1}(s,q) = -(q^2 - 2\text{Re}[s]q + |s|^2)^{-1}(q - \overline{s}).$$ (1.14)

The value of the integral depends neither on U, nor on the imaginary unit $I \in \mathbb{S}$. Moreover,

$$f^{(n)}(q) = \frac{n!}{2\pi} \int_{\partial(U \cap \mathbb{C}_I)} (q^2 - 2s_0 q + |s|^2)^{-n-1}(q - \overline{s})^{(n+1)*} ds_I f(s),$$

where

$$(q - \overline{s})^{n*} := \sum_{k=0}^{n} \frac{n!}{(n-k)!k!} q^{n-k} \overline{s}^k.$$ (1.15)

1.4 Preliminary results on quaternionic polynomials

Quaternionic polynomials are interesting for several reasons. First of all, polynomials with quaternionic coefficients may be written with coefficients on the left, on the right, or on both sides. In the literature, see, e.g., [154], all these situations are considered, however the majority of the results have been obtained for unilateral polynomials, i.e., with coefficients on one side, see [129]. In this section we will consider the case of coefficients written on the right, since they can be seen as a special case of (left) slice regular functions. Similarly, polynomials with coefficients written on the left can be seen as a special case of (right) slice regular functions.

From now on, by right polynomial or simply polynomial, we will mean an expression of the form

$$p(q) = q^n a_n + \cdots + q a_1 + a_0, \quad a_k \in \mathbb{H}, \ k = 0, \ldots, n.$$

Two such polynomials can be multiplied using the $*$-multiplication for slice regular functions, which coincides, in this case, with the multiplication in [129] and which is valid for polynomials with coefficients in a ring: given $p_1(q) = q^n a_n + \ldots + q a_1 + a_0$, $p_2(q) = q^m b_m + \cdots + q b_1 + b_0$ we have

$$(p_1 * p_2)(q) = \sum_{k=0}^{m+n} q^k c_k, \quad c_k = \sum_{i+j=k} a_i b_j.$$ (1.16)

Given a polynomial $p(q)$ as above, its evaluation at the point $q = \alpha$ is defined in the standard way as $p(\alpha) = \sum_{k=0}^{n} \alpha^k a_k$. As we noted already for slice regular functions, the notion of evaluation does not work well with respect to the $*$–multiplication, in fact it is not a ring homomorphism between the ring of quaternionic polynomials and $\mathbb{H}$: if $r(q) = (p_1 * p_2)(q)$, then in general $r(\alpha) \neq p(\alpha)q(\alpha)$. To see this, consider the following simple example: let $r(q) = (q - i) * (q - j) = q^2 - q(i + j) + k$; then $r(j) \neq 0$ but $(j - i)(j - j) = 0$.

We denote by $\mathbb{H}[q]$ the quaternionic right linear space of polynomials with quaternionic coefficients on the right, see [129].

To describe the zeros of polynomials in $\mathbb{H}[q]$, it is necessary to introduce an equivalence relation for quaternions: two elements $\alpha, \alpha' \in \mathbb{H}$ are equivalent and we write $\alpha \sim \alpha'$, if $\mathrm{Re}(\alpha) = \mathrm{Re}(\alpha')$ and $|\mathrm{Im}(\alpha)| = |\mathrm{Im}(\alpha')|$. It is immediate that $\alpha \sim \alpha'$ if and only if there exists $q \neq 0$ such that $q^{-1}\alpha q = \alpha'$ or, in other words, if α' can be obtained from α by a rotation. The set $[\alpha]$ is the equivalence class of α: it is the 2-dimensional sphere in $\mathbb{H}$ of the form $\mathrm{Re}(\alpha) + I|\mathrm{Im}(\alpha)|$, where $I \in \mathbb{S}$, see (1.1). When $\alpha \in \mathbb{R}$, the sphere reduces to the point α, i.e., its radius is equal to 0. We have the following well known results, see [111], [129], collected here as a theorem for the reader's convenience:

Theorem 1.4.1.   1. *A quaternion α is a zero of a nonzero polynomial $p \in \mathbb{H}[q]$ if and only if the polynomial $q - \alpha$ is a left divisor of $p(q)$.*

2. *If $p(q) = (q - \alpha_1) * \cdots * (q - \alpha_n) \in \mathbb{H}[q]$, where $\alpha_1, \ldots, \alpha_n \in \mathbb{H}$, then α_1 is a zero of p and every other zero of p is in the equivalence class of α_i, $i = 2, \ldots, n$.*

3. *If p has two distinct zeros in an equivalence class $[\alpha]$, then all the elements in $[\alpha]$ are zeros of p. In particular, if a polynomial has real coefficients, then its zeros are either real points or spheres.*

When all the elements in $[\alpha]$ are zeros of p we will say that $[\alpha]$ is a spherical zero of p. It is easy to verify that the elements belonging to the equivalence class $[\alpha]$ are roots of the following polynomial (sometimes called the minimal polynomial or companion polynomial of the equivalence class $[\alpha]$):

$$Q_\alpha(q) = (q - \alpha) * (q - \bar{\alpha}) = q^2 - 2\mathrm{Re}(\alpha)q + |\alpha|^2.$$

We also have (see [13], [154]):

Proposition 1.4.2. $Q_\alpha(q) = Q_{\alpha'}(q)$ *if and only if $[\alpha] = [\alpha']$. If Q_α divides a polynomial $p(q)$, then $p(\lambda) = 0$ for every $\lambda \in [\alpha]$. Otherwise, at most one element in $[\alpha]$ is a zero of p.*

Furthermore, we have the following result (see, e.g., [51, Corollary 4.3.6]):

Proposition 1.4.3. *The values of a slice regular function f and, in particular, of a polynomial at two points belonging to the same equivalence class $[\alpha]$ determine the values of f at all the other points in $[\alpha]$.*

In particular, here is an example that will be useful in the sequel; it concerns the roots of the polynomial $P(q) = q^n - 1$:

Proposition 1.4.4. *The quaternionic equation $q^n - 1 = 0$ has the real zero $q = 1$ and k spherical zeros if $n = 2k + 1$, while it has the real zeros $q = \pm 1$ and $k - 1$ spherical zeros if $n = 2k$.*

Proof. To solve the equation $q^n - 1 = 0$ over the quaternions, we first solve $z^n - 1 = 0$ over the complex field. This amounts to looking for solutions of quaternionic equations belonging to a complex plane $\mathbb{C}_I$. As it is well known, the complex equation has the real zero $+1$ and pairs of conjugate roots if $n = 2k + 1$, and it has real zeros ± 1 and $k - 1$ pairs of conjugate zeros if $n = 2k$. By Theorem 1.4.1, each pair of conjugate roots corresponds to a spherical zero. $\qquad\square$

Quaternionic polynomials display a peculiar behavior with respect to change of order of their linear factors. According to Theorem 1.4.1, if α and β are roots of a quaternionic polynomial P, such that $[\alpha] \neq [\beta]$, there exist $\alpha' \in [\alpha]$, $\beta' \in [\beta]$ such that $P(q) = (q - \alpha) * (q - \beta')$ or $P(q) = (q - \beta) * (q - \alpha')$. Let us consider the polynomial $P(q) = (q - \alpha) * (q - \beta)$ and assume that β belongs to the sphere of α. If $\beta = \overline{\alpha}$, then it is evident that we obtain the minimal polynomial of the sphere $[\alpha]$ and so the whole sphere consists of zeros of P. If $\beta \neq \overline{\alpha}$ there cannot be other zeros of P on $[\alpha]$ so α is the only zero of P with multiplicity 2.

To see this, we may also use the formula to find the zeros of $P(q) = (q - \alpha) * (q - \beta)$: besides α, also $(\beta - \overline{\alpha})^{-1}\beta(\beta - \overline{\alpha})$ is a zero.

This discussion shows that the notion of multiplicity of a zero differs from the standard one and we recall it below, see [99, 102]:

Definition 1.4.5. Let $P(q) = (q - \alpha_1) * \cdots * (q - \alpha_n) * g(q)$, $\alpha_{j+1} \neq \overline{\alpha}_j$, $j = 1, \ldots, n-1$, $g(q) \neq 0$ for $q \in [\alpha_1]$.

We say that $\alpha_1 \in \mathbb{H} \setminus \mathbb{R}$ is a zero of P of *multiplicity* 1, if $\alpha_j \notin [\alpha_1]$ for $j = 2, \ldots, n$.

We say that $\alpha_1 \in \mathbb{H} \setminus \mathbb{R}$ is a zero of P of *multiplicity* $n \geq 2$, if $\alpha_j \in [\alpha_1]$ for all $j = 2, \ldots, n$.

We say that $\alpha_1 \in \mathbb{R}$ is a zero of P of *multiplicity* $n \geq 1$, if $P(q) = (q - \alpha_1)^n g(q)$ with $g(\alpha_1) \neq 0$.

Assume now that $P(q)$ contains the factor $(q^2 + 2\mathrm{Re}(\alpha_j)q + |\alpha_j|^2)$ so that $[\alpha_j]$ is a zero of $P(q)$. We say that the *multiplicity of the spherical zero* $[\alpha_j]$ is m_j if m_j is the maximum of the integers m such that $(q^2 + 2\mathrm{Re}(\alpha_j)q + |\alpha_j|^2)^m$ divides $P(q)$.

Remark 1.4.6. Both polynomials $P_1(q) = (q - i) * (q - i) = (q - i)^{2*}$ and $P_2(q) = (q - i) * (q - j)$ have the root $q = i$ with multiplicity two. However, $P_1'(i) = 0$ while $P_2'(i) \neq 0$. This suggests that inequalities relating the size of a polynomial and its derivative may have peculiarities, see Chapter 6.

1.5 Slice monogenic functions

The notion of slice hyperholomorphic functions can be introduced also for functions with values in a Clifford algebra. In this case, we will say that a function is slice monogenic. We begin by recalling some basic facts about Clifford algebras.

Definition 1.5.1. The real (resp. complex) *Clifford algebra* $\mathbb{R}_n$ (resp. $\mathbb{C}_n$) is the associative algebra generated over $\mathbb{R}$ (resp. $\mathbb{C}$) by the n basis elements of $\mathbb{R}^n$ with the following presentation:

$$\mathbb{R}_n = \langle e_1, \dots e_m \mid e_i e_j + e_j e_i = -2\delta_{ij}, \ i, j = 1, \dots, n \rangle.$$

It is immediate that $\mathbb{R}_n$ is a non commutative algebra for $n \geq 2$. Its dimension as a real vector space is 2^n, in fact the set of elements

$$\{e_0 = 1\} \cup \{e_A = e_{a_1} \cdots e_{a_i} \mid 1 \leq i \leq n, \ 1 \leq a_1 < \cdots < a_i \leq n\}$$

forms a basis for $\mathbb{R}_n$ called the basis of *multivectors*. The length of a multivector e_A is the length $|A|$ of the multi-index A in the (minimal) representation $e_A = e_{a_1} \cdots e_{a_i}$ as above. Note that $\mathbb{R}_1$ is the algebra of complex numbers, the only commutative Clifford algebra, while $\mathbb{R}_2$ is the algebra of real quaternions. For $n \geq 3$, $\mathbb{R}_n$ contains zero divisors.

In the Clifford algebra $\mathbb{R}_n$, we can identify some elements with the vectors in the Euclidean space $\mathbb{R}^{n+1}$: specifically, an element $(x_0, x_1, \dots, x_n) \in \mathbb{R}^{n+1}$ can be identified with a so-called paravector in the Clifford algebra through the map

$$(x_0, x_1, \dots, x_n) \mapsto x = x_0 + x_1 e_1 + \cdots + x_n e_n.$$

Given an element $a = \sum_A a_A e_A \in \mathbb{R}_n$ we can define its norm as

$$|a| = \left(\sum_A a_A^2 \right)^{\frac{1}{2}}.$$

The norm is not multiplicative: in fact for any two elements $a, b \in \mathbb{R}_n$,

$$|ab| \leq C_n |a| \, |b|, \tag{1.17}$$

where C_n is a constant depending only on the dimension of the Clifford algebra $\mathbb{R}_n$. Moreover, we have $C_n \leq 2^{n/2}$. The norm is however multiplicative, for example, when a is a paravector or, in particular, a real number.

By $\mathbb{S}^{n-1}$ we denote the sphere of the unit 1-vectors in $\mathbb{R}^n$, i.e.,

$$\mathbb{S}^{n-1} = \{ \underline{x} = e_1 x_1 + \cdots + e_n x_n \mid x_1^2 + \cdots + x_n^2 = 1 \}.$$

Since each $I \in \mathbb{S}^{n-1}$ is such that $I^2 = -1$, the set $\mathbb{C}_I$ of elements of the form $u + Iv$, for $u, v \in \mathbb{R}$ can be identified with the complex plane. Any non-real paravector x belongs to a unique complex plane $\mathbb{C}_I$. Indeed, we can write $x = u + I_x v$, where $I_x = \underline{x}/|\underline{x}|$. A real number belongs to any $\mathbb{C}_I$. Similarly to the quaternionic case, we give the following

Definition 1.5.2. Let $U \subseteq \mathbb{R}^{n+1}$ be an open set and let $f : U \to \mathbb{R}_n$ be a real differentiable function. Let $I \in \mathbb{S}^{n-1}$ and let f_I be the restriction of f to the complex plane $\mathbb{C}_I$ and denote by $u + Iv$ an element of $\mathbb{C}_I$. We say that f is a left *slice monogenic* function if, for every $I \in \mathbb{S}^{n-1}$,

$$\frac{1}{2} \left(\frac{\partial}{\partial u} + I \frac{\partial}{\partial v} \right) f_I(u + Iv) = \frac{1}{2} \left(\frac{\partial}{\partial u} f_I(u + Iv) + I \frac{\partial}{\partial v} f_I(u + Iv) \right) = 0$$

on $U \cap \mathbb{C}_I$.

An analogous definition can be given in the case of right slice monogenicity. For the class of slice monogenic functions we can repeat, with suitable modifications, the results mentioned above for slice regular functions. We refer the reader to [51] for more details.

1.6 Other hypercomplex function theories

In the hypercomplex setting there are other notions, besides slice regularity, which generalize analyticity, and in fact the definition of slice regularity is rather recent compared with other notions of regularity. The most studied is regularity over the quaternions and monogenicity for functions with values in a Clifford algebra.

The idea behind the definition of regularity was to formalize formulas for the areolar derivative and for the mean derivative in the case of complex variable. The definition was introduced in the thirties by Moisil, see [146], and was further studied and developed by Fueter [72, 73]. This notion of regularity should be referred to as Cauchy–Moisil–Fueter regularity even though, in the literature, it is commonly known as Cauchy–Fueter regularity after the paper by Sudbery [182] was published.

To introduce the notion of regularity, we need two differential operators which generalize the Cauchy–Riemann operator to the quaternionic case:

$$\frac{\partial_l}{\partial \bar{q}} = \frac{1}{4} \left(\frac{\partial}{\partial x_0} + i \frac{\partial}{\partial x_1} + j \frac{\partial}{\partial x_2} + k \frac{\partial}{\partial x_3} \right),$$

$$\frac{\partial_r}{\partial \bar{q}} = \frac{1}{4} \left(\frac{\partial}{\partial x_0} + \frac{\partial}{\partial x_1} i + \frac{\partial}{\partial x_2} j + \frac{\partial}{\partial x_3} k \right).$$

These two operators are called the left and right Cauchy–Moisil–Fueter operator, and also Cauchy–Fueter operators, respectively.

Definition 1.6.1. Let $U \subseteq \mathbb{H}$ be an open set and let $f : U \to \mathbb{H}$ be a real differentiable function. We say that f is *left regular* on U if

$$\frac{\partial_l f}{\partial \bar{q}} = \frac{1}{4} \left(\frac{\partial f}{\partial x_0} + i \frac{\partial f}{\partial x_1} + j \frac{\partial f}{\partial x_2} + k \frac{\partial f}{\partial x_3} \right) = 0. \qquad (1.18)$$

We say that f is *right regular* on U if

$$\frac{\partial_r f}{\partial \bar{q}} = \frac{1}{4} \left(\frac{\partial f}{\partial x_0} + \frac{\partial f}{\partial x_1} i + \frac{\partial f}{\partial x_2} j + \frac{\partial f}{\partial x_3} k \right) = 0. \qquad (1.19)$$

In this chapter, we denote by $\mathscr{R}_l(U)$ the set of left regular functions on U and by $\mathscr{R}_r(U)$ the set of right regular functions on U.

The left and right regularity give rise to different classes of functions which are, however, basically equivalent.

Remark 1.6.2. We can also define the conjugate operators

$$\frac{\partial_l}{\partial q} = \frac{1}{4}\left(\frac{\partial}{\partial x_0} - i\frac{\partial}{\partial x_1} - j\frac{\partial}{\partial x_2} - k\frac{\partial}{\partial x_3}\right),$$

$$\frac{\partial_r}{\partial q} = \frac{1}{4}\left(\frac{\partial}{\partial x_0} - \frac{\partial}{\partial x_1}i - \frac{\partial}{\partial x_2}j - \frac{\partial}{\partial x_3}k\right).$$

Let Δ be the Laplace operator in $\mathbb{R}^4$. It is immediate to verify that

$$\Delta = \frac{\partial_l}{\partial q}\frac{\partial_l}{\partial \bar{q}} = \frac{\partial_l}{\partial \bar{q}}\frac{\partial_l}{\partial q} = \frac{\partial_r}{\partial q}\frac{\partial_r}{\partial \bar{q}} = \frac{\partial_r}{\partial \bar{q}}\frac{\partial_r}{\partial q}. \tag{1.20}$$

Cauchy–Moisil–Fueter regular functions constitute a nice class of functions to which almost all the classical results of complex analysis can be extended. However, an important difference with the case of holomorphic functions is that the powers q^k of the quaternionic variable are not regular, according to this definition of regularity. Hence, it is not possible to expand regular functions in terms of q^k; however it is possible to express them as series of suitable homogeneous polynomials. Let us consider a set ν of n integers

$$\nu = \{\lambda_1, \ldots, \lambda_n\}, \quad i = 1, \ldots, n, \quad 1 \le \lambda_i \le 3.$$

In other words, ν can be specified by giving three integers n_1, n_2, n_3, such that the numbers of 1's which appear in ν is n_1, the number of 2's is n_2, the number of 3's is n_3, and $n_1 + n_2 + n_3 = n$. Let us denote by σ_n the set of triples $\nu = [n_1, n_2, n_3]$; then for any $n > 0$, the set σ_n contains $\frac{1}{2}(n+1)(n+2)$ triples and when $n = 0$, we set $\nu = \emptyset$. We define

$$p_\nu(q) = \frac{1}{n!}\sum_{1 \le \lambda_1, \ldots, \lambda_n \le 3}(x_0 i_{\lambda_1} - x_{\lambda_1})\cdots(x_0 i_{\lambda_n} - x_{\lambda_n}), \tag{1.21}$$

where the sum is taken over the $\dfrac{n!}{n_1! n_2! n_3!}$ different arrangements of n_i elements equal to i, with $i = 1, 2, 3$. The polynomials $p_\nu(q)$ play the role of q^k in the Taylor expansion of a function regular at the origin, as shown in the following result:

Theorem 1.6.3. *Let* $f : U \subseteq \mathbb{H} \to \mathbb{H}$, $f \in \mathscr{R}_l(U)$, $q_0 \in U$. *Then there exists a ball* $|q - q_0| < \delta$ *of radius* $\delta < \text{dist}\,(q_0, \partial U)$ *in which* f *can be represented by a uniformly convergent series of the form*

$$f(q) = \sum_{n=0}^{+\infty}\sum_{\nu \in \sigma_n} p_\nu(q - q_0)a_\nu.$$

It is also possible to write Laurent expansions, as shown in the next result. To show this, we introduce the function

$$G(q) = \frac{q^{-1}}{|q|^2} = \frac{\bar{q}}{|q|^4} \tag{1.22}$$

which is the kernel used in the Cauchy formula. For any $\nu = [n_1, n_2, n_3] \in \sigma_n$, let

$$\partial_\nu = \frac{\partial^n}{\partial x_1^{n_1} \partial x_2^{n_2} \partial x_3^{n_3}} \quad \text{and} \quad G_\nu(q) = \partial_\nu G(q).$$

Theorem 1.6.4. *Let q_0 be a point in an open set U in $\mathbb{H}$ and let $f : U \backslash \{q_0\} \to \mathbb{H}$ be a left regular function. There exists a set of the form*

$$\{q \in \mathbb{H} : r_1 < |q - q_0| < r_2\} \quad \text{with} \quad 0 < r_1 < r_2 < \mathrm{dist}(q_0, \partial U)$$

on which f can be represented by a uniformly convergent series

$$f(q) = \sum_{n=0}^{+\infty} \sum_{\nu \in \sigma_n} [p_\nu(q - q_0)a_\nu + G_\nu(q - q_0)b_\nu],$$

where

$$a_\nu = \frac{1}{2\pi^2} \int_{|q-q_0|=r_2} G_\nu(q - q_0)Dqf(q),$$

$$b_\nu = \frac{1}{2\pi^2} \int_{|q-q_0|=r_1} p_\nu(q - q_0)Dqf(q).$$

The theory of Cauchy–Moisil–Fueter regular functions is widely studied; we refer the reader to [48, 117] and the references therein for more information. It is worthwhile mentioning that there is another important operator, called the Moisil–Teodorescu operator, introduced in [147], which acts on functions $f : U \subseteq \mathbb{R}^3 \to \mathbb{H}$ where U is an open set, by the rule

$$D_{MT}f = i\frac{\partial f}{\partial x_1} + j\frac{\partial f}{\partial x_2} + k\frac{\partial f}{\partial x_3}.$$

Functions in the kernel of D_{MT} are called Moisil–Teodorescu regular. They are relevant in the study of the div-rot system which describes an irrotational fluid without sources or sinks.

One can also consider classes of functions generalizing the holomorphic ones in the framework of Clifford algebras. There are several ways of defining such generalizations but the most popular and most interesting from the point of view of approximation theory is the one associated with the Dirac operator or with the so-called generalized Cauchy–Riemann operator started in [60, 124]. Classical references for this function theory are the books [20, 62, 117].

Definition 1.6.5. The operator $\partial_{\underline{x}} = \sum_{j=1}^{n} e_j \partial_{x_j}$, acting on functions $f(\underline{x}) = f(x_1, \ldots, x_n)$ defined on the n-dimensional Euclidean space and taking values in the real (or complex) Clifford algebra $\mathbb{R}_n$ (resp. $\mathbb{C}_n$), is called the *Dirac operator*.

Definition 1.6.6. A real differentiable function $f : U \subseteq \mathbb{R}^n \to \mathbb{R}_n$ is called (left) *monogenic* in the domain U if it satisfies $\partial_{\underline{x}} f(\underline{x}) = 0$ on U. An analogous definition can be given for a real differentiable function $f : \mathbb{R}^n \to \mathbb{C}_n$.

Left monogenic functions form a right module over $\mathbb{R}_n$ (resp. $\mathbb{C}_n$ if they are $\mathbb{C}_n$-valued). An analogous definition may be given for right monogenic functions.

Remark 1.6.7. A variation of the Dirac operator is the generalized Cauchy–Riemann operator

$$D = \partial_{x_0} + \partial_{\underline{x}};$$ (1.23)

real differentiable functions $f : U \subseteq \mathbb{R}^{n+1} \to \mathbb{R}_n$ lying in the kernel of D are still called (left) monogenic.

Another variation, which is more general and will not be treated in this work, is to consider functions from open sets in $\mathbb{R}^m$ or $\mathbb{R}^{m+1}$ with values in the Clifford algebra $\mathbb{R}_n$ or $\mathbb{C}_n$, $1 \leq m \leq n$ and the operators $\partial_{\underline{x}} = \sum_{j=1}^{m} e_j \partial_{x_j}$ and $D = \partial_{x_0} + \partial_{\underline{x}}$, respectively.

Remark 1.6.8. A remarkable property enjoyed by monogenic functions is that they are harmonic in n variables (or $n+1$ variables in the case when one considers the generalized Cauchy–Riemann operator). Thus the theory of monogenic functions is a refinement of the theory of harmonic functions. Note also that slice monogenic functions are harmonic in two variables, independently of the dimension of the Euclidean space.

It is immediate to verify that the identity function $f(x) = x$ is not monogenic and, similarly, none of the functions $f_k(x) = x^k$, $k > 1$ are monogenic.

Thus, monogenic functions do not admit power series expansions in term of the variable (1-vector or paravector), but instead in terms of symmetric polynomials, as it happens in the quaternionic case, see Theorem 1.6.3.

Let $\mathcal{A}_n$ denote either $\mathbb{R}_n$ or $\mathbb{C}_n$. The set of functions monogenic on an open set U with values in $\mathcal{A}_n$ is denoted by $\mathcal{M}(U, \mathcal{A}_n)$. For any $p \in \mathbb{N}$, consider $(\ell_1, \ldots, \ell_p)$, $\ell_i \in \{1, \ldots, n\}$, and the polynomials

$$V_{\ell_1 \ldots \ell_p}(x) = \frac{1}{p!} \sum_{\pi(\ell_1, \ldots, \ell_p)} z_{\ell_1} \cdots z_{\ell_p}$$ (1.24)

where $z_i = x_i - x_0 e_i$ and the sum is taken over all distinct permutations $\pi(\ell_1, \ldots, \ell_p)$ of $\ell_1, \ldots, \ell_p$. We set $V_0(x) = 1$. Symmetric polynomials of the form (1.24) are monogenic (both on the left and on the right). They form a basis for the set of the inner spherical monogenics of degree $p > 0$, i.e., homogeneous polynomials which are monogenic of degree $p > 0$. They play the role of the powers of a complex variable, in fact they allow to obtain the following expansion:

Theorem 1.6.9. *If f is monogenic in an open set U containing the origin, then there exists a neighborhood $U(0)$ of the origin in which f can be expanded in a normally convergent series of the form*

$$f(x) = \sum_{p=0}^{\infty} \left(\sum_{(\ell_1, \ldots, \ell_p)} V_{\ell_1 \ldots \ell_p}(x) \partial_{x_{\ell_1}} \cdots \partial_{x_{\ell_p}} f(0) \right).$$

More generally, if a function is monogenic in a neighborhood of a point $a = a_0 + \sum_{k=1}^{n} a_k e_k \in \mathbb{R}^{n+1}$, we introduce

$$z_k^{(a)} = (x_k - a_k) - (x_0 - a_0)e_k, \quad k = 1, \ldots, n,$$

and

$$V_{\ell_1 \ldots \ell_p}^{(a)}(x) = \frac{1}{p!} \sum_{\pi(\ell_1, \ldots, \ell_p)} z_{\ell_1}^{(a)} \cdots z_{\ell_p}^{(a)},$$

and we state:

Theorem 1.6.10. *Let* $a \in \mathbb{R}^{m+1}$, $U \subseteq \mathbb{R}^{n+1}$ *be an open set containing* a, *and* $f : U \to \mathcal{A}_n$ *be monogenic. Then there exists a neighborhood* $U(a) \subseteq U$ *of the point* a *such that* f *can be expanded in a unique, normally convergent Taylor series expansion of the form*

$$f(x) = \sum_{p=0}^{\infty} \left(\sum_{(\ell_1, \ldots, \ell_p)} V_{\ell_1 \ldots \ell_p}^{(a)}(x) \partial_{x_{\ell_1}} \cdots \partial_{x_{\ell_p}} f(a) \right). \tag{1.25}$$

It is well known that also all the derivatives of f converge normally in $U(a)$. It is important to mention that, besides the series (1.25), one may consider

$$f(x) = \sum_{p=0}^{\infty} \sum_{(\ell_1, \ldots, \ell_p)} V_{\ell_1 \ldots \ell_p}^{(a)}(x) \partial_{x_{\ell_1}} \cdots \partial_{x_{\ell_p}} f(a), \tag{1.26}$$

where the terms are not grouped together. As discussed in [20], Section 11.3, the two expansions (1.25) and (1.26) do not have the same domain of convergence, in fact in the latter case the domain is smaller. We will come back to the polynomials $V_{\ell_1 \ldots \ell_p}^{(a)}(x)$ and to Taylor expansions of monogenic functions in Chapter 7.

To write the Laurent expansion, we need some suitable functions, so-called outer spherical monogenics, which are obtained by means of the Cauchy kernel for monogenic functions

$$E(y - x) = \frac{1}{A_{n+1}} \frac{\bar{y} - \bar{x}}{|y - x|^{n+1}}, \quad y \neq x,$$

where $\bar{x} = x_0 - \underline{x}$ and $A_{n+1} = 2\pi^{(n+1)/2} \frac{1}{\Gamma((n+1)/2)}$.

Let us consider $(\ell_1, \cdots, \ell_p)$, $\ell_i \in \{1, \cdots, n\}$ and set $W_0(y) = E(y)$ and

$$W_{\ell_1, \ldots, \ell_p}(y) = (-1)^p \frac{\partial^p}{\partial y_{\ell_1} \ldots \partial y_{\ell_p}} E(y) = \left[\frac{\partial^p}{\partial u_{\ell_1} \ldots \partial u_{\ell_p}} E(y - u) \right]_{|u=0}. \tag{1.27}$$

The functions $W_{\ell_1, \ldots, \ell_p}(y)$ are left and right monogenic outside the origin and are homogeneous of degree $-(n + p)$. They form a basis for the set of the outer spherical monogenics of degree $-(n+p)$, i.e., homogeneous polynomials which are monogenic of degree $-(n+p)$. They play the role of the powers z^{-k} of a complex variable.

Theorem 1.6.11. *Let f be monogenic in the set*

$$A(0, r_1, r_2) = \{x \in \mathbb{R}^{n+1} \text{ such that } r_1 < |x| < r_2\}.$$

Then in $A(0, r_1, r_2)$ the function f can be expanded into a unique Laurent series

$$f(x) = \sum_{p=0}^{\infty} \sum_{(\ell_1,\ldots,\ell_p)} V_{\ell_1,\ldots,\ell_p}(x) a_{\ell_1,\ldots,\ell_p} + \sum_{p=0}^{\infty} \sum_{(\ell_1,\ldots,\ell_p)} W_{\ell_1,\ldots,\ell_p}(x) b_{\ell_1,\ldots,\ell_p}. \quad (1.28)$$

where both series in the right-hand side converge normally in the ball $B(0; r_2)$ centered at the origin with radius r_2 and $\mathbb{R}^{n+1} \setminus \overline{B(0, r_1)}$, respectively, and

$$a_{\ell_1,\ldots,\ell_p} = \int_{\partial B(0;R)} W_{\ell_1,\ldots,\ell_p}(y) d\sigma_y f(y), \quad R \in (r_1, r_2)$$

$$b_{\ell_1,\ldots,\ell_p} = \int_{\partial B(0;R)} V_{\ell_1,\ldots,\ell_p}(y) d\sigma_y f(y) \quad R \in (r_1, r_2).$$

1.7   A connection between theories of hyperholomorphy via the Fueter mapping theorem

As we have already pointed out, slice hyperholomorphic functions and functions in the kernel of the Dirac operator (or Cauchy–Moisil–Fueter operator, in the case of quaternionic-valued functions) are not in alternative and in fact they allow to solve different types of problems. However, they can still be compared via the Fueter mapping theorem.

This theorem was proved by Fueter in the thirties, see [72], in order to construct Cauchy–Moisil–Fueter regular functions starting from holomorphic functions. In his paper, Fueter considered a function f holomorphic in an open set of the complex upper half-plane,

$$f(x + iy) = u(x, y) + iv(x, y), \quad x, y \in \mathbb{R},$$

where u and v are real differentiable functions with values in $\mathbb{R}$. Let

$$q = x_0 + ix_1 + jx_2 + kx_3 := x_0 + \underline{q}, \quad x_0, \ldots, x_3 \in \mathbb{R}$$

be a quaternion and consider the function defined by

$$F(q) = u(x_0, |\underline{q}|) + \frac{\underline{q}}{|\underline{q}|} v(x_0, |\underline{q}|). \quad (1.29)$$

Fueter proved that the function

$$\Delta F(q) = \Delta \left(u(x_0, |\underline{q}|) + \frac{\underline{q}}{|\underline{q}|} v(x_0, |\underline{q}|) \right),$$

where Δ is the four-dimensional Laplace operator in $x_0, \ldots, x_3$, lies in the kernel of the Cauchy–Fueter operator, namely

$$\frac{\partial}{\partial \bar{q}}(\Delta F(q)) = \frac{\partial}{\partial \bar{q}}\left(\Delta\left(u(x_0, |\underline{q}|) + \frac{q}{|\underline{q}|}v(x_0, |\underline{q}|)\right)\right) = 0.$$

The above result was then generalized by Sce in [176] in order to obtain monogenic functions with values in a Clifford algebra. Specifically, consider the Euclidean space $\mathbb{R}^{n+1}$, n odd, whose variable is identified with a paravector $x_0 + \underline{x}$. Fueter's theorem in this setting states that, given a holomorphic function f as above and

$$F(x) = u(x_0, |\underline{x}|) + \frac{\underline{x}}{|\underline{x}|}v(x_0, |\underline{x}|), \tag{1.30}$$

the function

$$\Delta^{\frac{n-1}{2}}(F(x)) = \Delta^{\frac{n-1}{2}}\left(u(x_0, |\underline{x}|) + \frac{\underline{x}}{|\underline{x}|}v(x_0, |\underline{x}|)\right)$$

is monogenic, i.e., lies in the kernel of the operator

$$\frac{\partial}{\partial x_0} + \sum_{i=1}^{n} e_i \frac{\partial}{\partial x_i}.$$

The proof of this result for general n is due to Qian [162] and is based on Fourier multipliers in order to give meaning to the fractional powers of the Laplacian when n is even. The Fueter mapping theorem holds also when the holomorphic function f is defined in an open set Ω not necessarily chosen in the complex upper half-plane, see [162], [163], [164], provided that Ω is symmetric with respect to the real line, u is even in the variable y, while v is odd in y.

One immediately observes that the functions defined in (1.29) and (1.30) are slice regular and slice monogenic, respectively, by construction. Moreover, they are intrinsic since u, v are real valued. Since the Laplacian is a real operator, one may consider u, v as quaternionic valued or $\mathbb{R}_n$-valued, thus obtaining general slice hyperholomorphic functions.

Thus, the Fueter mapping theorem provides a mapping F_n between slice regular (resp. slice monogenic) functions and Cauchy–Moisil–Fueter regular (resp. monogenic) functions. By using the Cauchy formula in Theorem 1.3.14 and the analog formula in the case of slice monogenic functions, one can describe the map F_n in integral form, see [45]. Let us set

$$\mathcal{F}_n(s, x) := \Delta^{\frac{n-1}{2}} S^{-1}(s, x) = \gamma_n(s - \bar{x})(s^2 - 2\mathrm{Re}[x]s + |x|^2)^{-\frac{n+1}{2}},$$

where

$$\gamma_n := (-1)^{(n-1)/2} 2^{(n-1)/2}(n-1)!\left(\frac{n-1}{2}\right)!.$$

Then we have:

Theorem 1.7.1 (Theorem 3.9 in [45]). *Let n be an odd number. Let $W \subset \mathbb{R}^{n+1}$ be an axially symmetric open set and let f be slice monogenic in W. Let U be a bounded axially symmetric open set such that $\overline{U} \subset W$. Suppose that the boundary of $U \cap \mathbb{C}_I$ consists of a finite number of rectifiable Jordan curves for any $I \in \mathbb{S}^{n-1}$. Then, if $x \in U$, the function $\mathsf{F}_n[f](x)$ given by*

$$\mathsf{F}_n[f](x) = \Delta^{\frac{n-1}{2}} f(x)$$

is monogenic and it admits the integral representation

$$\mathsf{F}_n[f](x)(x) = \frac{1}{2\pi} \int_{\partial(U \cap \mathbb{C}_I)} \mathcal{F}_n(s, x) ds_I f(s), \quad ds_I = ds/I, \qquad (1.31)$$

where the integral does not depend neither on U nor on the imaginary unit $I \in \mathbb{S}^{n-1}$.

The map F_n is not surjective, in fact its image consists of Cauchy–Moisil–Fueter regular (resp. monogenic) functions of axial type. These are monogenic functions of the form

$$A(x_0, \underline{x}) + \underline{\omega} B(x_0, \underline{x}), \qquad (1.32)$$

where A, B satisfy a suitable Vekua-type system. Considering a suitable variation of slice monogenic functions one may obtain more general monogenic functions, the so-called axially monogenic functions of degree k that we will consider in Section 7.4. For a more general version of the Fueter mapping theorem one may consult [179], [127] and the references therein.

The next step is to ask whether the Fueter map admit an inverse from the subset of axially monogenic functions to the set of slice monogenic functions. The answer to this question is affirmative, in fact it can be proved that, given an axially monogenic function $\check{f}$, one can construct a slice monogenic function f such that $\mathsf{F}_n(f) = \check{f}$, see [46]. The approach can be generalized in order to find a Fueter primitive of an axially monogenic function of degree k, see [47], and in order to consider even dimensions n, see [64].

The Fueter mapping theorem is not the only way to exhibit a relation between slice monogenic functions and monogenic functions. In fact, the Radon transform for $\mathbb{R}_n$-valued functions maps monogenic functions into slice monogenic functions and, analogously, the dual Radon transform maps slice monogenic functions into monogenic functions. This description is based on the action of the dual Radon (resp. Radon) transform on the Taylor (resp. Laurent) part of a slice monogenic (resp. monogenic) function. For details, we refer the interested reader to [43].

Chapter 2

Approximation of Continuous Functions

In this chapter we present general density results of Stone–Weierstrass type and of Carleman type in the space of continuous quaternionic-valued functions. Also, Müntz type results are discussed and presented as open questions in the quaternionic setting.

2.1 Weierstrass–Stone Type Results

In this section we present some variants of the well known Stone–Weierstrass result in the case of quaternionic-valued functions. Note that the algebras and subalgebras considered in all the statements of this section are over the real numbers.

A generalization of the classical Weierstrass approximation theorem, concerning the possibility of uniform approximation of any continuous real-valued functions on a compact interval $[a, b]$ by algebraic polynomials was proved by M. H. Stone [181] and can be stated as follows.

Theorem 2.1.1. *For X a compact Hausdorff space, let us denote by $C(X; \mathbb{R})$ the space of all continuous real-valued functions on X. Let $A \subset C(X; \mathbb{R})$ be a closed subalgebra of $C(X; \mathbb{R})$ containing the constant functions and separating the points of X, that is, for any $x, y \in A$, $x \neq y$, there exists a function $f \in A$ such that $f(x) \neq f(y)$. Then A is dense in $C(X; \mathbb{R})$, i.e., $\overline{A} = C(X; \mathbb{R})$.*

Remark 2.1.2. It is known that if we replace $\mathbb{R}$ by $\mathbb{C}$, the set of complex numbers, then Theorem 2.1.1 does not remain, in general, valid. A counterexample is obtained by letting $X = \{z \in \mathbb{C} : |z| \leq 1\}$ and A be the subalgebra of functions which are analytic in $\{z \in \mathbb{C} : |z| < 1\}$.

However, it is interesting to note that if we replace $\mathbb{R}$ by the set of quaternions, $\mathbb{H}$, the above result remains valid. In fact, Holladay proved the following:

© Springer Nature Switzerland AG 2019

S. G. Gal, I. Sabadini, *Quaternionic Approximation*, Frontiers in Mathematics,
https://doi.org/10.1007/978-3-030-10666-9_2

Theorem 2.1.3 (Holladay [122]). *For X a compact Hausdorff space, let us denote by $C(X; \mathbb{H})$ the space of all continuous quaternionic-valued functions on X. Let $A \subset C(X; \mathbb{H})$ be a closed subalgebra of $C(X; \mathbb{H})$ containing the constant functions and separating the points of X. Then A is dense in $C(X; \mathbb{H})$, i.e., $A = C(X; \mathbb{H})$.*

Proof. Let $x \neq y$ be two distinct points of X and f such that $f(x) \neq f(y)$. We can multiply f by a suitable quaternion in order to obtain a function F such that its real part satisfies $\mathrm{Re}[F(x)] \neq \mathrm{Re}[F(y)]$. But the real part of F is $[F - iFi - jFj - kFk]/4$, which is an element of A. Therefore, A contains real-valued functions which separates the points of X. Since A is a closed subalgebra, it follows that the set of all real-valued functions in A also is a closed subalgebra. Theorem 2.1.1 implies that A contains all continuous real-valued functions on X, which evidently means that A contains all continuous quaternion-valued functions on X, since A contains the constant functions. $\square$

The Stone–Weierstrass theorem can be extended to two algebraic structures which contain or embed the quaternions, namely to Cayley–Dickson type algebras and to Clifford algebras.

Both the complex numbers and the quaternions are examples of Cayley–Dickson type algebras over the reals (see, e.g., [175]) of dimension 2^1 and 2^2, respectively. This facts suggest that the Stone–Weierstrass theorem is possibly valid for any real Cayley–Dickson type algebra $\mathcal{D}_n$ of dimension 2^n with $n > 1$.

The fact that the algebra of complex numbers as well as the one of real quaternions may also be seen as the first two cases of real Clifford algebras suggests that the Stone–Weierstrass result remains valid in the setting of Clifford algebras (possibly with some restriction on the dimension of the algebra).

A real Cayley–Dickson type algebra $\mathcal{D}_n$ is constructed over n imaginary units in an inductive way. For $n \geq 3$ it is not associative and for $n \geq 4$ it is not alternative. Denoting by $\mathbb{R}_n$ a Clifford algebra over n imaginary units, the following result holds.

Theorem 2.1.4 (Hausner [119]). *Let X be a compact Hausdorff space and A be a closed real subalgebra of $C(X; \mathcal{D}_n)$ with $n > 1$ or of $C(X; \mathbb{R}_n)$ with n even, respectively. If A contains all constant functions and separates the points in X, then $A = C(X; \mathcal{D}_n)$ or $A = C(X; \mathbb{R}_n)$, respectively.*

Remark 2.1.5. If $\mathbb{R}_n$ is a Clifford algebra with n odd, then for the validity of the Stone–Weierstrass theorem, the additional condition that A is stable with respect to the main involution is sufficient (see, e.g., [132]).

2.2   Carleman Type Results

This section contains the quaternionic version of some results proved by Carleman. The results that we present, except those where the authors are explicitly mentioned, were obtained in the paper of Gal and Sabadini [86].

We begin by recalling an approximation theorem in the complex setting which was proved by Carleman in [30] and so it bears his name. The result can be stated as follows.

Theorem 2.2.1. *Let $f : \mathbb{R} \to \mathbb{C}$ and $\varepsilon : \mathbb{R} \to (0, +\infty)$ be continuous functions on $\mathbb{R}$. Then there exists an entire function $G : \mathbb{C} \to \mathbb{C}$ such that*

$$|f(x) - G(x)| < \varepsilon(x), \quad \text{for all } x \in \mathbb{R}.$$

Carleman's theorem is a pointwise approximation result which can be considered a generalization of the Weierstrass result on uniform approximation by polynomials on compact intervals. In fact, on any compact interval of the real line $\mathbb{R}$, an entire function can be approximated uniformly by polynomials. More precisely, it can be approximated by the partial sums of its power series expansion, as we will show below in Remark 2.2.10. In this framework, a main result which we will prove in the setting of slice regular functions is the following.

Theorem 2.2.2. *Let $f : \mathbb{R} \to \mathbb{H}$ and $\varepsilon : \mathbb{R} \to (0, +\infty)$ be continuous functions on $\mathbb{R}$. Then there exists an entire function $G : \mathbb{H} \to \mathbb{H}$ such that*

$$|f(x) - G(x)| < \varepsilon(x), \quad \text{for all } x \in \mathbb{R}.$$

To prove Theorem 2.2.2, we need some preliminary results which are of independent interest. In their proofs, we follow the ideas in the complex case in Hoischen's paper [121], and in Burckel's book [21], pp. 273–276.

Lemma 2.2.3. *Let $f : \mathbb{R} \to \mathbb{H}$ be continuous on $\mathbb{R}$. There exists a zero free entire function $g : \mathbb{H} \to \mathbb{H}$ such that $g(x) \in \mathbb{R}$ for all $x \in \mathbb{R}$ and $g(x) > |f(x)|$, for all $x \in \mathbb{R}$.*

Proof. Let $n \in \mathbb{N}$ and set $M_n = \max\{|f(x)|; |x| \leq n+1\}$. Choose natural numbers $k_n \geq n$ such that $\left(n^2/(n+1)\right)^{k_n} > M_n$. If $q \in \mathbb{H}$ is such that $|q| \leq N$, then $|q^2/(n+1)| < 1/2$ for all $n \geq 2N^2$, so that the quaternionic power series

$$h(q) = M_0 + \sum_{n=1}^{\infty} \left(\frac{q^2}{n+1}\right)^{k_n}$$

converges uniformly in any closed ball $\overline{B(0; N)}$, for any $N > 0$. This implies that h is entire on $\mathbb{H}$. Furthermore, the coefficients in the series expansion of h are all real (and positive). By its definition, $h(x) \geq 0$ for all $x \in \mathbb{R}$ and for $|x| < 1$ we have

$$h(x) > M_0 > |f(x)|,$$

while for $1 \leq n \leq |x| < n + 1$ we have

$$h(x) > \left(\frac{x^2}{n+1}\right)^{k_n} \geq \left(\frac{n^2}{n+1}\right)^{k_n} > M_n \geq |f(x)|.$$

Thus, $h(x) \geq |f(x)|$, for all $x \in \mathbb{R}$. Finally, set $g(q) = \exp(h(q))$ to get the required entire function. Here it is important to note that, in general, the composition $f \circ h$ of two slice regular functions f and h is not slice regular. However, it is so when h is quaternionic intrinsic, as in the present case (see Remark 1.3.8).   □

Lemma 2.2.4. *Let $I = [a, b]$ be an interval in $\mathbb{R}$ and let $f : I \to \mathbb{H}$ be a continuous function. For any $k \in \mathbb{N}$ define*

$$f_k(x) = \frac{k}{C} \int_a^b e^{-k^2(x-t)^2} f(t) \, dt, \quad x \in \mathbb{R}, \tag{2.1}$$

where $C = \int_{-\infty}^{+\infty} e^{-x^2} \, dx$. Then for every $\varepsilon > 0$

$$\lim_{k \to +\infty} f_k(x) = \begin{cases} f(x), & \text{uniformly for } x \in [a + \varepsilon, b - \varepsilon], \\ 0, & \text{uniformly for } x \in \mathbb{R} \setminus [a + \varepsilon, b - \varepsilon]. \end{cases}$$

Proof. Fix a basis $\{1, i, j, k\}$ for $\mathbb{H}$. We can write $f(x) = f_0(x) + f_1(x)i + f_2(x)j + f_3(x)k = \varphi(x) + \psi(x)j$, where the functions $\varphi(x) = f_0(x) + f_1(x)i$, $\psi(x) = f_2(x) + f_3(x)i$ have values in the complex plane $\mathbb{C}_i$ whose imaginary unit is i. We define, for each $k \in \mathbb{N}$, the functions $\varphi_k(x)$ and $\psi_k(x)$ as in formula (2.1) by writing $\varphi(t)$, $\psi(t)$ instead of $f(t)$ in the integrand. Then for every $\varepsilon > 0$ we have that, uniformly, $\lim_{k \to +\infty} \varphi_k(x)$ is $\varphi(x)$ in $[a + \varepsilon, b - \varepsilon]$ and is 0 outside. In a similar way, we have that, uniformly, $\lim_{k \to +\infty} \psi_k(x)$ is $\psi(x)$ in $[a + \varepsilon, b - \varepsilon]$ and is 0 outside. Observe that the result we are proving holds true for the complex-valued functions φ, ψ, see, e.g., [21, Exercise 8.26 (ii)]. By setting $f_k(x) = \varphi_k(x) + \psi_k(x)j$ we obtain the statement.   □

Lemma 2.2.5. *Let $f : \mathbb{R} \to \mathbb{H}$ be continuous on $\mathbb{R}$. Then for each $n \in \mathbb{Z}$, there exists a continuous function $f_n : \mathbb{R} \to \mathbb{H}$ with support in $[-1, 1]$, such that for all $x \in \mathbb{R}$ we have $f(x) = \sum_{n=-\infty}^{+\infty} f_n(x - n)$.*

Proof. Fix a basis $\{1, i, j, k\}$ of $\mathbb{H}$ and, as in the proof of Lemma 2.2.4, write the function $f(x)$ as $\varphi(x) + \psi(x)j$, where the functions $\varphi(x) = f_0(x) + f_1(x)i$, $\psi(x) = f_2(x) + f_3(x)i$ have values in the complex plane $\mathbb{C}_i$.

Since the result is true for complex-valued functions, see, e.g., [21, Exercise 8.28 (i)], we have

$$\varphi(x) = \sum_{n=-\infty}^{+\infty} \varphi_n(x - n), \quad \psi(x) = \sum_{n=-\infty}^{+\infty} \psi_n(x - n)$$

and setting $f_n(x - n) = \varphi_n(x - n) + \psi_n(x - n)j$, the result follows.   □

Remark 2.2.6. In the sequel, it may be important to explicitly construct the functions $\varphi_n(x - n)$, $\psi_n(x - n)$ following [21]. Consider a piecewise linear function $\sigma(x)$ such that

$$\sigma(x) = \begin{cases} 1, & \text{on } (-1/2, 1/2), \\ 0, & \text{outside } (-1, 1), \end{cases}$$

and let

$$\Sigma(x) = \sum_{n=-\infty}^{\infty} \sigma(x-n), \quad x \in \mathbb{R}.$$

Then the functions $\varphi_n(x)$ can be defined as

$$\varphi_n(x) = \frac{\sigma(n)\varphi(x+n)}{\Sigma(x+n)}, \quad n \in \mathbb{N},$$

and similarly we can construct $\psi_n(x)$.

Lemma 2.2.7. *Let* $f : \mathbb{R} \to \mathbb{H}$ *be a continuous function on* $\mathbb{R}$ *with compact support in* $[-1, 1]$. *Set*

$$T = \{q \in \mathbb{H} : |\mathrm{Re}(q)| > 3 \text{ and } |\mathrm{Re}(q)| > 2|\mathrm{Im}(q)|\}.$$

For any number $\varepsilon > 0$, *there exists an entire function* $F : \mathbb{H} \to \mathbb{H}$, *such that* $|f(x) - F(x)| < \varepsilon$ *for all* $x \in \mathbb{R}$ *and* $|F(q)| < \varepsilon$ *for all* $q \in T$.

Proof. For $k \in \mathbb{N}$, set

$$f_k(q) = \frac{k}{C} \int_{-1}^{1} e^{-k^2(q-t)^2} f(t) dt, \quad q \in \mathbb{H},$$

where $C = \int_{-\infty}^{+\infty} e^{-x^2} dx$. The function $e^{-k^2(q-t)^2}$ is slice regular in the variable q and when we multiply it on the right by the quaternion-valued function $f(t)$ it remains slice regular, since slice regular functions form a right linear space over $\mathbb{H}$ (see Proposition 1.1.2). Moreover, since f has a compact support in $[-1, 1]$, so does $e^{-k^2(q-t)^2} f(t)$. If we expand the exponential in a power series, the uniform convergence allows us to exchange the series and the integral. Hence, $f_k(q)$ can be written as power series converging everywhere and so it is an entire slice regular function for all $k \in \mathbb{N}$.

Applying Lemma 2.2.4 to the function $f(x)$, by choosing $a = -2$, $b = 2$, we deduce that $f_k \to f$ uniformly in $[-3/2, 3/2]$ while, by choosing $a = -1$, $b = 1$ we have that $f \to 0$ uniformly in $\mathbb{R} \setminus [-3/2, 3/2]$. Thus $f_k \to f$ uniformly on $\mathbb{R}$. Let $q \in T$, $t \in [-1, 1]$, and let $q = x_0 + \mathrm{Im}(q)$. It is easy to show that

$$\mathrm{Re}(k^2(q-t)^2) = k^2((x_0 - t)^2 - |\mathrm{Im}(q)|^2) > \frac{3}{4}k^2.$$

Let us consider the function $f_k(x)$ written in real components as $f_k = f_{k0} + f_{k1}i + f_{k2}j + f_{k3}k$. On each interval $[a, b]$, f_k satisfies

$$\left| \int_a^b f_k(x)\, dx \right| \leq \sqrt{\sum_{n=0}^{3} \left(\int_a^b f_{kn}(x) dx \right)^2}$$

$$\leq \sum_{n=0}^{3} \int_a^b |f_{kn}(x)| dx$$

$$\leq 4 \int_a^b |f_k(x)|\, dx.$$

Then, for all $q \in T$, we have

$$
\begin{aligned}
|f_k(q)| &\leq 4\frac{k}{C} \int_{-1}^{1} |e^{-k^2(q-t)^2} f(t)| \, dt \\
&\leq 4\frac{k}{C} \int_{-1}^{1} e^{-\mathrm{Re}(-k^2(q-t)^2)} |f(t)| \, dt \\
&\leq 4\frac{k}{C} e^{-\frac{3}{4}k^2} \int_{-1}^{1} |f(t)| \, dt \\
&\leq \frac{k}{C} \frac{16}{3k^2} M,
\end{aligned}
\tag{2.2}
$$

where

$$
M = \int_{-1}^{1} |f(t)| \, dt.
$$

If we choose $F(q) = f_k(q)$, then for k large we have $|f(x) - F(x)| < \varepsilon$ for $x \in \mathbb{R}$ since $f_k \to f$ uniformly on $\mathbb{R}$. Moreover, estimate (2.2) implies $|F(q)| < \varepsilon$ for $q \in T$. $\qquad\square$

Lemma 2.2.8. *Let $f : \mathbb{R} \to \mathbb{H}$ be continuous on $\mathbb{R}$. There exists an entire function $F : \mathbb{H} \to \mathbb{H}$, such that $|f(x) - F(x)| < 1$ for all $x \in \mathbb{R}$.*

Proof. Let f_n be defined as in Lemma 2.2.5, for $n \in \mathbb{Z}$. By Lemma 2.2.7 we can associate to each f_n an entire function F_n such that

$$
|f_n(x) - F_n(x)| < 2^{-|n|-2}, \quad |F_n(x)| < 2^{-|n|}.
$$

Let $N \in \mathbb{N}$, and consider q such that $|q| \leq N$ and $n \in \mathbb{Z}$ such that $|n| > 3N + 3$. We have

$$
|\mathrm{Re}(q - n)| \geq |n| - |\mathrm{Re}(q)| > 3
$$

and

$$
|\mathrm{Im}(q - n)| = |\mathrm{Im}(q)| \leq N < \frac{1}{3}(|n| - N) \leq \frac{1}{2}|\mathrm{Re}(q - n)|.
$$

The above inequalities allow us to conclude that $q - n$ belongs to the set T defined in Lemma 2.2.7. Our assumption implies that

$$
|F_n(q - n)| < 2^{-|n|} \quad \text{for } |q| \leq N, \ |n| > 3N + 3,
$$

which in turn implies that the series $\sum_{n=-\infty}^{+\infty} F_n(q-n)$ converges uniformly for any q such that $|q| \leq N$, for any $N \in \mathbb{N}$. Thus the sequence $s_m(q) = \sum_{n=-m}^{m} F_n(q-n)$ converges uniformly to a function F. Moreover, the sequence of the restrictions of $s_m(q)$ to any complex plane $\mathbb{C}_I$ converges to the restriction of F to $\mathbb{C}_I$, for all $I \in \mathbb{S}$. Thus we have that

$$
\begin{aligned}
(\partial_x + I\partial_y)F(x + Iy) &= (\partial_x + I\partial_y) \lim_{m \to \infty} s_m(x + Iy) \\
&= \lim_{m \to \infty} (\partial_x + I\partial_y) s_m(x + Iy) = 0,
\end{aligned}
$$

for any q such that $|q| \leq N$, for any $N \in \mathbb{N}$, and so F is an entire function. Moreover, for any $x \in \mathbb{R}$,

$$|F(x) - f(x)| \leq |\sum_{n=-\infty}^{+\infty} F_n(x-n) - f_n(x-n)|$$

$$\leq \sum_{n=-\infty}^{+\infty} |F_n(x) - f_n(x)|$$

$$< \sum_{n=-\infty}^{+\infty} 2^{-|n|-2} < 1,$$

which completes the proof. □

With all these preliminary results at hand, we can now come back to the proof of Theorem 2.2.2.

Proof of Theorem 2.2.2. By Lemma 2.2.3, there exists an entire function $h : \mathbb{H} \to \mathbb{H}$ such that it has no zeros, all the coefficients in its series expansion are real numbers, and $h(x) > \frac{1}{\varepsilon(x)}$, for all $x \in \mathbb{R}$. Then, Lemma 2.2.8 provides an entire function $F : \mathbb{H} \to \mathbb{H}$ such that $|h(x)f(x) - F(x)| < 1$, for all $x \in \mathbb{R}$. The values of the function $h(x)$, $x \in \mathbb{R}$, are real since h has real coefficients; consequently

$$\left| f(x) - \frac{F(x)}{h(x)} \right| < \frac{1}{h(x)} < \varepsilon(x), \quad \text{for all } x \in \mathbb{R}.$$

Now the proof follows by choosing $G(q) = (h(q))^{-1} \cdot F(q)$. □

Remark 2.2.9. Note that in our proof we considered a function h quaternionic intrinsic. In general, if the function $h \notin \mathcal{N}(\mathbb{H})$ then, instead of setting $G(q) = (h(q))^{-1} \cdot F(q)$, we would have set $G(q) = (h(q))^{-*} * F(q)$.

Remark 2.2.10. It is important to note that the Weierstrass result on uniform approximation by polynomials on compact subintervals of $\mathbb{R}$ can be obtained as a consequence of Theorem 2.2.1. Indeed, take $[A, B] \subset \mathbb{R}$ and set $\varepsilon(x) := \varepsilon/2 > 0$, for all $x \in \mathbb{R}$, where ε is an arbitrary small constant. By Theorem 2.2.1, there exists an entire function $G(q) = \sum_{k=0}^{\infty} q^k a_k$, such that

$$|f(x) - G(x)| < \varepsilon/2, \quad \text{for all } x \in [A, B].$$

The uniform convergence of the series $G(q)$ in a closed ball $\overline{B(0; R)}$ that includes $[A, B]$ shows that there exists n_0, such that for all $n \geq n_0$ we have

$$|G(q) - \sum_{k=0}^{n} q^k a_k| < \varepsilon/2, \quad \text{for all } q \in \overline{B(0; R)},$$

which implies

$$|f(x) - \sum_{k=0}^{n} x^k a_k| \le |f(x) - G(x)| + |G(x) - \sum_{k=0}^{n} x^k a_k| < \varepsilon/2 + \varepsilon/2 = \varepsilon,$$

for all $x \in [A, B]$ and all $n \ge n_0$.

We now prove another Carleman-type result on simultaneous approximation. This result generalizes the corresponding result, see [126, Theorem 3], obtained by Kaplan in the complex case. We follow the proof in [126], which is is based on Lemma 1 and Lemma 2 in the same paper. Since Lemma 1 refers only to real-valued functions of real variable, it will remain unchanged and we state it below without proof:

Lemma 2.2.11. *Let $E(x)$ be a continuous function, positive for $x \in [0, +\infty)$. Then there exists a function $E_1(x)$ continuous for $x \in [0, +\infty)$, such that*

$$0 < E_1(x) < \frac{1}{2}E(x), \quad \int_x^{+\infty} E_1(t)\, dt < \frac{1}{2}E(x).$$

It remains to prove the analogue of Lemma 2 in the quaternionic setting, namely the following:

Lemma 2.2.12. *Let $E_1 : \mathbb{R} \to \mathbb{R}_+$ be a continuous function satisfying $E_1(x) = E_1(-x)$, for all $x \in \mathbb{R}$ and $k = \int_{-\infty}^{+\infty} E_1(t)dt < \infty$. Let $A, B \in \mathbb{H}$ be such that $|A - B| < 2k$. Then there exists an entire function $h : \mathbb{H} \to \mathbb{H}$, such that*

$$|h'(x)| < E_1(x), \ \ \text{for all } x \in \mathbb{R}, \ \ \text{and} \ \ \lim_{x \to -\infty} h(x) = A, \ \ \lim_{x \to +\infty} h(x) = B.$$

Proof. If $A = B$, then clearly we can choose $h(q) = A$, for all $q \in \mathbb{H}$. If $A \ne B$, denote $r = |A - B|/(2k)$ and $s = (1 - r)/(2(1 + r))$. By Theorem 2.2.1, there exists an entire function $G : \mathbb{H} \to \mathbb{H}$, such that for all $x \in \mathbb{R}$ we have

$$|G(x) - E_1(x)| < sE_1(x).$$

If $G(q) = \sum_{n=0}^{\infty} q^n a_n$ then

$$h_0(q) = \sum_{n=0}^{\infty} q^{n+1} \cdot \frac{a_n}{n+1}$$

is a convergent series with the same ray of convergence as G, therefore h_0 is also entire. Moreover, it is clear that $\partial_s h_0(q) = G(q)$ for all q. Thus, we get that there exists an entire function $h_0 : \mathbb{H} \to \mathbb{H}$, such that

$$|h_0'(x) - E_1(x)| < sE_1(x), \ \ \text{for all } x \in \mathbb{R}.$$

This last inequality implies

$$|h'(x)| \leq (1+s)E_1(x)$$

and therefore by the Leibniz–Newton formula $h_0(x) = \int_0^x h'(t)dt + h_0(0)$, we obtain that the following two limits exist in $\mathbb{H}$:

$$\lim_{x \to +\infty} h_0(x) = \int_0^{+\infty} h_0'(t)dt + h_0(0) := B_0,$$

$$\lim_{x \to -\infty} h_0(x) = \int_0^{-\infty} h_0'(t)dt + h_0(0) := A_0.$$

In addition, we easily get that

$$\mathrm{Re}[h'(x)] > (1-s)E_1(x), \quad \text{for all } x \in \mathbb{R}$$

and therefore

$$|A_0 - B_0| = \left| \int_{-\infty}^{+\infty} h'(x)dx \right| > \int_{-\infty}^{+\infty} \mathrm{Re}[h'(x)]dx > 2k(1-s).$$

Let us now choose the constants $a, b \in \mathbb{H}$ such that $aA_0 + b = A$, $aB_0 + b = B$ and put $h(q) = ah_0(q) + b$. Then

$$\lim_{x \to -\infty} h(x) = A, \quad \lim_{x \to +\infty} h(x) = B;$$

moreover,

$$|a| = \frac{|A - B|}{|A_0 - B_0|} < \frac{2rk}{2k(1-s)} = \frac{r}{1-s}.$$

Hence we conclude that

$$|h'(x)| < \frac{r}{1-s}|h_0'(x)| < \frac{r(1+s)}{1-s}E_1(x) < \frac{r^2 + 3r}{3r+1}E_1(x) < E_1(x),$$

and the statement follows. $\square$

We can now state and prove the announced Carleman-type result on simultaneous approximation:

Theorem 2.2.13. *Let $f : \mathbb{R} \to \mathbb{H}$ be a function having continuous derivative on $\mathbb{R}$ and let $E : \mathbb{R} \to (0, +\infty)$ be continuous on $\mathbb{R}$. Then there exists an entire function $G : \mathbb{H} \to \mathbb{H}$ such that*

$$|f(x) - G(x)| < E(x), \quad and \quad |f'(x) - G'(x)| < E(x), \quad for \ all \ x \in \mathbb{R}.$$

Proof. Without loss of generality, we can assume that $E(x) = E(-x)$, for all $x \in \mathbb{R}$, because for any positive function $E(x)$ on $\mathbb{R}$, we can define $E^*(x) = \min(E(x), E(-x))$, which is now an even function on $\mathbb{R}$. Let $E_1(x)$ (depending on $E(x)$ as in Lemma 2.2.11), so that E_1 is also an even function. By Theorem 2.2.1, there exists an entire function G_1 such that

$$|G_1(x) - f'(x)| < E_1(x), \quad \text{for all } x \in \mathbb{R}.$$

Set

$$g(x) = \int_0^x [G_1(t) - f'(t)]dt.$$

By the choice of $E_1(x)$, the limits $\lim_{x \to +\infty} g(x) = B$, $\lim_{x \to -\infty} g(x) = A$ exist (in $\mathbb{H}$) and

$$|A - B| < \int_{-\infty}^{+\infty} E_1(x)dx := 2k.$$

For these choices of A, B and $E_1(x)$, let h be the entire function given by the Lemma 2.2.12.

Define now $G(q) = \int_0^q G_1(t)dt + f(0) - h(q)$, $q \in \mathbb{H}$. We have:

$$|G'(x) - f'(x)| = |G_1(x) - f'(x) - h'(x)| < E_1(x) + E_1(x) < E(x)$$

and

$$\lim_{x \to +\infty} [G(x) - f(x)] = \lim_{x \to +\infty} \left[\int_0^x (G_1(t) - f'(t)) \, dt - h(x) \right] = B - B = 0.$$

Thus, for $x > 0$ we obtain:

$$G(x) - f(x) = -\int_x^{+\infty} (Q'(t) - f'(t)) \, dt,$$

$$|G(x) - f(x)| < \int_x^{+\infty} 2E_1(t) \, dt < E(x).$$

The inequality holds, by using a similar argument, for $x < 0$ and the conclusion of the theorem follows. $\square$

We now present some applications of the approximation results. A first application of Theorem 2.2.2 is the following.

Theorem 2.2.14. *Let $f : (-1, 1) \to \mathbb{H}$ and $\varepsilon : (-1, 1) \to (0, +\infty)$ be continuous functions. Then there exists a power series $P(t) = \sum_{n=0}^{\infty} t^n a_n$, with $a_n \in \mathbb{H}$, such that*

$$|f(t) - P(t)| < \varepsilon(t), \quad \text{for all } t \in (-1, 1).$$

Moreover, if f is real valued on $(-1, 1)$, then also P can be chosen real valued on $(-1, 1)$.

Proof. This is an immediate consequence of Theorem 2.2.2 by using the function w defined by

$$w(q) = \tan\left(\frac{\pi}{2}q\right)$$

$$= \sum_{n=1}^{\infty} q^{2n-1} \cdot \frac{(-1)^{n-1}2^{2n}(2^{2n}-1)B_{2n}}{(2n)!}$$

$$= q + q^3 \cdot \frac{1}{3} + q^5 \cdot \frac{2}{15} + q^7 \cdot \frac{17}{315} + \cdots,$$

where B_n denotes the nth Bernoulli number. Note that $w \in \mathcal{N}(B(0;R))$ for all $R > 0$, thus w is an entire function.

Let us define $F : \mathbb{R} \to \mathbb{H}$ by $F(x) = f((2/\pi)\arctan(x))$. Clearly, F is continuous on $\mathbb{R}$ and then, by Theorem 2.2.1, for the continuous function $E : \mathbb{R} \to \mathbb{R}_+$ given by $E(x) = \varepsilon((2/\pi)\arctan(x))$, there exists an entire function $G : \mathbb{H} \to \mathbb{H}$, such that $|F(x) - G(x)| < E(x)$, for all $x \in \mathbb{R}$, i.e.,

$$|f((2/\pi)\arctan(x)) - G(x)| < E(x)$$

for all $x \in \mathbb{R}$.

Setting $t = (2/\pi)\arctan(x)$ and substituting in the last inequality, we obtain

$$|f(t) - G(\tan(\pi t/2))| < E(\tan(\pi t/2)) = \varepsilon(t), \text{ for all } t \in (-1,1).$$

Since $w \in \mathcal{N}(B(0;R))$ for all $R > 0$, we can define $P(q) = G(w(q))$, which turns out to be an entire function on $\mathbb{H}$, being composition of two entire functions. Therefore, $P(q) = \sum_{n=0}^{\infty} q^n a_n$, for all $q \in \mathbb{H}$ and the statement follows. $\square$

The next result generalizes to this setting Theorem 7 in Kaplan [126].

Corollary 2.2.15. *Let $f : \partial(B(0;1)) \to \mathbb{R}$ be real valued and measurable. Then there exists a function $u : \overline{B(0;1)} \to \mathbb{H}$, harmonic in $B(0;1)$, that is, if $u(q) = u(x + Iy)$ then $\frac{\partial^2 u}{\partial x^2} + \frac{\partial^2 u}{\partial y^2} = 0$, for all $|q| < 1$, such that for any $I \in \mathbb{S}$ we have $u(re^{I\varphi}) \to f(e^{I\varphi})$ as $r \nearrow 1$, almost everywhere in φ.*

2.3 Notes on Müntz–Szász Type Results

As a generalization of the Weierstrass's approximation theorem, in [14] S. N. Bernstein asked which conditions on an increasing sequence

$$\Lambda = \{0 = \lambda_0 < \lambda_1 < \cdots\},$$

can guarantee that the vector space $\text{span}\{x^{\lambda_k}; \ k = 0, 1, \ldots\}$ is dense in

$$C[0,1] = \{f : [0,1] \to \mathbb{R}; \ f \text{ is continuous on } [0,1]\}.$$

The answer was given by Müntz in [152], who proved that a necessary and sufficient condition is that $\sum_{k=1}^{\infty} \lambda_k^{-1} = +\infty$.

An extension of the Müntz's approximation theorem to the complex setting was obtained by Szász in [185]:

Theorem 2.3.1. *Let us consider*

$$C([0,1]; \mathbb{C}) = \{f : [0,1] \to \mathbb{C}; \ f \ \text{is continuous on} \ [0,1]\}.$$

For $\Lambda = (\lambda_k)_{k=0}^{\infty} \subset \mathbb{C}$, let

$$\Pi_{\mathbb{C}}(\Lambda) = \text{span}_{\mathbb{C}}\{x^{\lambda_k}; \ k = 0, 1, \dots\}$$

be the space of generalized polynomials with complex coefficients and the complex exponents in Λ.

Suppose that $\lambda_0 = 0$ and $\text{Re}(\lambda_k) > 0$ for all $k \geq 1$. If

$$\sum_{k=1}^{\infty} \frac{\text{Re}\lambda_k}{1 + |\lambda_k|^2} = +\infty, \tag{2.3}$$

then $\Pi_{\mathbb{C}}(\Lambda)$ is dense in $C([0,1]; \mathbb{C})$.

Moreover, if

$$\sum_{k=1}^{\infty} \frac{1 + \text{Re}\lambda_k}{1 + |\lambda_k|^2} < +\infty, \tag{2.4}$$

then $\Pi_{\mathbb{C}}(\Lambda)$ is not a dense subset in $C([0,1]; \mathbb{C})$.

In particular, if $\liminf_{k \to \infty} \text{Re}(\lambda_k) > 0$, then $\Pi_{\mathbb{C}}(\Lambda)$ is dense in $C([0,1]; \mathbb{C})$ if and only if (2.3) holds.

The Müntz–Szász theorem was extended by Trent in [187], as follows. Denote $\mathbb{D} = \{z \in \mathbb{C}; \ |z| < 1\}$ and take a set $M \subset \mathbb{N} \times \mathbb{N}$. For each $k \in \mathbb{N}$, denote $M_k = \{k \in \mathbb{N}; \ (m, m+k) \in M\}$.

We have:

Theorem 2.3.2. *The set $\text{span}\{1, z^n, \overline{z}^m; \ (n, m) \in M\}$ is dense in $C(\overline{\mathbb{D}}) = \{f : \overline{\mathbb{D}} \to \mathbb{C}; \ f \ \text{is continuous in} \ \mathbb{D}\}$, if and only if for every $k \in \mathbb{N}$, $\sum_{m \in M_k} 1/m = +\infty$.*

It is natural to look for possible extensions of Theorems 2.3.1 and 2.3.2 to the quaternionic setting. Unfortunately, it seems that it is not possible to use the ideas of the original proof of Szász, which is based on some rather complicated determinants: in fact, in the quaternionic setting the situation becomes extremely complicated. Therefore, it seems to us that this would not be a way to construct the extension to quaternions.

We have also tried to use the simpler, more constructive idea of Golitschek in [109], but unfortunately this method, too, does not work in the quaternionic case.

Thus, the extensions of Theorems 2.3.1 and 2.3.2 to the quaternionic setting remain open questions.

Chapter 3

Approximation in Compact Balls by Bernstein and Convolution Type Operators

In this chapter we present the quaternionic counterparts, in the slice regular case, of several celebrated results in complex approximation. In particular, we discuss approximation by Bernstein polynomials and by convolution operators in the case of compact balls and Cassini cells.

3.1 Approximation by q-Bernstein Polynomials, $q \geq 1$, in Compact Balls

In the complex case, approximation using Bernstein polynomials is a widely studied topic. The interested reader may consult the book by Gal [77] and the one by Lorentz [135].

In this section we treat the quaternionic case, in the slice regular setting. The sources, except for the results where the authors are explicitly mentioned, are the papers of Gal [78], [79] and Section 4.1, pp. 295–299 in the book of Gal [77].

To prove our results, it is more convenient to work with the notion of W-analytic functions, see Definition 1.1.6, rather than that of slice regular function. Moreover, in order to keep the notation used in the complex case and at the same time to avoid any confusion with the standard notation for y as a quaternion, we point out that everywhere in this section q will denote a real number $q \geq 1$.

For $q \geq 1$ and for any $n = 1, 2, \ldots$, define the q-integer

$$[n]_q := 1 + q + \cdots + q^{n-1},$$

© Springer Nature Switzerland AG 2019

S. G. Gal, I. Sabadini, *Quaternionic Approximation*, Frontiers in Mathematics,
https://doi.org/10.1007/978-3-030-10666-9_3

$[0]_q := 0$, and the q-factorial

$$[n]_q! := [1]_q[2]_q \cdots [n]_q,$$

$[0]_q! := 1$. For $q = 1$ we obviously get $[n]_q = n$.

For integers $0 \leq k \leq n$, define

$$\binom{n}{k}_q := \frac{[n]_q!}{[k]_q![n-k]_q!}.$$

It is immediate that for $q = 1$ we get $[n]_1 = n$, $[n]_1! = n!$ and $\binom{n}{k}_1 = \binom{n}{k}$.

Due to the noncommutative setting, to each $\mathbb{H}$-valued function defined in $B(0; R)$ we can associate three polynomials as follows.

Definition 3.1.1. Let $q \geq 1$ and $R > 1$. For any given function $f : B(0; R) \to \mathbb{H}$, we define:

$$B_{n,q}(f)(w) = \sum_{l=0}^{n} f\left(\frac{[l]_q}{[n]_q}\right)\binom{n}{l}_q w^l \prod_{s=0}^{n-1-l}(1 - q^s w), \quad w \in \mathbb{H},$$

$$B_{n,q}^*(f)(w) = \sum_{l=0}^{n} \binom{n}{l}_q w^l \prod_{s=0}^{n-1-l}(1 - q^s w) f\left(\frac{[l]_q}{[n]_q}\right), \quad w \in \mathbb{H},$$

$$B_{n,q}^{**}(f)(w) = \sum_{l=0}^{n} \binom{n}{l}_q w^l f\left(\frac{[l]_q}{[n]_q}\right) \prod_{s=0}^{n-1-l}(1 - q^s w), \quad w \in \mathbb{H}.$$

We call $B_{n,q}(f)$ the left q-Bernstein polynomial, $B_{n,q}^*(f)$ the right q-Bernstein polynomial and $B_{n,q}^{**}(f)$ middle q-Bernstein polynomial, respectively.

In this section we deal with approximation using the q-Bernstein polynomials of quaternion variable just introduced.

It is easy to show with a simple example that these three kinds of q-Bernstein polynomials do not necessarily converge for all quaternionic-valued continuous functions f. Indeed, if we take $f(w) = iwi$, then we easily get

$$|B_{n,q}(f)(w) - iwi| = |B_{n,q}^*(f)(w) - iwi|$$

$$= |B_{n,q}^{**}(f)(w) - iwi|$$

$$= |-w - iwi| = |-iw + wi| > 0, \quad \text{for all } w \neq i.$$

However, for the W-analytic functions we can obtain convergence results with very good error estimates. To prove this, we first need an auxiliary result.

Theorem 3.1.2. *Let $q \geq 1$ be a real number. Suppose that $f : B(0; R) \to \mathbb{H}$ has the property that $f(w) \in \mathbb{R}$ for all $w \in [0, 1]$. Then we have the representation formula*

$$B_{n,q}(f)(w) = \sum_{m=0}^{n} \binom{n}{m}_q [\Delta_{1/[n]_q}^m f(0)]_q w^m, \quad \text{for all } w \in \mathbb{H},$$

where $[\Delta_h^p f(0)]_q = \sum_{k=0}^{p}(-1)^k q^{k(k-1)/2}\binom{p}{k}_q f([p-k]_q h)$.

Proof. Our assumption on f implies that the real values $f\left(\frac{l}{[n]_q}\right)$ commute with the other terms in the expression of $B_n(f)(w)$. Hence, since $w^{n+m} = w^n w^m = w^m w^n$ (and by associativity) we have

$$w^l \prod_{s=0}^{n-1-l}(1 - q^s w) = \prod_{s=0}^{n-1-l}(1 - q^s w)w^l,$$

and $\alpha w = w\alpha$, for all $\alpha \in \mathbb{R}$, $w \in \mathbb{H}$; moreover in the product $\prod_{s=0}^{n-1-l}(1 - q^s w)$ we can interchange the order of the terms. Using these facts and reasoning exactly as in the case of q-Bernstein polynomials of a real variable (see Phillips [159], proof of Theorem 1) we obtain that the coefficient of w^m in the expression of $B_{n,q}(f)(w)$ is

$$\sum_{r=0}^{m} f([m-r]_q/[n]_q)\binom{n}{m-r}_q(-1)^r q^{r(r-1)/2}\binom{n-m+r}{r}_q$$

$$= \binom{n}{m}_q \sum_{r=0}^{m}(-1)^r q^{r(r-1)/2}\binom{m}{r}_q f([m-r]_q/[n]_q),$$

which immediately proves the statement. $\qquad \square$

Remark 3.1.3. An analog of Theorem 3.1.2 holds also for the right and middle q-Bernstein polynomials, $B_{n,q}^*(f)(w)$ and $B_{n,q}^{**}(f)(w)$, as similar arguments show.

In the case $q = 1$ we can state and prove the following approximation result.

Theorem 3.1.4. *Suppose that $f : B(0;R) \to \mathbb{H}$ is left W-analytic in $B(0;R)$, i.e., $f(w) = \sum_{p=0}^{\infty} c_p w^p$, for all $w \in B(0;R)$, where $c_p \in \mathbb{H}$ for all $p = 0, 1, 2, \ldots$. Then for all $1 \leq r < R$, $|w| \leq r$ and $n \in \mathbb{N}$,*

$$|B_n(f)(w) - f(w)| \leq \frac{2}{n}\sum_{p=2}^{\infty}|c_p|p(p-1)r^p.$$

Proof. It is immediate to verify that $B_n(f)(0) = f(0)$ and $B_n(f)(1) = f(1)$. Then, applying Theorem 3.1.2 to $f(w) = e_p(w) = w^p$ we get

$$B_n(e_p)(w) = \sum_{l=0}^{n}\binom{n}{l}\Delta_{1/n}^l e_p(0)w^l,$$

and

$$B_n(e_p)(1) = 1 = \sum_{l=0}^{n}\binom{n}{l}\Delta_{1/n}^l e_p(0),$$

where, since e_p is convex of any order, one has that

$$\binom{n}{l}\Delta_{1/n}^l e_p(0) \geq 0$$

for all $0 \leq l \leq n$.

Further, using the formula relating the finite differences and the divided differences, we can write

$$
\begin{aligned}
B_n(e_p)(w) &= \sum_{l=0}^{n} \binom{n}{l} l! \frac{1}{n^l} [0, 1/n, \ldots, l/n; \; e_p] w^l \\
&= \sum_{l=0}^{n} \frac{n(n-1)\cdots(n-l+1)}{n^l} [0, 1/n, \ldots, l/n; \; e_p] w^l.
\end{aligned}
$$

On the other hand, $B_n(f)$ satisfies the properties

$$
\begin{aligned}
B_n(f+g) &= B_n(f) + B_n(g), \\
B_n(\alpha f) &= \alpha B_n(f), \quad \alpha \in \mathbb{H}, \\
B_n(f\alpha) &\neq \alpha B_n(f), \quad \alpha \in \mathbb{H} \setminus \mathbb{R}.
\end{aligned}
$$

Now let us prove that

$$
B_n(f)(w) = \sum_{k=0}^{\infty} c_k B_n(e_k)(w).
$$

Denoting $f_m(w) = \sum_{j=0}^{m} c_j w^j$, $|w| \leq r$, $m \in \mathbb{N}$, and using the above properties of B_n, we obviously have

$$
B_n(f_m)(w) = \sum_{k=0}^{m} c_k B_n(e_k)(w).
$$

Thus it suffices to prove that for any fixed $n \in \mathbb{N}$ and $|w| \leq r$ with $r \geq 1$, we have

$$
\lim_{m \to \infty} B_n(f_m)(w) = B_n(f)(w).
$$

But this is immediate from $\lim_{m \to \infty} \|f_m - f\|_r = 0$, where

$$
\|f\|_r := \sup\{|f(w)|; \; |w| \leq r\} \tag{3.1}
$$

and from the inequality

$$
\begin{aligned}
|B_n(f_m)(w) - B_n(f)(w)| &\leq \sum_{k=0}^{n} \binom{n}{k} |w^k (1-w)^{n-k}| \cdot \|f_m - f\|_r \\
&\leq M_{r,n} \|f_m - f\|_r,
\end{aligned}
$$

valid for all $|w| \leq r$.

This immediately yields

$$
|B_n(f)(w) - f(w)| \leq \sum_{p=0}^{\infty} |c_p| \cdot |B_n(e_p)(w) - e_p(w)|.
$$

To estimate $|B_n(e_p) - e_p|$ we have to consider two cases: 1) $0 \le p \le n$, and 2) $p > n$.

Case 1). We get

$$B_n(e_p)(w) = \sum_{l=0}^{p} \frac{n(n-1)\cdots(n-l+1)}{n^l}[0, 1/n, \ldots, l/n; \ e_p]w^l$$

and denoting

$$C_{n,l} = \frac{n(n-1)\cdots(n-l+1)}{n^l}[0, 1/n, \ldots, l/n; \ e_p]$$

it follows

$$B_n(e_p)(w) - e_p(w) = \left(1 - \frac{n(n-1)\cdots(n-(p-1))}{n^p}\right)e_p(w) + \sum_{l=0}^{p-1} C_{n,l}w^l.$$

Passing to the modulus $|\cdot|$ with $|w| \le r$ and using the obvious inequality

$$1 - \frac{n(n-1)\cdots(n-(p-1))}{n^p} = 1 - (1 - 1/n)\cdots\cdots(1 - (p-1)/n)$$

$$\le \frac{p(p-1)}{2n},$$

we obtain

$$|B_n(e_p)(w) - e_p(w)| \le \left|1 - \frac{n(n-1)\cdots(n-(p-1))}{n^p}\right|r^p + \sum_{l=0}^{p-1} C_{n,l}r^p$$

$$= 2\left|1 - \frac{n(n-1)\cdots(n-(p-1))}{n^p}\right|r^p$$

$$\le \frac{p(p-1)}{n}r^p.$$

Case 2). For all $|w| \le r$, $r \ge 1$, $w \in \mathbb{H}$, $p > n \ge 1$ we get

$$|B_n(e_p)(w)| \le \sum_{l=0}^{p} \frac{n(n-1)\cdots(n-l+1)}{n^l}[0, 1/n, \ldots, l/n; \ e_p] \cdot |w^l| \le r^n,$$

and therefore

$$|B_n(e_p)(w) - e_p(w)| \le r^n + r^p \le 2r^p \le 2r^p n \le 2r^p \frac{p(p-1)}{n}.$$

Combining Case 1) and 2), we conclude that

$$|B_n(f)(w) - f(w)| \le \sum_{p=0}^{\infty} |c_p| \cdot |B_n(e_p)(w) - e_p(w)|$$

$$\le \frac{2}{n}\sum_{p=0}^{\infty} |c_p|p(p-1),$$

which completes the proof of the theorem. $\square$

In a similar way we obtain the following.

Corollary 3.1.5. *Let* $f : B(0; R) \to \mathbb{H}$ *be right* W*-analytic in* $B(0; R)$*, i.e.,* $f(w) = \sum_{p=0}^{\infty} w^p c_p$*, for all* $w \in B(0; R)$*, where* $c_p \in \mathbb{H}$ *for all* $p = 0, 1, 2, \ldots$*. Then for all* $1 \leq r < R$*,* $|w| \leq r$ *and* $n \in \mathbb{N}$ *we have*

$$|B_n^*(f)(w) - f(w)| \leq \frac{2}{n} \sum_{p=2}^{\infty} |c_p| p(p-1) r^p.$$

Remark 3.1.6. It is not difficult to see that in the case of Bernstein-type polynomials $B_n^{**}(f)(w)$, an estimate of the form in Theorem 3.1.4 cannot be obtained. In fact, in general, we cannot write a formula of the type $B_n^{**}(f)(w) = \sum_{p=0}^{\infty} c_p B_n^{**}(e_p)(w)$ for f left W-analytic or of the type $B_n^{**}(f)(w) = \sum_{p=0}^{\infty} B_n^{**}(e_p)(w) c_p$ for f right W-analytic.

In the case $q > 1$, we have the following approximation result.

Theorem 3.1.7. *Let* $1 < q < R$ *and let* $f : B(0; R) \to \mathbb{H}$ *be left* W*-analytic in* $B(0; R)$*, i.e.,* $f(w) = \sum_{k=0}^{\infty} a_k w^k$*, for all* $w \in B(0; R)$*, where* $a_k \in \mathbb{H}$*,* $k = 0, 1, 2, \ldots$*. Then for all* $1 < r < \frac{R}{q}$*,* $|w| \leq r$ *and* $n \in \mathbb{N}$ *we have*

$$|B_{n,q}(f)(w) - f(w)| \leq \frac{2}{(q-1)[n]_q} \cdot \sum_{k=1}^{\infty} |a_k| k (qr)^k.$$

Proof. Let $e_k(w) = w^k$. We will prove that

$$B_{n,q}(f)(w) = \sum_{k=0}^{\infty} a_k B_{n,q}(e_k)(w), \quad \text{for all } |w| \leq r.$$

Denote by $f_m(w) = \sum_{k=0}^{m} a_k e_k(w)$, $m \in \mathbb{N}$, the partial sum of the power series expansion of f.

By the linearity of $B_{n,q}$ we easily get

$$B_{n,q}(f_m)(w) = \sum_{k=0}^{m} a_k B_{n,q}(e_k)(w), \quad \text{for all } |w| \leq r.$$

To obtain the result, it suffices to prove that

$$\lim_{m \to \infty} B_{n,q}(f_m)(w) = B_{n,q}(f)(w),$$

for all $|w| \leq r$ and $n \in \mathbb{N}$.

By Theorem 3.1.2, we have

$$B_{n,q}(f_m)(w) = \sum_{p=0}^{n} \binom{n}{p}_q [\Delta_{1/[n]_q}^p f_m(0)]_q e_p(w).$$

Hence, for all $n, m \in \mathbb{N}$ and $|w| \leq R$,

$$|B_{n,q}(f_m)(w) - B_{n,q}(f)(w)| \leq \sum_{p=0}^{n} \binom{n}{p}_q |[\Delta_{1/[n]_q}^p (f_m - f)(0)]_q| \cdot |e_p(w)|$$

$$\leq \sum_{p=0}^{n} \binom{n}{p}_q \sum_{j=0}^{p} q^{j(j-1)/2} \binom{p}{j}_q |(f_m - f)([p-j]_q/[n]_q)| \cdot |e_p(w)|$$

$$\leq \sum_{p=0}^{n} \binom{n}{p}_q \sum_{j=0}^{p} q^{j(j-1)/2} \binom{p}{j}_q C_{j,p,\beta} |||f_m - f|||_r \cdot |e_p(w)|$$

$$\leq M_{n,p,r,q} |||f_m - f|||_r,$$

which because $\lim_{m \to \infty} |||f_m - f|||_r = 0$ implies the desired conclusion where, as customary, $\|f_m - f\|_r = \max\{|f_m(w) - f(w)|; \ |w| \leq r\}$.

As a consequence, we obtain

$$|B_{n,q}(f)(w) - f(w)| \leq \sum_{k=0}^{\infty} |a_k| \cdot |B_{n,q}(e_k)(w) \quad c_k(w)|$$

$$= \sum_{k=0}^{n} |a_k| \cdot |B_{n,q}(e_k)(w) - e_k(w)| + \sum_{k=n+1}^{\infty} |a_k| \cdot |B_{n,q}(e_k)(w) - e_k(w)|.$$

Therefore it remains to estimate $|B_{n,q}(e_k)(w) - e_k(w)|$, first for all $0 \leq k \leq n$, and then for $k \geq n+1$, where

$$B_{n,q}(e_k)(w) = \sum_{p=0}^{n} \binom{n}{p}_q [\Delta_{1/[n]_q}^p e_k(0)]_q \cdot e_p(w).$$

Set

$$D_{n,p,k}^{(q)} = \binom{n}{p}_q [\Delta_{1/[n]_q}^p e_k(0)]_q.$$

By relation (12), p. 513 in Phillips [159], we can write

$$D_{n,p,k}^{(q)} = q^{p(p-1)/2} \binom{n}{p}_q [0, [1]_q/[n]_q, \dots, [p]_q/[n]_q; \ e_k] \cdot ([p]_q!)/[n]_q^p,$$

where $[0, [1]_q/[n]_q, \dots, [p]_q/[n]_q; \ e_k]$ denotes the divided difference of $e_k(w) = w^k$.

To explain in more detail, let us recall the definition of divided difference of a function f on the knots $x_0, x_1, \dots, x_p$:

$$[x_0, x_1, \dots, x_p; \ f] = \sum_{i=0}^{p} \frac{f(x_i)}{(x_i - x_0) \cdots (x_i - x_{i-1})(x_i - x_{i+1}) \cdots (x_i - x_p)}.$$

So the above formula follows immediately from the relation between the finite q-difference Δ and the divided difference (which can be easily proved by mathematical induction on p)

$$[0, [1]_q/[n]_q, \ldots, [p]_q/[n]_q; \ f] = q^{-p(p-1)/2}[n]_q^p \cdot \frac{[\Delta_{1/[n]_q}^p f(0)]_q}{[p]_q!},$$

by taking here $f = e_k$.

Therefore,

$$B_{n,q}(e_k)(w) = \sum_{p=0}^{n} D_{n,p,k}^{(q)} \cdot e_p(w).$$

By the relationship (7), p. 236 in Ostrovska [156], we get the formula (easily proved by direct calculation)

$$\binom{n}{p}_q \cdot \frac{[p]_q!}{[n]_q^k} \cdot q^{p(p-1)/2} = \left(1 - \frac{[1]_q}{[n]_q}\right) \cdots \left(1 - \frac{[p-1]_q}{[n]_q}\right),$$

which combined with the above relationship for $D_{n,p,k}^{(q)}$, yields

$$D_{n,p,k}^{(q)} = \left(1 - \frac{[1]_q}{[n]_q}\right) \cdots \left(1 - \frac{[p-1]_q}{[n]_q}\right) [0, [1]_q/[n]_q, \ldots, [p]_q/[n]_q; \ e_k].$$

Since each e_k is convex of any order and $B_{n,q}(e_k)(1) = e_k(1) = 1$ for all k, it follows that all $D_{n,p,k}^{(q)} \geq 0$ and $\sum_{p=0}^{n} D_{n,p,k}^{(q)} = 1$, for all k and n.

Also, note that

$$D_{n,k,k}^{(q)} = \left(1 - \frac{[1]_q}{[n]_q}\right) \cdots \left(1 - \frac{[k-1]_q}{[n]_q}\right),$$

for all $k \geq 1$ and that $D_{n,0,0}^{(q)} = 1$.

In the estimate of $|B_{n,q}(e_k)(w) - e_k(w)|$ we distinguish two cases: 1) $0 \leq k \leq n$ and 2) $k > n$.

Case 1). We have

$$|B_{n,q}(e_k)(w) - e_k(w)| \leq |e_k(w)| \cdot |1 - D_{n,k,k}^{(q)}| + \sum_{p=0}^{k-1} D_{n,p,k}^{(q)} \cdot |e_p(w)|.$$

Since $|e_p(w)| \leq r^p$, for all $|w| \leq r$ and $p \geq 0$, by the relationships in the proof of Theorem 5, p. 247 and by Corollary 6, p. 244, both in Ostrovska [156], we immediately get

$$|B_{n,q}(e_k)(w) - e_k(w)| \leq 2[1 - D^{(q)}_{n,k,k}]r^k$$

$$\leq 2\frac{(k-1)[k-1]_q}{[n]_q}r^k$$

$$\leq 2\frac{kq^k}{(q-1)[n]_q}r^k$$

$$= 2\frac{k(qr)^k}{(q-1)[n]_q},$$

for all $|w| \leq r$.

Case 2). We have

$$|B_{n,q}(e_k)(w) - e_k(w)| \leq |B_{n,q}(e_k)(w)| + |e_k(w)| \leq 2r^k$$

$$\leq 2\frac{(k-1)[k-1]_q}{[n]_q}r^k$$

$$\leq 2\frac{(k-1)[k-1]_q}{[n]_q}(qr)^k$$

$$\leq 2\frac{k(qr)^k}{(q-1)[n]_q}.$$

In conclusion, the estimates in the Cases 1) and 2) yield

$$|B_{n,q}(f)(w) - f(w)| \leq \frac{2}{(q-1)[n]_q} \cdot \sum_{k=0}^{\infty} |a_k|k(qr)^k, |w| \leq r, \quad n \in \mathbb{N},$$

which completes the proof of the theorem. $\qquad\qquad\qquad\qquad\qquad\qquad\square$

In a similar way we can establish the following.

Corollary 3.1.8. *Let $1 < q < R$ and let $f : B(0; R) \to \mathbb{H}$ be right W-analytic in $B(0; R)$, i.e., $f(w) = \sum_{p=0}^{\infty} w^p a_p$, for all $w \in B(0; R)$, where $c_p \in \mathbb{H}$ for all $p = 0, 1, 2, \ldots$. Then for all $1 \leq r < \frac{R}{q}$, $|w| \leq r$ and $n \in \mathbb{N}$ we have*

$$|B^*_{n,q}(f)(w) - f(w)| \leq \frac{2}{(q-1)[n]_q} \cdot \sum_{k=1}^{\infty} |a_k|k(qr)^k.$$

Remark 3.1.9. Since $\frac{1}{[n]_q} \leq q \cdot q^{-n}$ for all $n \in \mathbb{N}$, it follows that the order of approximation in Theorem 3.1.7 and in Corollary 3.1.8 is in fact q^{-n}.

Remark 3.1.10. It is not difficult to see that in the case of middle q-Bernstein-type polynomials $B^{**}_{n,q}(f)(w)$ an estimate of the form in Theorem 3.1.7 cannot be obtained, because in general, for f left W-analytic, we cannot write a formula of the type

$$B_{n,q}^{**}(f)(w) = \sum_{p=0}^{\infty} c_p B_{n,q}^{**}(e_p)(w).$$

A similar remark holds for the case when f is right W-analytic.

Remark 3.1.11. In the case of functions of a complex variable, the estimate in Theorem 3.1.7 was obtained in Ostrovska [156].

We now prove some Voronovskaya-type results for $B_{n,q}(f)(w)$ and $B_{n,q}^*(f)(w)$, $q \geq 1$. Then, as an application, we obtain that for functions which are not left polynomials of degree ≤ 1, the orders of approximation in Theorems 3.1.4 and 3.1.7 are exactly $1/n$ and q^{-n}, respectively. The results obtained are extensions to the quaternionic case of analogous results on the approximation by q-Bernstein operators ($q \geq 1$) in the complex case in Gal [76], [79] and Wang and Wu [194].

Theorem 3.1.12. *Suppose that $1 \leq q < R$ and that $f : B(0; R) \to \mathbb{H}$ is left W-analytic in $B(0; R)$, i.e., $f(w) = \sum_{p=0}^{\infty} a_p w^p$, for all $w \in B(0; R)$, where $a_p \in \mathbb{H}$ for all $p = 0, 1, 2, \ldots$. Set*

$$S_k^{(q)} = [1]_q + \cdots + [k-1]_q, \quad k \geq 2.$$

(i) *If $q = 1$, then for any $1 \leq r < R$, $|w| \leq r$ and $n \in \mathbb{N}$, the estimate*

$$\left\| B_{n,1}(f) - f - \sum_{k=2}^{\infty} a_k \cdot \frac{S_k^{(1)}}{n} [e_{k-1} - e_k] \right\|_r \leq \frac{1}{n^2} \cdot \sum_{k=2}^{\infty} |a_k|(k-1)^4 r^k,$$

holds, where $e_p(w) = w^p$ and $\sum_{k=2}^{\infty} |a_k|(k-1)^4 r^k < \infty$.

(ii) *If $q > 1$, $1 < r < R/q^2$, $|w| \leq r$ and $n \in \mathbb{N}$, then the following upper estimate*

$$\left\| B_{n,q}(f) - f - \sum_{k=2}^{\infty} a_k \cdot \frac{S_k^{(q)}}{[n]_q} [e_{k-1} - e_k] \right\|_r \leq \frac{C_{r,q}(f)}{[n]_q^2}$$

holds, where

$$C_{r,q}(f) = 4 \cdot \max \left\{ \frac{1}{(q-1)^2}, \frac{1}{(q-1)^3} \right\} \cdot \sum_{k=2}^{\infty} |a_k|(k-1)^2 (q^2 r)^k.$$

(iii) *If $q \geq 1$, then for any $1 < r < R/q$ we have*

$$\lim_{n \to \infty} [n]_q (B_{n,q}(f)(w) - f(w)) = A_q(f)(w),$$

uniformly in $\overline{B(0; r)}$, where

$$A_q(f)(w) = \sum_{k=2}^{\infty} a_k \cdot S_k^{(q)} \cdot [w^{k-1} - w^k], \quad w \in \mathbb{H}.$$

Proof. Let $1 < r < R/q$. We begin by recalling the most important relations which will be used in our proof. Simple calculations show that

$$
\begin{aligned}
S_k^{(q)} &= \frac{k(k-1)}{2}, \quad \text{for } q = 1 \text{ and} \\
S_k^{(q)} &= \frac{q^k - k(q-1) - 1}{(q-1)^2}, \quad \text{for } q > 1.
\end{aligned}
\tag{3.2}
$$

Moreover, reasoning as in the proof of Theorem 2.3 in [78], we easily get

$$
B_{n,q}(f)(w) = \sum_{k=0}^{\infty} a_k B_{n,q}(e_k)(w).
\tag{3.3}
$$

Here, note that by Theorem 2.1 in [78] we have

$$
\begin{aligned}
B_{n,q}(e_k)(w) &= \sum_{p=0}^{n} \binom{n}{p}_q [\Delta^p_{1/[n]_q} e_k(0)]_q e_p(w) \\
&= \sum_{p=0}^{n} \binom{n}{p}_q [\Delta^p_{1/[n]_q} e_k(0)]_q e_p(w),
\end{aligned}
$$

where

$$
[\Delta^p_h f(0)]_q = \sum_{k=0}^{p} (-1)^q q^{k(k-1)/2} \binom{p}{k}_q f([p-k]_q h).
$$

By (5), (6), and (7) in Ostrovska [156], p. 236 (see also relation (1), proof of Theorem 2.3 in [78]), it follows that

$$
B_{n,q}(e_k)(w) = \sum_{p=0}^{n} D_{n,p,k}^{(q)} e_p(w),
$$

where

$$
\begin{aligned}
D_{n,p,k}^{(q)} &= \binom{n}{p}_q \cdot q^{p(p-1)/2} \cdot [0, 1/[n]_q, \ldots, p/[n]_q; \ e_k] \cdot \frac{[p]_q!}{[n]_q^p} \\
&= \lambda_{n,p}^{(q)} \cdot [0, 1/[n]_q, \ldots, p/[n]_q; \ e_k],
\end{aligned}
\tag{3.4}
$$

with

$$
\lambda_{n,p}^{(q)} = \left(1 - \frac{[1]_q}{[n]_q}\right) \cdot \ldots \cdot \left(1 - \frac{[p-1]_q}{[n]_q}\right).
$$

Note that by (3.4) (see also Lemma 3, p. 245 in [156]), we obtain

$$
\begin{aligned}
D_{n,p,k}^{(q)} &\geq 0 \quad \text{for all } 0 \leq p \leq n, \ k \geq 0, \\
\sum_{p=0}^{n} D_{n,p,k}^{(q)} &= 1 \quad \text{for all } 0 \leq k \leq n
\end{aligned}
\tag{3.5}
$$

and

$$D_{n,k,k}^{(q)} = \prod_{i=1}^{k-1} \left(1 - \frac{[i]_q}{[n]_q} \right)$$

$$D_{n,k-1,k}^{(q)} = \frac{S_k^{(q)}}{[n]_q} \cdot \prod_{i=1}^{k-2} \left(1 - \frac{[i]_q}{[n]_q} \right), \quad k \le n. \tag{3.6}$$

We now prove that the function $A_q(f)(w)$ given by

$$A_q(f)(w) = \sum_{k=2}^{\infty} a_k \cdot S_k^{(q)} \cdot [w^{k-1} - w^k],$$

is left W-analytic in $\overline{B(0;r)}$, for $1 < r < R/q$.
     Indeed, since

$$|A_q(f)(w)| \le \sum_{k=0}^{\infty} |a_k| \cdot S_k^{(q)} \cdot [|w^{k-1}| + |w^k|],$$

and since, by (3.2), $S_k^{(q)} \le \frac{q^k}{(q-1)^2}$ for $q > 1$, it immediately follows that

$$|A_q(f)(w)| \le \frac{2}{(q-1)^2} \cdot \sum_{k=0}^{\infty} |a_k|(qr)^k < \infty, \quad \text{if } q > 1,$$

and

$$|A_q(f)(w)| \le \sum_{k=0}^{\infty} |a_k|k(k-1)(qr)^k < \infty, \quad \text{if } q = 1,$$

for all $w \in \overline{B(0;r)}$. These facts immediately show that for $q \ge 1$, the function $A_q(f)$ is well defined and left W-analytic in $\overline{B(0;r)}$.
     Now, by (3.3), we obtain

$$\left| B_{n,q}(f)(w) - f(w) - \sum_{k=0}^{\infty} a_k \cdot \frac{S_k^{(q)}}{[n]_q}[w^{k-1} - w^k] \right| \le \sum_{k=0}^{\infty} |a_k| \cdot |E_{k,n}^{(q)}(w)|,$$

where

$$E_{k,n}^{(q)}(w) = B_{n,q}(e_k)(w) - w^k - \frac{S_k^{(q)}}{[n]_q}[w^{k-1} - w^k].$$

Since simple calculations and (3.2) imply that

$$E_{0,n}^{(q)}(w) = E_{1,n}^{(q)}(w) = E_{2,n}^{(q)}(w) = 0,$$

in fact we have to estimate the expression

$$\sum_{k=3}^{\infty} |a_k| \cdot |E_{k,n}^{(q)}(w)| = \sum_{k=3}^{n} |a_k| \cdot |E_{k,n}^{(q)}(w)| + \sum_{k=n+1}^{\infty} |a_k| \cdot |E_{k,n}^{(q)}(w)|.$$

To estimate $|E_{k,n}^{(q)}(w)|$, we distinguish two cases: 1) $3 \leq k \leq n$ and 2) $k \geq n+1$.
    Case 1). We obtain

$$[n]_q |E_{k,n}^{(q)}(w)| = |[n]_q (B_{n,q}(e_k)(w) - w^k) - S_k^{(q)} \cdot (w^{k-1} - w^k)|$$

$$\leq r^k [n]_q \sum_{i=1}^{k-2} D_{n,i,k}^{(q)} + |[n]_q D_{n,k-1,k}^{(q)} - S_k^{(q)} |r^k + |[n]_q (1 - D_{n,k,k}^{(q)}) - S_k^{(q)} |r^k.$$

Taking now into account (3.5) and (3.6) and following exactly the reasoning in the proof of Lemma 3, p. 747 in [194], we arrive at

$$|E_{k,n}^{(q)}(w)| \leq \frac{4(k-1)^2 [k-1]_q^2}{[n]_q^2} \cdot r^k, \quad \text{for all } w \in \overline{B(0;r)}. \tag{3.7}$$

By (3.7), it follows that

$$\sum_{k=3}^{n} |a_k| \cdot |E_{k,n}^{(q)}(w)| \leq \frac{4}{[n]_q^2} \cdot \sum_{k=3}^{n} |a_k| (k-1)^2 [k-1]_q^2 r^k, \tag{3.8}$$

for all $w \in \overline{B(0;r)}$ and $n \in \mathbb{N}$.
    Case 2). We get

$$\sum_{k=n+1}^{\infty} |a_k| \cdot |E_{k,n}^{(q)}(w)| \leq \sum_{k=n+1}^{\infty} |a_k| \cdot |B_{n,q}(e_k)(w)| + \sum_{k=n+1}^{\infty} |a_k| \cdot |w^k|$$

$$+ \frac{1}{[n]_q} \sum_{k=n+1}^{\infty} |a_k| \cdot S_k^{(q)} \cdot |w^{k-1}| + \frac{1}{[n]_q} \sum_{k=n+1}^{\infty} |a_k| \cdot S_k^{(q)} \cdot |w^k|$$

$$=: L_{1,q}(w) + L_{2,q}(w) + L_{3,q}(w) + L_{4,q}(w). \tag{3.9}$$

We distinguish two subcases: 2(i)) $q = 1$ and 2(ii)) $q > 1$.
    Subcase (2(i)). Formula (3.5) immediately yields

$$L_{1,1}(w) \leq \sum_{k=n+1}^{\infty} |a_k| \cdot \sum_{p=0}^{n} D_{n,p,k}^{(1)} |e_p(w)|$$

$$\leq \frac{1}{n^2} \sum_{k=n+1}^{\infty} |a_k| (k-1)^2 r^k,$$

and, similarly,

$$L_{2,1}(w) \le \frac{1}{n^2} \sum_{k=n+1}^{\infty} |a_k|(k-1)^2 r^k.$$

for all $w \in \overline{B(0;r)}$.

Reasoning as above and using (3.2) and (3.5), we obtain

$$L_{3,1}(w) \le \frac{1}{n} \sum_{k=n+1}^{\infty} |a_k| \cdot \frac{k(k-1)}{2} r^k$$

$$\le \frac{1}{n^2} \sum_{k=n+1}^{\infty} |a_k| \cdot (k-1)^3 r^k,$$

and

$$L_{4,1}(w) \le \frac{1}{n^2} \sum_{k=n+1}^{\infty} |a_k| \cdot (k-1)^3 r^k.$$

The above inequalities for $L_{1,1}(w), \dots, L_{4,1}(w)$ and (3.9) imply that

$$\sum_{k=n+1}^{\infty} |a_k| \cdot |E_{k,n}^{(1)}(w)| \le \frac{1}{n^2} \sum_{k=n+1}^{\infty} |a_k|(k-1)^3 r^k,$$

for all $w \in \overline{B(0;r)}$ and $n \in \mathbb{N}$.

Now note that the sequence

$$\{a_n = \sum_{k=n+1}^{\infty} |a_k|(k-1)^3 r^k, \quad n \in \mathbb{N}\}$$

converges to zero (and thus it is bounded by a positive constant $M > 0$ independent of n), since it is the remainder of the convergent series $\sum_{k=0}^{\infty} |a_k|(k-1)^3 r^k$. This implies that

$$\sum_{k=n+1}^{\infty} |a_k| \cdot |E_{k,n}^{(1)}(w)| \le \frac{M}{n^2} \le \frac{\sum_{k=1}^{\infty} |a_k|(k-1)^3 r^k}{n^2},$$

for all $w \in \overline{B(0;r)}$ and $n \in \mathbb{N}$.

Next, set $q = 1$ in (3.8) and use that the series $\sum_{k=3}^{\infty} |a_k|(k-1)^4 r^k$ is convergent to obtain

$$\sum_{k=3}^{n} |a_k| \cdot |E_{k,n}^{(1)}(w)| \le \frac{4}{n^2} \cdot \sum_{k=3}^{n} |a_k|(k-1)^4 r^k.$$

$$\le \frac{4}{n^2} \cdot \sum_{k=3}^{\infty} |a_k|(k-1)^4 r^k,$$

which combined with the previous estimate gives

$$\sum_{k=0}^{\infty} |a_k| \cdot |E_{k,n}^{(1)}(w)| \leq \frac{4 \sum_{k=2}^{\infty} |a_k|(k-1)^4 r^k}{n^2},$$

for all $w \in \overline{B(0;r)}$ and $n \in \mathbb{N}$.

This proves (i) in the statement of Theorem 3.1.12.

Subcase $(2(ii))$. By (3.5), for all $w \in \overline{B(0;r)}$

$$L_{1,q}(w) \leq \sum_{k=n+1}^{\infty} |a_k| \cdot \sum_{p=0}^{n} D_{n,p,k}^{(q)} |w^p| \leq \sum_{k=n+1}^{\infty} |a_k| \cdot r^k$$

$$\leq \frac{1}{[n]_q^2} \sum_{k=n+1}^{\infty} |a_k| \cdot [k-1]_q^2 r^k \leq \frac{1}{(q-1)^2 [n]_q^2} \sum_{k=n+1}^{\infty} |a_k| \cdot (q^2 r)^k$$

and similarly

$$L_{2,q}(w) \leq \sum_{k=n+1}^{\infty} |a_k| \cdot r^k$$

$$\leq \frac{1}{[n]_q^2} \sum_{k=n+1}^{\infty} |a_k| \cdot [k-1]_q^2 r^k$$

$$\leq \frac{1}{(q-1)^2 [n]_q^2} \sum_{k=n+1}^{\infty} |a_k| (q^2 r)^k.$$

Moreover, from (3.2) we get $S_k^{(q)} \leq \frac{q^k}{(q-1)^2}$, for all $w \in \overline{B(0;r)}$ and so we have

$$L_{3,q}(w) \leq \frac{1}{(q-1)^2 [n]_q} \sum_{k=n+1}^{\infty} |a_k| \cdot q^k r^k$$

$$= \frac{1}{(q-1)^2 [n]_q} \sum_{k=n+1}^{\infty} |a_k| \cdot (qr)^k$$

$$\leq \frac{1}{(q-1)^3 [n]_q^2} \sum_{k=n+1}^{\infty} |a_k| (q^2 r)^k,$$

and similarly

$$L_{4,q}(w) \leq \frac{1}{[n]_q} \sum_{k=n+1}^{\infty} |a_k| \cdot q^k r^k \leq \frac{1}{(q-1)^3 [n]_q^2} \sum_{k=n+1}^{\infty} |a_k| \cdot (q^2 r)^k.$$

From (3.9), it immediately follows that

$$\sum_{k=n+1}^{\infty} |a_k| \cdot |E_{k,n}^{(q)}(w)| \leq \frac{C_{r,q}(f)}{[n]_q^2},$$

for all $w \in \overline{B(0;r)}$, where

$$C_{r,q}(f) = \max\left\{\frac{1}{(q-1)^2}, \frac{1}{(q-1)^3}\right\} \cdot \sum_{k=n+1}^{\infty} |a_k|(q^2 r)^r.$$

Since

$$[k-1]_q^2 \leq [k]_q^2 \leq \frac{q^{2k}}{(q-1)^2},$$

for $w \in \overline{B(0;r)}$ with $1 < r < R/q^2$, by (3.8) we easily obtain

$$\sum_{k=3}^{n} |a_k| \cdot |E_{k,n}^{(q)}| \leq \frac{4}{(q-1)^2[n]_q^2} \cdot \sum_{k=2}^{\infty} |a_k|(k-1)^2(q^2 r)^k.$$

Collecting these results, we get the upper estimate in (ii).

(iii) The case $q = 1$ follows directly by multiplying the estimate in (i) by n and letting $n \to \infty$.

The case $q > 1$ and $1 < r < R/q^2$ follows by multiplying (ii) by $[n]_q$ and letting $n \to \infty$.

What remains to be proved, for $q > 1$, is that the limit in (iii) still holds under the more general condition $1 < r < \frac{R}{q}$.

Since $R/q^{1+t} \nearrow R/q$ as $t \searrow 0$, it is evident that given $1 < r < R/q$, there exists a $t \in (0,1)$, such that $q^{1+t}r < R$. The fact that f is left W-analytic in $B(0;R)$ implies that

$$\sum_{k=2}^{\infty} |a_k| k^4 q^{(1+t)k} r^k = \sum_{k=2}^{\infty} |a_k| \cdot k^4 (q^{1+t}r)^k < \infty,$$

for all $w \in \overline{B(0;r)}$. Moreover, the convergence of the above series implies that for arbitrary $\varepsilon > 0$, there exists n_0, such that

$$\sum_{k=n_0+1}^{\infty} |a_k| \cdot k^2 q^k r^k < \varepsilon.$$

By using (3.7), for all $w \in \overline{B(0;r)}$ and $n > n_0$ we get

$$|[n]_q(B_{n,q}(f)(w) - f(w)) - A_q(f)(w)|$$

$$\leq \sum_{k=2}^{n_0} |a_k| \cdot \left|[n]_q(B_{n,q}(e_k)(w) - e_k(w)) - S_k^{(q)}[w^{k-1} - w^k]\right|$$

$$+ \sum_{k=n_0+1}^{\infty} |a_k| \cdot ([n]_q |B_{n,q}(e_k)(w) - w^k| + S_k^{(q)}|w^{k-1} - w^k|)$$

$$\leq \sum_{k=2}^{n_0} |a_k| \cdot \frac{4(k-1)^2[k-1]_q^2}{[n]_q} \cdot r^k$$

$$+ \sum_{k=n_0+1}^{\infty} |a_k| \cdot ([n]_q |B_{n,q}(e_k)(w) - w^k| + S_k^{(q)} |w^{k-1} - w^k|).$$

From the proof of Theorem 2.3, Case 1 in [78], for $k \leq n$ we have

$$|B_{n,q}(e_k)(w) - w^k| \leq \frac{k[k-1]_q}{[n]_q} \cdot r^k,$$

while for $k > n$ and using (3.5), we get

$$|B_{n,q}(e_k)(w) - w^k| \leq |B_{n,q}(e_k)(w)| + |w^k|$$

$$\leq \sum_{p=0}^{n} D_{n,p,k}^{(q)} |w^p| + |w^k|$$

$$\leq r^n + r^k \leq 2r^k \leq 2\frac{k[k-1]_q}{[n]_q} \cdot r^k,$$

for all $w \in \overline{B(0;r)}$.

Furthermore, since $S_k^{(q)} \leq (k-1)[k-1]_q$, it is immediate that

$$S_k^{(q)} \cdot |w^{k-1} - w^k| \leq S_k^{(q)} \cdot [|w^{k-1}| + |w^k|] \leq 2(k-1)[k-1]_q r^k.$$

Therefore

$$\sum_{k=n_0+1}^{\infty} |a_k| \cdot \left([n]_q |B_{n,q}(e_k)(w) - w^k| + S_k^{(q)} |w^{k-1} - w^k| \right)$$

$$\leq 2 \sum_{k=n_0+1}^{\infty} |a_k| \cdot (k-1)[k-1]_q r^k,$$

for all $w \in \overline{B(0;r)}$.

Thus, for all $w \in \overline{B(0;r)}$ and $n > n_0$ we have

$$|[n]_q(B_{n,q}(f)(w) - f(w)) - A_q(f)(w)|$$

$$\leq \sum_{k=2}^{n_0} |a_k| \cdot \frac{4(k-1)^2[k-1]_q^2}{[n]_q} \cdot r^k + 2 \sum_{k=n_0+1}^{\infty} |a_k| \cdot (k-1)[k-1]_q r^k$$

$$\leq \frac{4}{[n]_q^t} \cdot \sum_{k=2}^{n_0} |a_k| \cdot k^2[k-1]_q^{1+t} \cdot r^k + 2 \sum_{k=n_0+1}^{\infty} |a_k| \cdot k^2 q^k r^k$$

$$\leq \frac{4}{[n]_q^t} \cdot \sum_{k=2}^{\infty} |a_k| \cdot k^4 q^{(1+t)k} \cdot r^k + 2\varepsilon.$$

Now, since $4/[n]_q^t \to 0$ as $n \to \infty$ and $\sum_{k=2}^{\infty} |a_k| \cdot k^4 q^{(1+t)k} \cdot r^k < \infty$, for the given $\varepsilon > 0$, there exists an index n_1, such that

$$\frac{4}{[n]_q^t} \cdot \sum_{k=2}^{\infty} |a_k| \cdot k^4 q^{(1+t)k} \cdot r^k < \varepsilon$$

for all $n > n_1$.

To conclude, for all $n > \max\{n_0, n_1\}$ and $w \in \overline{B(0;r)}$, we get

$$|[n]_q(B_{n,q}(f)(w) - f(w)) - A_q(f)(w)| \le 3\varepsilon,$$

which shows that

$$\lim_{n \to \infty} [n]_q(B_{n,q}(f)(w) - f(w)) = A_q(f)(w), \text{ uniformly in } \overline{B(0;r)}.$$

The theorem is proved. $\square$

As a consequence of Theorem 3.1.12 we get the exact order of approximation, in the following result.

Corollary 3.1.13. *Under the hypotheses of Theorem 3.1.12, suppose that $R > q \ge 1$. If $1 < r < R/q$ and f is not of the form $f(w) = a_0 + a_1 w$, then*

$$\|B_{n,q}(f) - f\|_r \sim \frac{1}{[n]_q}, \quad n \in \mathbb{N},$$

holds, where $\|f\|_r = \sup\{|f(w)|; \ |w| \le r\}$ and the constants in the equivalence depend on f, r and q but are independent of n.

Proof. Suppose that f is such that the approximation order by $B_{n,q}(f)$ is less than $1/[n]_q$, that is, suppose that

$$\|B_{n,q}(f) - f\|_r \le M \frac{s_n}{[n]_q},$$

for all $n \in \mathbb{N}$, where $s_n \to 0$ as $n \to \infty$. This would imply that

$$\lim_{n \to \infty} [n]_q \|B_{n,q}(f) - f\|_r = 0.$$

Then, by Theorem 3.1.12, (iii), we would have $A_q(f) = 0$ for all $w \in \overline{B(0;r)}$, where $A_q(f)(w)$ is defined in the statement of Theorem 3.1.12, (iii). But $A_q(f)(w) = 0$ for all $w \in \overline{B(0;r)}$ and simple calculations show that

$$2a_2 S_2^{(q)} w + \sum_{k=2}^{\infty} [S_{k+1}^{(q)} a_{k+1} - S_k^{(q)} a_k] w^k = 0, \quad w \in \overline{B(0;r)}.$$

By the uniqueness of power series (which follows from Theorem 2.7 in [101]), since by (3.2) it is clear that $S_k^{(q)} > 0$ for all $k \geq 2$, we would get that $a_2 = 0$ and

$$S_{k+1}^{(q)} a_{k+1} - S_k^{(q)} a_k = 0, \quad \text{for all } k = 2, 3, \ldots.$$

For $k = 2$ we easily get $a_3 = 0$ and, iterating for $k = 3, 4, \ldots$, we obtain that $a_k = 0$ for all $k \geq 2$. Thus $f(w) = a_0 + a_1 w$ for all $w \in \overline{B(0;r)}$, which contradicts our assumption.

So, if f is not a left-polynomial of degree ≤ 1, then Theorem 3.1.4 and Theorem 3.1.7 imply that the approximation order is exactly $1/[n]_q$, which proves the corollary. $\qquad\square$

Remark 3.1.14. The obvious inequalities

$$\frac{q-1}{q^n} \leq \frac{1}{[n]_q} \leq \frac{q}{q^n},$$

for all $n \in \mathbb{N}$ and $q > 1$, show that $B_{n,q}(f)$ approximates f in the closed ball $\overline{B(0;r)}$, with the exact order $1/q^n$.

Remark 3.1.15. In the complex variable case, Corollary 3.1.13 becomes Corollary 1 in [194].

Remark 3.1.16. If f is right W-analytic, then replacing $A_q(f)(w)$ by

$$A_q^*(f)(w) = \sum_{k=2}^{\infty} S_k^{(q)} [w^{k-1} - w^k] a_k$$

and $B_{n,q}(f)(w)$ by $B_{n,q}^*(f)(w)$, and reasoning as above, one can prove that the estimates and conclusions in Theorem 3.1.12 and Corollary 3.1.13 remain valid. In Corollary 3.1.13, the concept of left-polynomial must be replaced by the one of right-polynomial.

3.2 Approximation in Compact Balls and in Cassini Cells

Except for the results where we mention the authors, the sources of this section are the papers of Gal and Sabadini [85] and [93].

As we have already discussed, on balls centered at real points the concept of right W-analytic function coincides with the one of left slice regular function, and the concept of left W-analytic function coincides with the one of right slice regular function. However, also in this section (as in the previous one) the terminology of W-analytic function is more suitable, even though in our proofs we will use some general results on slice regular functions.

In Chapter 3 of the book [77], the author introduced complex convolution operators acting on analytic functions in compact disks centered at the origin and studied their approximation properties.

The operators considered are of the form (note that i is the imaginary unit of the complex numbers, and so $i^2 = -1$)

$$T_n(f)(w) = \alpha_n \int_a^b f(we^{iu})K_n(u)du, \tag{3.10}$$

where usually $a = -\pi$, $b = \pi$ (or equivalently $a = 0$, $b = 2\pi$) and $K_n(u)$ is a positive, even, trigonometric kernel (i.e., a trigonometric polynomial), or $a = -\infty$, $b = +\infty$ and $K_n(u)$ is a positive, continuous kernel. Here $\alpha_n > 0$ is a constant that may depend on n, but it is independent of f and is chosen so that $T_n(e_0)(w) = 1$, for all w, where $e_0(w) = 1$, for all w.

The purpose of this section is to carry out a similar study for the approximation of W-analytic functions by convolution operators of a quaternion variable.

To introduce the convolution operators of a quaternion variable that will be used in the sequel, we need a suitable exponential function of quaternion variable. For any $I \in \mathbb{S}$, we adopt the following well-known definition for the exponential:

$$e^{It} = \cos(t) + I\sin(t), \qquad t \in \mathbb{R},$$

see [117]. The Euler formula holds:

$$(\cos(t) + I\sin(t))^k = \cos(kt) + I\sin(kt),$$

and therefore we can write $(e^{It})^k = e^{Ikt}$.

For any $q \in \mathbb{H} \setminus \mathbb{R}$, let $r := |q|$; then there exists a unique $a \in (0, \pi)$ such that $\cos(a) := x_1/r$ and a unique $I_q \in \mathbb{S}$, such that $q = re^{I_q a}$, with

$$I_q = iy + jv + ks, \quad y = \frac{x_2}{r\sin(a)}, \quad v = \frac{x_3}{r\sin(a)}, \quad s = \frac{x_4}{r\sin(a)}.$$

If $q \in \mathbb{R}$, then we set $a = 0$, if $q > 0$ and $a = \pi$ if $q < 0$, and as I_q we can fix any $I \in \mathbb{S}$. Thus, if $q \in \mathbb{R} \setminus \{0\}$, then we can write $q = |q|(\cos(a) + I\sin(a))$, but with a non-unique choice of I. The above is called the trigonometric form of the quaternion number $q \neq 0$. For $q = 0$ we do not have a trigonometric form for q (exactly as in the complex case). As in the case of a complex variable, we can introduce an analog of (3.10) as follows:

Definition 3.2.1. Let $a = -\pi$, $b = \pi$ or $a = 0$, $b = 2\pi$ and let $K_n(u)$ be a positive, even, trigonometric kernel, or let $a = -\infty$, $b = \infty$ and $K_n(u)$ be a positive, continuous kernel.

If $f : \mathbb{B}_R \to \mathbb{H}$ is (right) W-analytic on $B(0; R)$, then we can define the right *convolution operator* of quaternion variable as

$$T_{n,r}(f)(q) = \alpha_n \int_a^b f(qe^{I_q u})K_n(u)du, \quad \text{if } q \in \mathbb{H} \setminus \mathbb{R}, \quad q = re^{I_q t} \in B(0; R),$$

$$T_{n,r}(f)(q) = \alpha_n \int_a^b f(qe^{Iu})K_n(u)du, \text{ if } q \in \mathbb{R} \setminus \{0\}, \quad q = |q|e^{It} \in B(0;R),$$

$$t = 0 \text{ or } \pi,$$

(3.11)

where $I \in \mathbb{S}$ is arbitrary but fixed, and

$$T_{n,r}(f)(0) = \alpha_n f(0) \int_a^b K_n(u)du,$$

where $\alpha_n > 0$ is a constant independent of f such that $T_n(e_0)(w) = 1$, for all w where $e_0(w) = 1$. If $f : B(0;R) \to \mathbb{H}$ is left W-analytic on $B(0;R)$, then we can define in an analogous way the left convolution operator of quaternion variable by taking $f(e^{I_q u}q)$ instead of $f(qe^{I_q u})$ in the integrals (3.11).

The integral in (3.11), for example, is understood in the Riemann sense and is of the form

$$T_{n,r}(f)(q) = \int_a^b P_n du + i \int_a^b Q_n du + j \int_a^b R_n du + k \int_a^b S_n du,$$

where $P_n, Q_n R_n, S_n$ are real-valued functions of the variables x_1, x_2, x_3, x_4, u, at least continuous, and

$$\alpha_n f(qe^{I_q u})K_n(u) := P_n(x_1, x_2, x_3, x_4, u) + iQ_n(x_1, x_2, x_3, x_4, u)$$
$$+ jR_n(x_1, x_2, x_3, x_4, u) + kS_n(x_1, x_2, x_3, x_4, u).$$

We have the following result (an analogous statement holds when f is left W-analytic):

Theorem 3.2.2. *Let $K_n(u)$ be a positive and even trigonometric kernel, $a = -\pi$, $b = +\pi$ and $f : B(0;R) \to \mathbb{H}$. If f is right W-analytic, that is, $f(q) = \sum_{k=0}^{\infty} q^k c_k$, then $T_{n,r}(f)(q)$ is a right W-analytic function given by the formula*

$$T_{n,r}(f)(q) = \sum_{k=0}^{\infty} q^k c_k A_{k,n}, \quad q \in B(0;R),$$

where $A_{k,n} = \alpha_n \int_{-\pi}^{\pi} \cos(ku)K_n(u)du \in \mathbb{R}$, $k = 0, 1, \ldots$.

Proof. Assume that q is not real. Since the kernel $K_n(u)$ is real valued, using the trigonometric form $q = r(\cos(\alpha) + I_q \sin(\alpha))$ we easily get

$$\alpha_n f(qe^{I_q u})K_n(u) = \alpha_n \sum_{k=0}^{\infty} [r(\cos(\alpha + u) + I_q \sin(\alpha + u)]^k c_k K_n(u)$$

$$= \alpha_n \sum_{k=0}^{\infty} r^k \cos(k(\alpha + u))K_n(u)c_k + I_q \alpha_n \sum_{k=0}^{\infty} r^k \sin(k(\alpha + u))K_n(u)c_k.$$

Since the last two series are uniformly and absolutely convergent with respect to the real variable u, they can be integrated term by term, so that it easily follows that

$$T_{n,r}(f)(q) = \sum_{k=0}^{\infty} r^k \left[\alpha_n \int_{-\pi}^{\pi} \cos(k(\alpha+u))K_n(u)du \right] c_k$$

$$+ I_q \sum_{k=0}^{\infty} r^k \left[\alpha_n \int_{-\pi}^{\pi} \sin(k(\alpha+u))K_n(u)du \right] c_k.$$

Since $\int_{-\pi}^{\pi} \sin(ku)K_n(u)du = 0$, we get

$$T_{n,r}(f)(q) = \sum_{k=0}^{\infty} r^k \cos(k\alpha)A_{k,n}c_k + I_q \sum_{k=0}^{\infty} r^k \sin(k\alpha)A_{n,k}c_k$$

$$= \sum_{k=0}^{\infty} q^k c_k A_{k,n},$$

which proves the formula for a non-real q.

If $q = 0$, then $f(0) = c_0$ and

$$T_{n,r}(f)(0) = \alpha_n c_0 A_{0,n} = \alpha_n c_0 \int_a^b K_n(u)du.$$

We now assume that $q \in \mathbb{R} \setminus \{0\}$. Then the proof follows as in the complex case. Thus the representation for $T_{n,r}(f)(q)$ is valid for all $q \in B(0; R)$.   □

Remark 3.2.3. Similar formulas hold for convolution operators of quaternion variable, in the case where $a = -\infty$, $b = +\infty$ and $K_n(u)$ is a continuous, positive, even and bounded kernel on $(-\infty, +\infty)$.

3.2.1   Approximation by quaternionic polynomial convolutions

It is well known that the classical de la Vallée Poussin kernel

$$K_n(u) = \frac{(n!)^2}{(2n)!} \left(2\cos\frac{u}{2} \right)^{2n} = 1 + 2\sum_{j=1}^{n} \frac{(n!)^2}{(n-j)!(n+j)!} \cos(ju), \quad u \in \mathbb{R},$$

and the de la Vallée Poussin convolution trigonometric polynomials of real variable associated to a 2π-periodic real function g, defined by

$$V_n(g)(x) = \frac{1}{2\pi} \int_{-\pi}^{\pi} g(x-u)K_n(u)du, \quad x \in \mathbb{R}, \ n \in \mathbb{N},$$

were introduced in [189] in order to provide a constructive solution to the second approximation theorem of Weierstrass. In [189] it is proved that

$\lim_{n\to\infty} V_n(f)(x) = f(x)$, uniformly on $\mathbb{R}$. A quantitative upper estimate of $|V_n(f)(x) - f(x)|$ in terms of the second order modulus of smoothness $\omega_2(f; 1/\sqrt{n})$ was obtained in [22].

Let f be function analytic in the disk D_R centered at the origin and of radius $R > 0$. Replacing in the integral expressing $V_n(f)$ the translation $x - u \in \mathbb{R}$ by the rotation $ze^{-iu} = ze^{iu} \in \mathbb{C}$, one gets the complex convolution polynomials defined by

$$V_n(f)(z) = \frac{1}{2\pi} \int_{-\pi}^{\pi} f(ze^{it}) K_n(t) dt, \quad z \in D_R, \ n \in \mathbb{N}.$$

These polynomials were introduced and studied in [160], including the proof of the fact that all $V_n(f)(z)$, $n \in \mathbb{N}$, preserve the spirallikeness and convexity of f in the unit disk. These nice shape preservation properties do not hold for the partial sums of the Taylor expansion of f. We discuss this fact in the following Remark 3.2.11. On the other hand, it was also natural to study the approximation properties of the complex de la Vallée Poussin polynomials, see [77], pp. 182-187 for the details.

At this point, we should mention that the approximation by the partial sums of the Taylor expansion provides a better upper estimate (of geometrical order) than the approximation by the de la Vallée Poussin complex polynomials. However, for the latter, the exact order of approximation and a Voronovskaya-type result can be obtained.

The first goal of this subsection is to extend the approximation properties of the de la Vallée Poussin polynomials of a complex variable, to the case of a quaternion variable. Thus, taking in Definition 3.2.1 $\alpha_n = \frac{1}{2\pi}$ and for $K_n(u)$ the above kernel and using Theorem 3.2.2, we see that the de la Vallée Poussin convolution operator of a quaternion variable acts on a function $f(q) = \sum_{k=0}^{\infty} q^k c_k$ which is W-analytic by the rule

$$P_{n,r}(f)(q) = \sum_{k=0}^{n} q^k c_k \frac{(n!)^2}{(n-k)!(n+k)!}, \quad q \in B(0; R). \tag{3.12}$$

We note that the subscript "r" indicates that we are considering polynomials with coefficients written on the right. Note also that in Theorem 3.2.2, we used the formula

$$A_{k,n} = \begin{cases} \dfrac{(n!)^2}{(n-k)!(n+k)!}, & \text{for } 0 \le k \le n, \\ 0, & \text{for } 0 \le k > n, \end{cases}$$

which, in the complex case, can be found in [77], p. 182.

The first result establishes upper estimates in the approximation of f by $P_{n,r}(f)$ with explicit constants.

Theorem 3.2.4. *Let $R > 1$, and let $f : B(0; R) \to \mathbb{H}$ be right W-analytic in $B(0; R)$, that is, $f(q) = \sum_{k=0}^{\infty} q^k c_k$, for all $q \in B(0; R)$. Let $\| \cdot \|_d$ as in (3.1). Then:*

(i) *Denoting $M_d(f) = \sum_{k=1}^{\infty} |c_k| k^2 d^k < \infty$, for any $d \in [1, R)$, we have*

$$\|P_{n,r}(f) - f\|_d \leq \frac{M_d(f)}{n}, \quad n \in \mathbb{N}.$$

(ii) *If $1 \leq d < r_1 < R$ and $p \in \mathbb{N}$, then*

$$\|\partial_s^p P_{n,r}(f) - \partial_s^p f\|_d \leq \frac{r_1 p! M_{r_1}(f)}{(r_1 - d)^{p+1} n}, \quad n \in \mathbb{N},$$

where ∂_s^p denotes the slice derivative of order p.

Proof. (i) Let $e_k(q) = q^k$. We have $e_k(q)c_k = \sum_{j=0}^{\infty} q^j c_k a_j$ with $a_k = 1$ and $a_j = 0$ for all $j \neq k$. Hence, by (3.12), it is immediate that

$$P_{n,r}(e_k c_k)(q) = e_k(q)c_k \frac{(n!)^2}{(n-k)!(n+k)!}$$

$$= P_{n,r}(e_k)(q)c_k, \quad \text{for all } 0 \leq k \leq n,$$

and $P_{n,r}(e_k) = 0$ for $k > n$. This implies $P_{n,r}(f)(q) = \sum_{k=0}^{\infty} P_{n,r}(e_k)(q)c_k$ and for all $\|q\| \leq d$ we get

$$|P_{n,r}(f)(q) - f(q)| \leq \sum_{k=1}^{\infty} |c_k| |P_{n,r}(e_k)(q) - e_k(q)|$$

$$\leq \sum_{k=1}^{n} |c_k| |P_{n,r}(e_k)(q) - e_k(q)| + \sum_{k=n+1}^{\infty} |c_k| \frac{k^2}{n} d^k.$$

Since

$$P_{n,r}(e_k)(q) = e_k(q) \frac{(n!)^2}{(n-k)!(n+k)!}$$

$$= e_k(q) \prod_{j=1}^{k} \left(1 - \frac{k}{n+j}\right), \quad \text{if } k \leq n,$$

for all $0 \leq k \leq n$ and $|q| \leq d$ we get

$$|P_{n,r}(e_k)(q) - e_k(q)| \leq \left|1 - \prod_{j=1}^{k} \left(1 - \frac{k}{n+j}\right)\right| d^k$$

$$\leq k d^k \sum_{j=1}^{k} \frac{1}{n+j} \leq \frac{k^2 d^k}{n}.$$

Note that we used the inequality

$$1 - \prod_{i=1}^{k} x_i \leq \sum_{i=1}^{k} (1 - x_i),$$

for $0 \leq x_i \leq 1$, $i = 1, \ldots, k$.

In conclusion, for all $|q| \leq d$ we have

$$|P_{n,r}(e_k)(q) - e_k(q)| \leq \frac{k^2}{n} d^k, \quad \text{for all } k, n \in \mathbb{N},$$

which implies the estimate in (i).

(ii) Let q be such that $|q| \leq d$. We can use the Cauchy formula (see Theorem 1.3.14) and integrate on a specific complex plane $\mathbb{C}_I$, for example we can choose $I = I_q$. Let γ be the circle of radius $r_1 > d$ and center 0 in the plane $\mathbb{C}_{I_q}$. For any $v \in \gamma$, we have $|v - q| \geq r_1 - d$, and by the Cauchy's formula it follows that for all $|q| \leq d$ and $n \in \mathbb{N}$, we have

$$|\partial_s^p P_{n,r}(f)(q) - \partial_s^p f(q)|$$

$$= \frac{p!}{2\pi} \left| \int_\gamma [S^{-1}(v,q)(q - \overline{v})^{-1}]^{p+1} (q - \overline{v})^{(p+1)*} dv_{I_q}(P_{n,r}(f)(v) - f(v)) \right|.$$

Since v and q commute, we get

$$[S^{-1}(v,q)(q - \overline{v})^{-1}]^{p+1} (q - \overline{v})^{(p+1)*} = (v - q)^{-(p+1)},$$

whence

$$|\partial_s^p P_{n,r}(f)(q) - \partial_s^p f(q)| = \frac{p!}{2\pi} \left| \int_\gamma (v - q)^{-(p+1)} dv_{I_q}(P_{n,r}(f)(v) - f(v)) \right|$$

$$\leq \frac{M_{r_1}(f)}{n} \frac{p!}{2\pi} \frac{2\pi r_1}{(r_1 - d)^{p+1}},$$

which proves (ii) and the theorem. $\qquad \square$

We now present a Voronovskaya-type theorem for the quaternionic case.

Theorem 3.2.5. *Let $R > 1$, and let $f : B(0; R) \to \mathbb{H}$ be right W-analytic in $B(0; R)$, that is, we can write $f(q) = \sum_{k=0}^\infty q^k c_k$, for all $q \in B(0; R)$. Then for any $d \in [1, R)$ we have*

$$\left\| P_{n,r}(f) - f + \frac{e_2 \partial_s^2 f}{n} + \frac{e_1 \partial_s f}{n} \right\|_d \leq \frac{A_d(f)}{n^2}, \quad n \in \mathbb{N}$$

where $A_d(f) = \sum_{k=1}^\infty |c_k| k^4 d^k < \infty$, $e_k(q) = q^k$.

Proof. Since the left slice derivatives of a power series representing a W analytic function coincide with its derivatives, we can write

$$\partial_s f(q) = \sum_{k=1}^\infty q^{k-1} c_k k \quad \text{and} \quad \partial_s^2 f(q) = \sum_{k=2}^\infty q^{k-2} c_k k(k-1).$$

Therefore, setting

$$|E_{k,n}(q)| = \left| P_{n,r}(e_k)(q) - e_k(q) + \frac{q^k k(k-1)}{n} + \frac{q^k k}{n} \right|,$$

for all $|q| \le d$ we obtain

$$\left| P_{n,r}(f)(q) - f(q) + \frac{e_2(q)\partial_s^2 f(q)}{n} + \frac{e_1(q)\partial_s f(q)}{n} \right|$$

$$\le \sum_{k=0}^{\infty} |E_{k,n}(q)|\,|c_k| = \sum_{k=1}^{n} |E_{k,n}(q)|\,|c_k| + \sum_{k=n+1}^{\infty} |E_{k,n}(q)|\,|c_k|$$

$$= \sum_{k=1}^{n} |E_{k,n}(q)|\,|c_k| + \sum_{k=n+1}^{\infty} \left| -q^k + \frac{q^k k(k-1)}{n} + \frac{q^k k}{n} \right| |c_k|.$$

But for $|q| \le d$ we have

$$\sum_{k=n+1}^{\infty} |c_k| \left| -q^k + \frac{q^k k(k-1)}{n} + \frac{q^k k}{n} \right|$$

$$= \sum_{k=n+1}^{\infty} |c_k| d^k \left| -1 + \frac{k(k-1)}{n} + \frac{k}{n} \right|$$

$$\le \sum_{k=n+1}^{\infty} |c_k| d^k \frac{k^2}{n} \le \sum_{k=n+1}^{\infty} |c_k| \frac{k}{n} d^k \frac{k^2}{n}$$

$$= \frac{1}{n^2} \sum_{k=n+1}^{\infty} |c_k| k^3 d^k \le \frac{1}{n^2} \sum_{k=n+1}^{\infty} |c_k| k^4 d^k. \tag{3.13}$$

Therefore, it remains to estimate $|E_{k,n}(q)|$ for $|q| \le d$ and $0 \le k \le n$. Since it is immediate that $E_{0,n}(q) = 0$, it suffices to consider $1 \le k \le n$. We obtain

$$|E_{k,n}(q)| = \left| q^k \frac{(n!)^2}{(n-k)!(n+k)!} - q^k + \frac{q^k k(k-1)}{n} + \frac{q^k k}{n} \right|$$

$$= |q|^k \left| \frac{(n!)^2}{(n-k)!(n+k)!} - 1 + \frac{k^2}{n} \right|. \tag{3.14}$$

By induction, one can prove (see (3.1), p. 184 in [77]) that

$$0 \le \frac{(n!)^2}{(n-k)!(n+k)!} - 1 + \frac{k^2}{n} \le \frac{k^4}{n^2}, \text{ for all } k = 1, 2, \ldots, n \text{ and } n \in \mathbb{N}. \tag{3.15}$$

Using this inequality in (3.14), and then (3.13), we obtain the statement.   □

Now we are in position to obtain the exact degree of approximation by $P_{n,r}(f)(q)$. To start with, we present a lower estimate for the approximation error obtained in Theorem 3.2.4, (i).

Theorem 3.2.6. *Let $R > 1$ and let $f : B(0; R) \to \mathbb{H}$ be right W-analytic, that is, $f(q) = \sum_{k=0}^{\infty} q^k c_k$, for all $q \in B(0; R)$. If f is not a constant function, then for any $d \in [1, R)$ we have*

$$\|P_{n,r}(f) - f\|_d \geq \frac{C_d(f)}{n}, \quad n \in \mathbb{N},$$

where the constant $0 < C_d(f) < \infty$ depends only on f and d.

Proof. For all $q \in B(0; R)$ and $n \in \mathbb{N}$,

$$P_{n,r}(f)(q) - f(q) = -\frac{1}{n}[q^2 \partial_s^2 f(q) + q \partial_s f(q)]$$

$$+ \frac{1}{n^2} \left[n^2 \left(P_{n,r}(f)(q) - f(q) + \frac{q^2 \partial_s^2 f(q)}{n} + \frac{q \partial_s f(q)}{n} \right) \right].$$

Since

$$\|F + G\|_d \geq |\, \|F\|_d - \|G\|_d \,| \geq \|F\|_d - \|G\|_d,$$

it follows that

$$\|P_{n,r}(f) - f\|_d$$

$$\geq \frac{1}{n} \left\{ \|e_2 \partial_s^2 f + e_1 \partial_s f\|_d - \frac{1}{n} \left[n^2 \|P_{n,r}(f) - f + \frac{e_2 \partial_s^2 f}{n} + \frac{e_1 \partial_s f}{n} \|_d \right] \right\}.$$

Since, by hypothesis, f is not a constant function in $B(0; R)$, we get

$$\|e_2 \partial_s^2 f + e_1 \partial_s f\|_d > 0.$$

In fact, if we suppose the contrary, it follows that $q^2 \partial_s^2 f(q) + q \partial_s f(q) = 0$ for all $q \in B(0; r)$. Using the uniqueness of the power series expansion of f and identifying the like coefficients in the above equation, it easily follows that $c_k = 0$ for all $k \geq 1$. This implies that $f(q) = c_0$, for all $q \in \overline{B(0; d)}$, which contradicts the hypothesis. Theorem 3.2.5 implies that

$$n^2 \|P_{n,r}(f) - f + \frac{e_2}{n} \partial_s^2 f + \frac{e_1}{n} \partial_s f\|_d \leq A_d(f).$$

Therefore, there exists an index n_0, depending only on f and d, such that for all $n \geq n_0$,

$$\|e_2 \partial_s^2 f + e_1 \partial_s f\|_d - \frac{1}{n} \left[n^2 \|P_{n,r}(f) - f + \frac{e_2}{n} \partial_s^2 f + \frac{e_1}{n} \partial_s f\|_d \right]$$

$$\geq \frac{1}{2} \|e_2 \partial_s^2 f + e_1 \partial_s f\|_d,$$

whence

$$\|P_{n,r}(f) - f\|_d \geq \frac{1}{2n} \|e_2 \partial_s^2 f + e_1 \partial_s f\|_d, \quad \text{for all } n \geq n_0.$$

For $n \in \{1, \ldots, n_0 - 1\}$ we have

$$\|P_{n,r}(f) - f\|_d \geq \frac{M_{d,n}(f)}{n},$$

with

$$M_{d,n}(f) = n\|P_{n,r}(f) - f\|_d > 0.$$

This yields

$$\|P_{n,r}(f) - f\|_d \geq \frac{C_d(f)}{n} \quad \text{for all } n \in \mathbb{N},$$

where

$$C_d(f) = \min\{M_{d,1}(f), \ldots, M_{d,n_0-1}(f), \frac{1}{2}\|e_2\partial_s^2 f + e_1\partial_s f\|_d\}.$$

This completes the proof.   □

Definition 3.2.7. We write $a_n \sim b_n$, $n \in \mathbb{N}$, if there exist two constants $c_1, c_2 > 0$ independent of n, such that

$$c_1 b_n \leq a_n \leq c_2 b_n$$

for all $n \in \mathbb{N}$.

It is easy to verify that $\sim$ is an equivalence relation.
We immediately get the following result:

Theorem 3.2.8. Let $R > 1$, and let $f : B(0; R) \to \mathbb{H}$ be right W-analytic in $B(0; R)$. If f is not a constant function in $B(0; R)$, then for any $d \in [1, R)$ we have

$$\|P_{n,r}(f) - f\|_d \sim \frac{1}{n}, \quad n \in \mathbb{N},$$

where the constants in the equivalence $\sim$ depend only on f and d.

Proof. This follows from Theorem 3.2.6 and Theorem 3.2.4, (i).   □

In the case of approximation by the slice derivatives of $P_{n,r}(f)(q)$ we have

Theorem 3.2.9. Let $R > 1$ and let $f : B(0; R) \to \mathbb{H}$ be right W-analytic in $B(0; R)$, i.e., $f(q) = \sum_{k=0}^{\infty} q^k c_k$, for all $q \in B(0; R)$. Also, let $1 \leq d < r_1 < R$ and $p \in \mathbb{N}$ be fixed. If f is not a polynomial of degree $\leq p - 1$, then

$$\|\partial_s^p P_{n,r}(f) - \partial_s^p f\|_d \sim \frac{1}{n},$$

where the constants in the equivalence $\sim$ depend on f, d, r_1 and p.

Proof. We reason as in the proof of Theorem 3.2.4. Let $q \neq 0$ be such that $|q| \leq d$ and consider the complex plane $\mathbb{C}_{I_q^*}$. Let γ be the circle of radius $r_1 > d \geq 1$ and center 0 in $\mathbb{C}_{I_q^*}$. For any $v \in \gamma$, we have $|v - q| \geq r_1 - d$. The Cauchy formula implies that for all $|q| \leq d$ and $n \in \mathbb{N}$, we have

$$\partial_s^p P_{n,r}(f)(q) - \partial_s^p f(q)$$
$$= \frac{p!}{2\pi} \int_\gamma [S^{-1}(v,q)(q-\bar{v})^{-1}]^{p+1}(q-\bar{v})^{(p+1)*} dv_{I_q^*}(P_{n,r}(f)(v) - f(v)),$$

where $I_q^* := I_q$ for q not real, and $I_q^* = I$ arbitrary in $\mathbb{S}$ for $q \in \mathbb{R} \setminus \{0\}$.

Taking into account Theorem 3.2.4, (ii), we only have to prove the lower estimate for $\|\partial_s^p P_n(f) - \partial_s^p f\|_d$.

To this end, we follow the proof of Theorem 3.2.6, and we note that for all $v \in \gamma$ and $n \in \mathbb{N}$,

$$P_{n,r}(f)(v) - f(v) = -\frac{1}{n}[v^2 \partial_s^2 f(v) + v \partial_s f(v)]$$
$$+ \frac{1}{n^2}\left[n^2\left(P_{n,r}(f)(v) - f(v) + \frac{v^2}{n}\partial_s^2 f(v) + \frac{v\partial_s f(v)}{n}\right)\right],$$

which can be substituted in the above Cauchy formula. Since v and q belong to the same complex plane, we have

$$\partial_s^p P_{n,r}(f)(q) - \partial_s^p f(q) = \frac{p!}{2n\pi}\int_\gamma -(v-q)^{p+1} dv_{I_q^*}[v^2 \partial_s^2 f(v) + v \partial_s^2 f(v)]$$
$$+ \frac{1}{n^2}\frac{p!}{2\pi}\int_\gamma (v-q)^{p+1} dv_{I_q^*}\left[n^2\left(P_{n,r}(f)(v) - f(v) + \frac{v^2}{n}\partial_s^2 f(v) + \frac{v}{n}\partial_s f(v)\right)\right]$$
$$= \frac{1}{n}\partial_s^p\left[-q^2\partial_s^2 f(q) - q\partial_s f(q)\right]$$
$$+ \frac{1}{n^2}\frac{p!}{2\pi}\int_\gamma (v-q)^{p+1} dv_{I_q^*}\left[n^2\left(P_{n,r}(f)(v) - f(v) + \frac{v^2}{n}\partial_s^2 f(v) + \frac{v}{n}\partial_s f(v)\right)\right].$$

Taking the norm $\|\cdot\|_d$, we get

$$\|\partial_s^p P_{n,r}(f) - \partial_s^p f\|_d \geq \frac{1}{n}\|\partial_s^p [e_2 \partial_s^2 f + e_1 \partial_s f]\|_d$$
$$- \frac{1}{n^2}\left\|\frac{p!}{2\pi}\int_\gamma (v-q)^{p+1} dv_{I_q^*}\left[n^2\left(P_n(f)(v) - f(v) + \frac{v^2}{n}\partial_s^2 f(v) + \frac{v}{n}\partial_s f\right)\right]\right\|_d$$

and using Theorem 3.2.2 we finally obtain

$$\left\|\frac{p!}{2\pi}\int_\gamma (v-q)^{p+1} dv_{I_q^*}\left[n^2\left(P_{n,r}(f)(v) - f(v) + \frac{v^2}{n}\partial_s^2 f(v) + \frac{v}{n}\partial_s^2 f\right)\right]\right\|_d$$
$$\leq \frac{p!}{2\pi}\frac{2\pi r_1 n^2}{(r_1 - d)^{p+1}}\left\|P_{n,r}(f) - f + \frac{e_2}{n}\partial_s^2 f + \frac{e_1}{n}\partial_s^2 f\right\|_{r_1} \leq \frac{A_{r_1}(f)p!r_1}{(r_1 - d)^{p+1}}.$$

The assumption on f implies

$$\| \partial_s^p \left[e_2 \partial_s^2 f + e_1 \partial_s f \right] \|_d > 0.$$

In fact, if we assume the contrary, then we have

$$q^2 \partial_s^2 f(q) + q \partial_s f(q) = Q_{p-1}(q), \qquad (3.16)$$

for all $q \in \overline{B(0;r)}$, where $Q_{p-1}(q) = \sum_{j=1}^{p-1} A_j q^j$ is a polynomial of degree $\leq p-1$, vanishing at $q = 0$. Let us set $\partial_s f(q) = g(q)$. Then (3.16) becomes

$$q^2 \partial_s g(q) + q g(q) = Q_{p-1}(q),$$

for all $q \in \overline{B(0;d)}$. Let us now look for a W-analytic solution of the form $g(q) = \sum_{j=0}^{\infty} q^j \alpha_j$: by replacing in the differential equation, and by the identification of coefficients we easily conclude that $g(q)$ necessarily is a polynomial of degree $\leq p-2$. Thus, $f(q)$ is a polynomial of degree $\leq p-1$, contradicting the hypothesis. Finally, reasoning exactly as in the proof of Theorem 3.2.6, we get the desired conclusion. □

Remark 3.2.10. Similar results could be easily adapted for the left convolution operator of the de la Vallée Poussin type, $P_{n,l}(f)(q)$ associated to left W-analytic functions. Moreover, one can obtain approximation results for other choices of the trigonometric kernel $K_n(u)$, like for example for the Fejér, Riesz–Zygmund, Jackson and Beatson kernels (see, e.g., Chapter 3 of the book [77], where the corresponding complex convolutions were studied).

Remark 3.2.11. It is not difficult to prove, and we will not do the explicit computations, that for quaternionic W-analytic functions with real coefficients (namely intrinsic functions), the de la Vallée Poussin polynomials of quaternion variable given by (3.12) preserve some geometric properties, as it happens in the complex case. More precisely, the de la Vallée Poussin quaternion polynomials given by (3.12) preserve the starlikeness and the convexity of $f \in \mathcal{N}$, where for $f: B(0;1) \to \mathbb{H}$ normalized by $f(0) = 0$ and $\partial_s(f)(0) \neq 0$, the starlikeness (convexity) is understood in the sense that for all $0 < r \leq 1$, $f(B(0;r))$ are starlike (convex, respectively) sets in $\mathbb{R}^4$. For more information on these geometric properties of slice regular functions we refer the reader to [82], [83], [196]. We recall that $A \subset \mathbb{R}^4$ is called starlike with respect to the origin 0, if for any point $p \in A$, the Euclidean segment determined by 0 and p entirely lies in A. Moreover, $A \subset \mathbb{R}^4$ is called convex, if for all $p, q \in A$, the Euclidean segment joining p and q entirely lies in A. Thus, we can easily find many examples of starlike (or convex) quaternionic intrinsic functions f by simply replacing $z \in \mathbb{C}$ with $q \in \mathbb{H}$ in the Taylor expansion (which has real coefficients) of a starlike respectively, convex function of a complex variable. Moreover, we can easily construct polynomials of quaternion variable with nice geometric properties, if these polynomials are interpreted as transformations from $\mathbb{R}^4$ to $\mathbb{R}^4$.

We next turn to approximation results by convolution polynomials in two cases: for compact balls centered at real numbers and for the so-called Cassini cells. We will obtain quantitative estimates in terms of moduli of smoothness.

In the next two results, instead of W-analytic functions we will use the terminology of slice regular functions (which better suits the Mergelyan theorem).

Keeping the notation $P_{n,r}$ for the de la Vallée Poussin polynomials studied above, we prove the following result.

Theorem 3.2.12 (Quantitative approximation on compact balls). *Let $R > 0$, $K = \overline{B(0;R)}$ and $f : K \to \mathbb{H}$ be continuous on K and (left) slice regular on the interior of K, $int(K) = B(0;R)$. Then for any $\varepsilon > 0$ there exists a polynomial P such that $|f(q) - P(q)| < \varepsilon$ for all $q \in K$.*

In fact, for all $q \in \overline{B(0;R)}$ and $n \in \mathbb{N}$ we have

$$|P_{n,r}(q) - f(q)| \le 3(R+1)\,\omega_1(f; 1/\sqrt{n}),$$

where $\omega_1(f;\delta) = \sup\{|f(u) - f(v)|;\ u,v \in \overline{B(0;R)},\ |u - v| \le \delta\}$.

Proof. Let $q \in \overline{B(0;R)}$. Here we follow the complex case in [77], p. 427, and get

$$|P_{n,r}(f)(q) - f(q)| \le \frac{1}{2\pi} \int_{-\pi}^{\pi} |f(qe^{I_q u}) - f(q)| K_n(u) du$$

$$\le \frac{1}{2\pi} \int_{-\pi}^{\pi} \omega_1(f; |q| \cdot |e^{I_q u} - 1|) K_n(u) du$$

$$\le \frac{1}{2\pi} \int_{-\pi}^{\pi} \omega_1(f; R|u|) K_n(u) du$$

$$\le (R+1)\frac{1}{2\pi} \int_{-\pi}^{\pi} \omega_1\left(f; \frac{1}{\sqrt{n}}|u|\sqrt{n}\right) K_n(u) du$$

$$\le (R+1)\omega_1(f; 1/\sqrt{n})\frac{1}{2\pi} \int_{-\pi}^{\pi} (|u|\sqrt{n} + 1) K_n(u) dt$$

$$\le 3(R+1)\omega_1(f; 1/\sqrt{n}),$$

where

$$\omega_1(f;\delta) = \sup\{|f(u) - f(v)|;\ u,v \in \overline{B(0;R)},\ |u - v| \le \delta\}.$$

Let $\varepsilon > 0$ be arbitrary. By the continuity of f on $\overline{B(0;R)}$ we have $\lim_{n\to\infty} \omega_1(f; 1/\sqrt{n}) = 0$, so that there exists n_0 such that for all $n \ge n_0$ we have

$$|P_{n,r}(f)(q) - f(q)| \le \varepsilon, \text{ for all } q \in \overline{B(0;R)}.$$

The formula expressing $P_{n,r}$, for $q \in B(0;R)$ shows that $P_{n,r}(f)(q)$ is a (right slice) polynomial, and

$$|P_{n,r}(q) - f(q)| \le \varepsilon, \text{ for all } q \in B(0;R). \tag{3.17}$$

Now, let $q_0 \in \overline{B(0; R)}$. There exists a sequence $q_m \in B(0; R)$, $m \in \mathbb{N}$, such that $|q_m - q_0| \to 0$ as $m \to \infty$. Replacing in (3.17) q by q_m, letting $m \to \infty$ and using the continuity of $P_{n,r}$ and f, we deduce that $|P_{n,r}(q_0) - f(q_0)| \leq \varepsilon$, which proves the statement. $\qquad\square$

Remark 3.2.13. Theorem 3.2.12 can be easily extended to functions defined on balls $\overline{B(x_0; R)}$ centered at a real point x_0, by taking $Q_n(q) = P_n(q - x_0)$, $q \in \overline{B(x_0; R)}$, with $P_n(q)$ as in the proof of Theorem 3.2.12.

If we consider balls centered at points in $\mathbb{H} \setminus \mathbb{R}$, then we cannot immediately extend Theorem 3.2.12. However, using the new series expansion (1.5), originally given in [180], the results extend to compact sets more general than closed balls centered at real points.

Recall that

$$B(x_0 + y_0\mathbb{S}; R) = \{q \in \mathbb{H};\ |(q - x_0)^2 + y_0^2| < R^2\},$$

where $q_0 = x_0 + Iy_0 \in \mathbb{H}$, with $x_0, y_0 \in \mathbb{R}$, $y_0 > 0$, $I \in \mathbb{S}$ and $R > 0$ and formula (1.5) that allows to write the slice regular function $f : B(x_0 + y_0\mathbb{S}; R) \to \mathbb{H}$ as

$$f(q) = \sum_{k=1}^{\infty} [(q - x_0)^2 + y_0^2]^k [c_{2k} + qc_{2k+1}], \quad \text{for all } q \in B(x_0 + y_0\mathbb{S}; R),$$

where $c_{2k}, c_{2k+1} \in \mathbb{H}$, for all $k \in \mathbb{N}$.

For functions f of the above form, we define the convolution

$$V_{n,l}(f)(q) = \frac{1}{2\pi} \int_{-\pi}^{\pi} f(qe^{I_q u}) K_n(u) du,$$

for $q \in \mathbb{H} \setminus \mathbb{R}$, $q = re^{I_q t} \in \overline{B(x_0 + y_0\mathbb{S}; R)}$, and

$$V_{n,l}(f)(q) = \frac{1}{2\pi} \int_{-\pi}^{\pi} f(qe^{Iu}) K_n(u) du,$$

for $q \in \mathbb{R} \setminus \{0\}, q = |q|e^{It} \in \overline{B(x_0 + y_0\mathbb{S}; R)}, t = 0$ or π, where I is an arbitrary element in $\mathbb{S}$, and

$$V_{n,l}(f)(0) = \frac{1}{2\pi} f(0) \int_{-\pi}^{\pi} K_n(u) du = f(0).$$

Here the subscript "l" refers to "left". We have:

Theorem 3.2.14 (Quantitative approximation on Cassini cells)**.** *Let* $q_0 = x_0 + Iy_0 \in \mathbb{H}$, *with* $x_0, y_0 \in \mathbb{R}$, $y_0 > 0$, $I \in \mathbb{S}$ *and* $R > 0$. *If* $f : \overline{B(x_0 + y_0\mathbb{S}; R)} \to \mathbb{H}$ *is continuous in* $\overline{B(x_0 + y_0\mathbb{S}; R)}$ *and (left) slice regular in* $B(x_0 + y_0\mathbb{S}; R)$, *then for* any $\varepsilon > 0$ *there exists a polynomial* P *such that* $|f(q) - P(q)| < \varepsilon$ *for all* $q \in \overline{B(x_0 + y_0\mathbb{S}; R)}$.

In fact, for all $q \in \overline{B(x_0 + y_0 \mathbb{S}; R)}$ and $n \in \mathbb{N}$ we have

$$|V_{n,l}(f)(q) - f(q)| \leq 3(R+1)\omega_1(f; 1/\sqrt{n}),$$

where $V_{n,l}(f)(q)$ is a polynomial.

Proof. Reasoning as in the proof of Theorem 3.2.12, for any $q \in \overline{B(x_0 + y_0 \mathbb{S}; R)}$ we obtain

$$
\begin{aligned}
|V_{n,l}(f)(q) - f(q)| &\leq \frac{1}{2\pi} \int_{-\pi}^{\pi} |f(qe^{I_q u}) - f(q)|K_n(u)du \\
&\leq \frac{1}{2\pi} \int_{-\pi}^{\pi} \omega_1(f; |q| \cdot |e^{I_q u} - 1|)K_n(u)du \\
&\leq \frac{1}{2\pi} \int_{-\pi}^{\pi} \omega_1(f; (M_{R,q_0})|u|)K_n(u)du \\
&\leq (M_{R,q_0} + 1)\frac{1}{2\pi} \int_{-\pi}^{\pi} \omega_1\left(f; \frac{1}{\sqrt{n}}|u|\sqrt{n}\right)K_n(u)du \\
&\leq 3(M_{R,q_0} + 1)\omega_1(f; 1/\sqrt{n}).
\end{aligned}
$$

Here we have used that if $q \in \overline{B(x_0 + y_0 \mathbb{S}; R)}$, then

$$
\begin{aligned}
|q|^2 = |q^2| &= |(q - x_0 + x_0)^2 + y_0^2 - y_0^2| \\
&= |(q - x_0)^2 + y_0^2 + 2x_0(q - x_0) + x_0^2 - y_0^2| \\
&\leq |(q - x_0)^2 + y_0^2| + 2|x_0| \cdot |q - x_0| + |x_0^2 + y_0^2| \\
&\leq R^2 + 2|x_0| \cdot \sqrt{R^2 + y_0^2} + |q_0|^2 := M_{R,q_0}^2,
\end{aligned}
$$

whence $|q| \leq M_{R,q_0}$.

We claim that $V_{n,l}(f)(q)$ is a polynomial.

Indeed, by (1.5) we can write

$$
\begin{aligned}
f(q) &= \sum_{k=1}^{\infty} \left(\sum_{j=0}^{k} \binom{k}{j}(q - x_0)^{2j} \cdot y_0^{2(k-j)} \right) c_{2k} \\
&+ \sum_{k=1}^{\infty} \left(\sum_{j=0}^{k} \binom{k}{j}(q - x_0)^{2j} \cdot q \cdot y_0^{2(k-j)} \right) c_{2k+1} \\
&= \sum_{k=1}^{\infty} \left(\sum_{j=0}^{k} \binom{k}{j} y_0^{2(k-j)} \sum_{p=0}^{2j} \binom{2j}{p} q^p (-1)^{2j-p} x_0^{2j-p} \right) c_{2k} \\
&+ \sum_{k=1}^{\infty} \left(\sum_{j=0}^{k} \binom{k}{j} y_0^{2(k-j)} \sum_{p=0}^{2j} \binom{2j}{p} q^p (-1)^{2j-p} x_0^{2j-p} \right) qc_{2k+1},
\end{aligned}
$$

which implies

$$V_{n,l}(f)(q) = \frac{1}{2\pi} \int_{-\pi}^{\pi} f(qe^{I_q u}) K_n(u) du$$

$$= \sum_{k=1}^{\infty} \left(\sum_{j=0}^{k} \binom{k}{j} y_0^{2(k-j)} \sum_{p=0}^{2j} \binom{2j}{p} q^p \right.$$

$$\times \left[\frac{1}{2\pi} \int_{-\pi}^{\pi} e^{I_q pu} K_n(u) du \right] (-1)^{2j-p} x_0^{2j-p} \right) c_{2k}$$

$$+ \sum_{k=1}^{\infty} \left(\sum_{j=0}^{k} \binom{k}{j} y_0^{2(k-j)} \sum_{p=0}^{2j} \binom{2j}{p} q^p \right.$$

$$\times \left[\frac{1}{2\pi} \int_{-\pi}^{\pi} e^{I_q pu} K_n(u) du \right] (-1)^{2j-p} x_0^{2j-p} \right) q c_{2k+1}.$$

But

$$\int_{-\pi}^{\pi} e^{I_q pu} K_n(u) du = \int_{-\pi}^{\pi} e^{I_q pu} \left(1 + \sum_{s=1}^{n} \frac{(n!)^2}{(n-s)!(n+s)!} (e^{I_q su} + e^{-I_q su}) \right) du$$

$$= \int_{-\pi}^{\pi} e^{I_q pu} du + \sum_{s=1}^{n} \frac{(n!)^2}{(n-s)!(n+s)!} \left(\int_{-\pi}^{\pi} e^{I_q(p+s)u} du + \int_{-\pi}^{\pi} e^{I_q(p-s)u} du \right).$$

Now, taking into account that

$$\int_{-\pi}^{\pi} e^{I_q pu} du = 0, \quad \text{if } p \neq 0, \qquad \int_{-\pi}^{\pi} e^{I_q pu} du = 2\pi, \quad \text{if } p = 0,$$

and

$$\int_{-\pi}^{\pi} e^{I_q(p+s)u} du = 0, \quad \text{for all } s \geq 1, \ p \geq 0,$$

$$\int_{-\pi}^{\pi} e^{I_q(p-s)u} du \neq 0, \quad \text{only for } p = s,$$

since $s \in \{1, \ldots, n\}$, it immediately follows that the infinite sum which expresses $V_{n,l}(f)(q)$ reduces to a finite sum containing powers of q less than n and with all the coefficients written on the right. In conclusion, $V_{n,l}(f)(q)$ is a polynomial, which ends the proof of the theorem. $\square$

Remark 3.2.15.   1. If in the definitions of $P_{n,r}(f)(q)$ and $V_{n,l}(f)(q)$ and in the statements of Theorem 3.2.12 and Theorem 3.2.14, instead of the de la Vallée Poussin kernel $\frac{1}{2\pi} K_n(u)$, we use the Jackson kernel

$$J_n(u) = \frac{1}{\pi} \cdot \frac{3}{2n(2n^2 + 1)} \left(\frac{\sin(nu/2)}{\sin(u/2)} \right)^4.$$

Then, reasoning as in the complex case, (see, e.g., [75], pp. 422–423), we get a better quantitative estimate in terms of $\omega_2(f; 1/n)$.

2. If, instead of the polynomial operators $P_{n,r}(f)(q)$ and $V_{n,l}(f)(q)$ in Theorem 3.2.12 and Theorem 3.2.14, we consider for any fixed $p \in \mathbb{N}$ the polynomial operators of the form

$$L_{n,l,p}(f)(q) = -\int_{-\pi}^{\pi} K_{n,r}(u) \sum_{k=1}^{p+1} \binom{p+1}{k} f(qe^{I_q ku}) du,$$

where r is the smallest integer for which $r \geq (p+2)/2$ and

$$K_{n,s}(u) = \frac{1}{\lambda_{n,s}} \left(\frac{\sin(nu/2)}{\sin(u/2)} \right)^{2s}$$

is the normalized generalized Jackson kernel. Reasoning as in the complex variable case in [75], p. 424, better quantitative estimates in terms of $\omega_{p+1}(f; 1/n)$ are obtained.

3. In the most general case, we can consider the Fejér kernel

$$F_n(u) = \frac{1}{2} \left(\frac{\sin(nu/2)}{\sin(u/2)} \right)^2,$$

set

$$L_n(f)(q) = \frac{1}{n\pi} \int_{-\pi}^{\pi} F_n(u) f(qe^{I_q u}) du,$$

and define the new polynomial operators

$$P_n(f)(q) = 2L_{2n}(f)(q) - L_n(f)(q).$$

Following the complex case, see [75], pp. 424–425, we obtain the results in Theorem 3.2.12 and Theorem 3.2.14 with quantitative estimates in terms of $E_n(f)$, where

$$E_n(f) = \inf\{\|p - f\|_R; \ p \text{ polynomial of degree } \leq n\}$$

is the best approximation of f by polynomials of degree $\leq n$.

3.2.2 Approximation by nonpolynomial quaternion convolutions

In this subsection we deal in detail with the approximation properties of the convolution based on the classical Gauss–Weierstrass kernel

$$K_t(u) = e^{-u^2/(2t)}, \quad u \in \mathbb{R}.$$

Note that t is a real, strictly positive parameter which replaces the natural parameter $n \in \mathbb{N}$ in the definition of the trigonometric kernels.

In the formulas (3.11) for the convolution operator in Definition 3.2.1 one may use $a = -\infty$, $b = +\infty$, and replace α_n by $\alpha_t = 1/\sqrt{2\pi t}$, and $K_n(u)$ by

$K_t(u) = e^{-u^2/(2t)}$, and then acting on W-analytic functions f on $B(0; R)$ we have the right Gauss–Weierstrass convolution operator of a quaternionic variable

$$W_{t,r}(f)(q) = \frac{1}{\sqrt{2\pi t}} \int_{-\infty}^{+\infty} f(qe^{I_q u}) e^{-u^2/(2t)} du, \quad q \in \mathbb{H} \setminus \mathbb{R}, \ q = re^{I_q a} \in B(0; R),$$

$$W_{t,r}(f)(q) = \frac{1}{\sqrt{2\pi t}} \int_{-\infty}^{+\infty} f(qe^{Iu}) e^{-u^2/(2t)} du, \quad q \in \mathbb{R} \setminus \{0\},$$

$$q = re^{Ia} \in B(0; R), \ a = 0 \text{ or } \pi,$$

$$\tag{3.18}$$

$$W_{t,r}(f)(0) = f(0),$$

where $I \in \mathbb{S}$ is fixed (but arbitrary).

Set $f(q) = \sum_{k=0}^{\infty} q^k c_k$. Reasoning as in the proof of Theorem 3.2.2, and taking into account that

$$A_{k,t} = \frac{1}{\sqrt{2\pi t}} \int_{-\infty}^{+\infty} e^{-u^2/(2t)} \cos(ku) du = e^{-k^2 t/2},$$

we have

$$W_{t,r}(f)(q) = \sum_{k=0}^{\infty} q^k c_k A_{k,t} = \sum_{k=0}^{\infty} q^k c_k e^{-k^2 t/2}, \quad q \in B(0; R), \ t > 0. \tag{3.19}$$

Theorem 3.2.16. *Let* $f : B(0; R) \to \mathbb{H}$ *be* W-*analytic in* $B(0; R)$, *for* $R > 1$, *so that* $f(q) = \sum_{k=0}^{\infty} q^k c_k$, *for all* $q \in B(0; R)$. *Then:*

(i) *For any* $d \in [1, R)$

$$\|W_{t,r}(f) - f\|_d \le \frac{t}{2} M_d(f), \quad t > 0,$$

where

$$M_d(f) = \sum_{k=1}^{\infty} |c_k| k^2 d^k < \infty.$$

(ii) *If* $1 \le d < R$ *and* $p \in \mathbb{N}$, *then for all* $I \in \mathbb{S}$

$$\|\partial_s^p W_{t,r}(f) - \partial_s^p f\|_d \le \frac{t}{2} M_{d,p}(f), \quad t > 0,$$

where

$$M_{d,p}(f) = \sum_{k=p}^{\infty} d^{k-p} |c_k| k^3 (k-1) \cdots (k-p+1) < \infty.$$

Proof. (i) Since $|\cdot|$ is multiplicative, (3.19) shows that for all $|q| \leq d$

$$|W_{t,r}(f)(q) - f(q)| \leq \sum_{k=0}^{\infty} |q^k| \, |c_k| \left| e^{-k^2 t/2} - 1 \right| \leq \sum_{k=0}^{\infty} d^k \, |c_k| \left| e^{-k^2 t/2} - 1 \right|.$$

If we set $h(t) = e^{-k^2 t/2}$ and note that $h(0) = 1$, then the mean value theorem shows that there exists a point $\xi \in (0, t)$ such that

$$\left| e^{-k^2 t/2} - 1 \right| = |h'(\xi)| t = \frac{t}{2} k^2 e^{-k^2 \xi/2} \leq \frac{t}{2} k^2,$$

which in conjunction with the above inequality, yield

$$|W_{t,r}(f)(q) - f(q)| \leq \frac{t}{2} \sum_{k=0}^{\infty} r^k |c_k| k^2.$$

Taking the supremum for $|q| \leq d$, we get the desired estimate.

(ii) Using the formula for the slice derivative of the quaternionic power series, and reasoning as before we have

$$|\partial_s^p W_{t,r}(f)(q) - \partial_s^p f(q)| = |\sum_{k=p}^{\infty} q^{k-p} c_k k(k-1) \cdots (k-p+1)(e^{-k^2 t/2} - 1)|$$

$$\leq \frac{t}{2} \sum_{k=p}^{\infty} d^{k-p} k^3 (k-1) \cdots (k-p+1),$$

which proves the second part of the theorem. $\qquad\square$

We also have the following exact estimate:

Theorem 3.2.17. *Let $f : B(0; R) \to \mathbb{H}$ be W-analytic in $B(0; R)$, for $R > 1$, so that $f(q) = \sum_{k=0}^{\infty} q^k c_k$, for all $q \in B(0; R)$. Consider $\partial_s^j W_{t,r}(f) - \partial_s^j f$. Then, if f is not a constant function for $j = 0$ and not a polynomial of degree $\leq j - 1$ for $j \in \mathbb{N}$, then for all $I \in \mathbb{S}$ we have*

$$\|\partial_s^j W_{t,r}(f) - \partial_s^j f\|_d \sim t,$$

where the constants in the equivalence $\sim$ depend only on f, d and j. Here $a(t) \sim b(t)$ means that there exists two absolute constants $C_1 > 0$ and $C_2 > 0$ such that $0 \leq C_1 a(t) \leq b(t) \leq C_2 a(t)$, for all $t > 0$.

Proof. Having the upper estimate in Theorem 3.2.16, (i), it remains to derive a lower estimate for $\|\partial_s^j W_{t,r}(f) - \partial_s^j f\|_d$.

To this end, we write a nonzero quaternion q in the trigonometric form $q := de^{I_q \varphi}$ and take $p \in \mathbb{N} \bigcup \{0\}$ and $j \in \mathbb{N} \bigcup \{0\}$. We get

$$\frac{1}{2\pi} e^{-I_q p \varphi} [\partial_s^j f(q) - \partial_s^j W_{t,r}(f)(q)]$$

$$= \frac{1}{2\pi} \sum_{k=j}^{\infty} d^{k-j} e^{I_q \varphi (k-j-p)} c_k k(k-1) \cdots (k-j+1)[1 - e^{-k^2 t/2}].$$

Integrating from $-\pi$ to π, and after some computations, we obtain

$$\frac{1}{2\pi}\int_{-\pi}^{\pi}e^{-I_q p\varphi}[\partial_s^j f(q)-\partial_s^j W_{t,r}(f)(q)]d\varphi$$

$$=\frac{1}{2\pi}\sum_{k=j}^{\infty}d^{k-j}\int_{-\pi}^{\pi}e^{I_q\varphi(k-j-p)}d\varphi c_k k(k-1)\cdots(k-j+1)[1-e^{-k^2 t/2}]$$

$$=d^p c_{j+p}(j+p)(j+p-1)\cdots(p+1)[1-e^{-(j+p)^2 t/2}].$$

Since $|e^{I_q\varphi}|=1$, by taking the modulus $|\cdot|$, we easily obtain

$$|a_{j+p}|(j+p)(j+p-1)\cdots(p+1)d^p[1-e^{-(j+p)^2 t/2}]\le \|\partial_s^j f-\partial_s^j(f)\|_d.$$

Let us first consider the case $j=0$ and set

$$V_t=\inf_{1\le p}(1-e^{-p^2 t/2})=1-e^{-t/2}.$$

By the mean value theorem applied to $h(x)=e^{-x/2}$ on $[0,t]$, there exists $\eta\in(0,t)$ such that for all $t\in(0,1]$

$$V_t=h(0)-h(t)=(-t)h'(\eta)=(t/2)e^{-\eta/2}$$

$$\ge (t/2)e^{-t/2}\ge \frac{e^{-1/2}}{2}t\ge t/4.$$

From the above lower estimate for $\|W_{t,r}(f)-f\|_d$, for all $p\ge 1$ and $t\in(0,1]$ it follows that

$$\frac{4\|W_{t,r}(f)-f\|_d}{t}\ge \frac{\|W_{t,r}(f)-f\|_d}{V_t}\ge \frac{\|W_{t,r}(f)-f\|_d}{1-e^{-p^2 t/2}}\ge |c_p|d^p.$$

But this implies that if there exists a subsequence $(t_k)_k$ in $(0,1]$ with $\lim_{k\to\infty}t_k=0$ and such that

$$\lim_{k\to\infty}\frac{\|W_{t_k,r}(f)-f\|_d}{t_k}=0,$$

then $c_p=0$ for all $p\ge 1$, that is, f is constant on $\overline{\mathbb{B}_d}$.

Thus, if f is not a constant then

$$\inf_{t\in(0,1]}\frac{\|W_{t,r}(f)-f\|_d}{t}>0,$$

which implies that there exists a constant $C_d(f)>0$ such that

$$\frac{\|W_{t,r}(f)-f\|_d}{t}\ge C_d(f),$$

for all $t \in (0, 1]$, that is,

$$\|W_{t,r}(f) - f\|_d \geq C_d(f)t, \text{ for all } t \in (0, 1].$$

Now take $j \geq 1$ and set

$$V_{j,t} = \inf_{p \geq 0} (1 - e^{-(p+j)^2 t/2}).$$

It is immediate that

$$V_{j,t} \geq \inf_{p \geq 1} (1 - e^{-p^2 t/2}) \geq t/4.$$

Reasoning as in the case $j = 0$, we obtain

$$\frac{4\|\partial_s^j W_{t,r}(f) - \partial_s^j f\|_d}{t} \geq \frac{\|\partial_s^j W_{t,r}(f) - \partial_s^j f\|_d}{V_{j,t}} \geq |c_{j+p}| \frac{(j+p)!}{p!} d^p,$$

for all $p \geq 0$ and $t \in (0, 1]$. Thus, if there exists a subsequence $(t_k)_k$ in $(0, 1]$ with $\lim_{k \to \infty} t_k = 0$ and such that

$$\lim_{k \to \infty} \frac{\|\partial_s^j W_{t_k,r}(f) - \partial_s^j f\|_d}{t_k} = 0,$$

then $c_{j+p} = 0$ for all $p \geq 0$, that is, f is a polynomial of degree $\leq j-1$ on $\overline{B(0; d)}$. Therefore,

$$\inf_{t \in (0,1]} \frac{\|\partial_s^j W_{t,r}(f) - \partial_s^j f\|_d}{t} > 0$$

when f is not a polynomial of degree $\leq j - 1$. This implies that there exists a constant $C_{d,j}(f) > 0$ such that

$$\frac{\|\partial_s^j W_{t,r}(f) - \partial_s^j f\|_d}{t} \geq C_{d,j}(f),$$

for all $t \in (0, 1]$, that is,

$$\|\partial_s^j W_{t,r}(f) - \partial_s^j f\|_d \geq C_{d,j}(f)t, \text{ for all } t \in (0, 1],$$

which concludes the proof. $\qquad \square$

Definition 3.2.18. Denote by $\mathcal{A}_r(B(0; R))$ the right quaternionic Banach space of the W-analytic functions on $B(0; R)$, that are continuous on $\overline{B(0; R)}$.

The space $\mathcal{A}_r(B(0; R))$, which coincides with the set of (left) slice regular functions on $B(0; R)$, is endowed with the uniform norm

$$\|f\|_R = \max\{|f(u)|; \ u \in \overline{B(0; R)}\}.$$

We now prove that the right Gauss–Weierstrass convolution of quaternion variable defines a contraction semigroup on the right quaternionic Banach space $\mathcal{A}_r(B(0; R))$ with respect to the uniform norm.

Theorem 3.2.19. *Let $f \in \mathcal{A}_r(B(0;R))$, $f(q) = \sum_{k=0}^{\infty} q^k c_k$, $q \in B(0;R)$. Then:*

(i) *For all $t > 0$, $W_{t,r}(f) \in \mathcal{A}_r(B(0;R))$ and*

$$W_{t,r}(f)(q) = \sum_{k=0}^{\infty} q^k c_k e^{-k^2 t/2}, \quad \text{for all } q \in B(0;R).$$

(ii) *For all $q \in \overline{B(0;R)}$, $t > 0$, the following estimate holds:*

$$|W_{t,r}(f)(q) - f(q)| \leq C_R \omega_1(f; \sqrt{t})_{\overline{B(0;R)}},$$

where

$$\omega_1(f;\delta)_{\overline{B(0;R)}} = \sup\{|f(u) - f(v)|; \ |u - v| \leq \delta, \ u, v \in \overline{B(0;R)}\}.$$

and $C_R > 0$ is a constant independent of t and f.

(iii) *For all $q \in \overline{B(0;R)}$, $t \in V_s \subset (0, +\infty)$ we have*

$$|W_{t,r}(f)(q) - W_{s,r}(f)(q)| \leq C_s|\sqrt{t} - \sqrt{s}|,$$

where $C_s > 0$ is a constant depending on f, and independent of q and t, and V_s is any neighborhood of s.

(iv) *The operator $W_{t,r} : \mathcal{A}_r(B(0;R)) \to \mathcal{A}_r(B(0;R))$ is contractive, that is*

$$\|W_{t,r}(f)\|_R \leq \|f\|_R, \quad \text{for all } t > 0, \ f \in \mathcal{A}_r(B(0;R)).$$

(v) *$(W_{t,r}, t \geq 0)$ is a C_0-contraction semigroup of linear operators on the real Banach space $(\mathcal{A}_r(B(0;R)), \|\cdot\|_R)$ and the unique solution $u(t,q)$ of the Cauchy problem for a heat-type equation in t and φ,*

$$\frac{\partial u}{\partial t}(t, q) = \frac{1}{2}\frac{\partial^2 u}{\partial \varphi^2}(t, q), \quad (t, q) \in (0, +\infty) \times B(0;R), \ q = he^{I_q\varphi} \neq 0,$$

$$u(0, q) = f(q), \quad q \in \overline{B(0;R)}, \ f \in \mathcal{A}_r(B(0;R)),$$

is given by $u(t, q) = W_{t,r}(f)(q)$. This unique solution belongs to $\mathcal{A}_r(B(0;R))$ for each fixed $t > 0$.

Proof. (i) The fact that $W_{t,r}(f)$ is W-analytic in $B(0;R)$ follows from (3.19). We have to prove the continuity in $\overline{B(0;R)}$. To this end, we take q_0 and a sequence $q_n \in \overline{B(0;R)}$ such that $\lim_{n \to \infty} q_n = q_0$, i.e., $\lim_{n \to \infty} |q_n - q_0| = 0$.

Let q_0 be a non-real quaternion. Then, without loss of generality, we may assume that q_n are non-real quaternions, for all $n \in \mathbb{N}$. In this case, if we set $q_n = r_n e^{I_{q_n} a_n}$ and $q_0 = r_0 e^{I_{q_0} a_0}$, then the definition of trigonometric form readily shows that $a_n \to a_0$ and $|I_{q_n} - I_{q_0}| \to 0$ as $n \to \infty$.

Since

$$|q_n e^{I_{q_n} u} - q_0 e^{I_{q_0} u}| = |(q_n - q_0)\cos(u) + (q_n I_{q_n} - q_0 I_{q_0})\sin(u)|$$
$$\leq |q_n - q_0| + |q_n I_{q_n} - q_0 I_{q_0}|$$
$$= |q_n - q_0| + |q_n I_{q_n} - q_n I_{q_0} + q_n I_{q_0} - q_0 I_{q_0}|$$
$$\leq |q_n - q_0| + |q_n||I_{q_n} - I_{q_0}| + |q_n - q_0||I_{q_0}|$$
$$= 2|q_n - q_0| + |q_n||I_{q_n} - I_{q_0}|,$$

we easily get

$$|W_{t,r}(f)(q_n) - W_{t,r}(f)(q_0)\|$$
$$\leq \frac{1}{\sqrt{2\pi t}} \int_{-\infty}^{+\infty} |f(q_n e^{I_{q_n} u}) - f(q_0 e^{I_{q_0} u})| e^{-u^2/(2t)}\, du$$
$$\leq \frac{1}{\sqrt{2\pi t}} \int_{-\infty}^{+\infty} \omega_1(f; |q_n e^{I_{q_n} u} - q_0 e^{I_{q_0} u}|)_{\overline{B(0;R)}} e^{-u^2/(2t)}\, du$$
$$\leq \frac{1}{\sqrt{2\pi t}} \int_{-\infty}^{+\infty} \omega_1(f; 2|q_n - q_0| + |q_n||I_{q_n} - I_{q_0}|)_{\overline{B(0;R)}} e^{-u^2/(2\pi t)}\, du$$
$$= \omega_1(f; 2|q_n - q_0| + |q_n||I_{q_n} - I_{q_0}|)_{\overline{B(0;R)}}.$$

We now note that $\omega_1(f;\delta)_{\overline{B(0;R)}}$ enjoys all the usual properties of a modulus of continuity for real-valued functions of a real variable, including the property that if f is continuous on its compact domain of definition, then

$$\lim_{\delta \searrow 0} \omega_1(f;\delta)_{\overline{B(0;R)}} = 0.$$

Using this fact, we have that $W_{t,r}(f)(q_n)$ converges to $W_{t,r}(f)(q_0)$, as $n \to \infty$.

We now suppose that $q_0 \in \mathbb{R} \setminus \{0\}$ and that $q_0 > 0$.

If all q_n are real quaternions, then we can assume that $q_n > 0$ for all $n \in \mathbb{N}$ and writing $q_0 = |q_0|(\cos(0) + I\sin(0))$, $q_n = |q_n|(\cos(0) + I\sin(0))$, with arbitrary $I \in \mathbb{S}$, we immediately obtain

$$|q_n e^{Iu} - q_0 e^{Iu}| = |(q_n - q_0)\cos(u)| = |q_n - q_0|,$$

which implies

$$|W_{t,r}(f)(q_n) - W_{t,r}(f)(q_0)|$$
$$\leq \frac{1}{\sqrt{2\pi t}} \int_{-\infty}^{+\infty} |f(q_n e^{Iu}) - f(q_0 e^{Iu})| e^{-u^2/(2t)}\, du$$
$$\leq \frac{1}{\sqrt{2\pi t}} \int_{-\infty}^{+\infty} \omega_1(f; |q_n e^{Iu} - q_0 e^{Iu}|)_{\overline{B(0;R)}} e^{-u^2/(2t)}\, du$$
$$= \frac{1}{\sqrt{2\pi t}} \int_{-\infty}^{+\infty} \omega_1(f; |q_n - q_0|)_{\overline{B(0;R)}} e^{-u^2/(2\pi t)}\, du$$
$$= \omega_1(f; |q_n - q_0|)_{\overline{B(0;R)}}.$$

This again yields that

$$\lim_{n\to\infty} W_{t,r}(f)(q_n) = W_{t,r}(f)(q_0).$$

If all the q_n are non-real quaternions, we write $q_n = r_n e^{I_{q_n} a_n}$ and $q_0 = q_0(\cos(0) + I \sin(0))$, with arbitrary $I \in \mathbb{S}$. Therefore, we can choose $I = I_{q_n}$ and write $q_0 = q_0(\cos(0) + I_{q_n} \sin(0)) = q_0 e^{I_{q_n} 0}$ and also in the definition of $W_{t,r}(f)(q_0)$ we can set $I = I_{q_n}$, which implies

$$|q_n e^{I_{q_n} u} - q_0 e^{I_{q_n} u}| = |q_n - q_0| |e^{I_{q_n} u}| = |q_n - q_0|.$$

Reasoning as above, we obtain

$$|W_{t,r}(f)(q_n) - W_{t,r}(f)(q_0)|$$
$$\leq \frac{1}{\sqrt{2\pi t}} \int_{-\infty}^{+\infty} |f(q_n e^{I_{q_n} u}) - f(q_0 e^{I_{q_n} u})| e^{-u^2/(2t)} \, du \leq \omega_1(f; |q_n - q_0|)_{\overline{B(0;R)}}.$$

Therefore, we obtain also in this case that $\lim_{n\to\infty} W_{t,r}(f)(q_n) = W_{t,r}(f)(q_0)$.

If $q_0 < 0$, we reason exactly as above, with the only difference that we write $q_0 = |q_0|(\cos(\pi) + I \sin(\pi))$, with arbitrary $I \in \mathbb{S}$.

Thus, we conclude that $W_{t,r}(f)$ is continuous at any $q_0 \in \overline{B(0;R)}$, since f is continuous on $\overline{B(0;R)}$.

(ii) If $|q| \leq R$, q is a non real quaternion, we get

$$|W_{t,r}(f)(q) - f(q)| \leq \frac{1}{\sqrt{2\pi t}} \int_{-\infty}^{+\infty} |f(q e^{I_q u}) - f(q)| e^{-u^2/(2t)} \, du$$

$$\leq \frac{1}{\sqrt{2\pi t}} \int_{-\infty}^{\infty} \omega_1(f; R|1 - e^{I_q u}|)_{\overline{B(0;R)}} e^{-u^2/(2t)} \, du$$

$$= \frac{1}{\sqrt{2\pi t}} \int_{-\infty}^{+\infty} \omega_1\left(f; 2R\left|\sin(u/2)\right|\right)_{\overline{B(0;R)}} e^{-u^2/(2t)} \, du$$

$$\leq \frac{2R+1}{\sqrt{2\pi t}} \int_{-\infty}^{+\infty} \omega_1(f; |u|)_{\overline{B(0;R)}} e^{-u^2/(2t)} \, du$$

$$\leq \frac{2R+1}{\sqrt{2\pi t}} \int_{-\infty}^{+\infty} \omega_1(f; \sqrt{t})_{\overline{B(0;R)}} \left(|u|/\sqrt{t} + 1\right) e^{-u^2/(2t)} \, du$$

$$= (2R+1) \left[\omega_1(f; \sqrt{t})_{\overline{B(0;R)}} + \frac{\omega_1(f; \sqrt{t})_{\overline{B(0;R)}}}{\sqrt{t}\sqrt{2\pi t}} \int_0^{\infty} 2u e^{-u^2/(2t)} \, du \right].$$

Since

$$\int_0^{\infty} 2u e^{-u^2/(2t)} \, du = 2t \int_0^{\infty} e^{-v} \, dv = 2t,$$

we infer that

$$|W_{t,r}(f)(q) - f(q)| \leq (2R+1) \left[\omega_1(f; \sqrt{t})_{\overline{B(0;R)}} + \left(\omega_1(f; \sqrt{t})_{\overline{B(0;R)}}\right) \frac{2t}{t\sqrt{2\pi}} \right]$$
$$\leq C_R \omega_1(f; \sqrt{t})_{\overline{B(0;R)}}.$$

Now, for $|q| \leq R$, $q \in \mathbb{R} \setminus \{0\}$, fix an arbitrary $I \in \mathbb{S}$. We have

$$|W_{t,r}(f)(q) - f(q)| \leq \frac{1}{\sqrt{2\pi t}} \int_{-\infty}^{+\infty} |f(qe^{Iu}) - f(q)| e^{-u^2/(2t)} \, du$$

and reasoning exactly as in the case of q non real, we obtain the same upper estimate. Finally, for $q = 0$ we get $|W_{t,r}(f)(q) - f(q)| = 0$, which all together imply the estimate in (ii) for all $q \in B(0; R)$.

(iii) From the definition of $W_{t,r}(f)(q)$ in (3.18), for all $q \in B(0; R)$ we get

$$|W_{t,r}(f)(q) - W_{s,r}(f)(q)| \leq \frac{\|f\|_R}{\sqrt{2\pi}} \int_{-\infty}^{+\infty} \left| \frac{e^{-u^2/t}}{\sqrt{t}} - \frac{e^{-u^2/s}}{\sqrt{s}} \right| du.$$

Set $\sqrt{t} = a$, $\sqrt{s} = b$. Applying the mean value theorem, we deduce that there exists $c \in (a, b)$, such that

$$\left| \frac{e^{-u^2/a^2}}{a} - \frac{e^{-u^2/b^2}}{b} \right| = |a - b| e^{-u^2/c^2} \left[\frac{2u^2}{c^4} - \frac{1}{c^2} \right],$$

and since

$$\int_{-\infty}^{+\infty} e^{-u^2/(2c)} < \infty, \qquad \int_{-\infty}^{+\infty} u^2 e^{-u^2/(2c)} < \infty,$$

we immediately deduce the desired inequality for $W_{t,r}$.

(iv) From

$$\frac{1}{\sqrt{2\pi t}} \int_{-\infty}^{+\infty} e^{-u^2/(2t)} \, du = 1,$$

we deduce

$$|W_{t,r}(f)(q)| \leq \frac{1}{\sqrt{2\pi t}} \int_{-\infty}^{+\infty} |f(qe^{I_q^* u})| e^{-u^2/(2t)} \, du$$

$$\leq \|f\|_R, \quad q \in \overline{B(0; R)},$$

where $I_q^* := I_q$ if q is not real, and $I_q^* := I \in \mathbb{S}$ is arbitrary if $q \in \mathbb{R} \setminus \{0\}$. Together with $W_{t,r}(f)(0) = f(0)$ all these relations easily yield that $\|W_{t,r}(f)\|_R \leq \|f\|_R$.

(v) Let $f \in \mathcal{A}_r(B(0; R))$, so that $f(q) = \sum_{k=0}^{\infty} q^k a_k$, $q \in B(0; R)$. If $q \in B(0; R)$, $q = de^{I_q \varphi}$, $0 < d < R$, then by (i) we can write

$$W_{t,r}(f)(q) = \sum_{k=0}^{\infty} d^k e^{kI_q \varphi} c_k e^{-k^2 t/2}.$$

It easily follows that

$$W_{t+s,r}(f)(q) = W_{s,r}[W_{t,r}(f)](q), \quad \text{for all } t, \ s > 0.$$

If q is on the boundary of $B(0;R)$, then we can take a sequence of points $(q_n)_{n \in \mathbb{N}}$ in $B(0;R)$ such that $\lim_{n \to \infty} q_n = q$ and apply the above relation and the continuity property from (i). Furthermore, denoting $W_{t,r}(f)(q)$ by $T(t)(f)(q)$, it is easy to check that the property $\lim_{t \searrow 0} T(t)(f) = f$, the continuity of $T(\cdot)$ and its contractiveness follow from (ii), (iii) and (iv), respectively. To conclude, we note that all these facts together show that $(W_{t,r}, t \geq 0)$ is a C_0-contraction semigroup of linear operators on $\mathcal{A}_r(B(0;R))$.

The above series representation for $W_{t,r}(f)(q)$ is uniformly and absolutely convergent in any compact ball included in $B(0;R)$, so it can be differentiated term by term, with respect to t and φ. We then easily get

$$\frac{\partial W_{t,r}(f)(q)}{\partial t} = \frac{1}{2} \frac{\partial^2 W_{t,r}(f)(q)}{\partial \varphi^2}.$$

From the same series representation it is immediate that

$$W_{0,r}(f)(q) = f(q), \qquad q \in \overline{B(0;R)}.$$

This completes the proof of the theorem.

We point out that in the differential equation we must take $q \neq 0$ simply because $q = 0$ has no polar representation, and so $q = 0$ cannot be expressed as a function of φ. $\square$

Remark 3.2.20. Similar results can be adapted for the left convolution operator of Gauss–Weierstrass type $W_{n,l}(f)(q)$ associated to left W-analytic functions. Moreover, similar results can be obtained by choosing different kernels, like the Picard kernel

$$K_t(u) = e^{-|u|t},$$

the Poisson–Cauchy kernel

$$K_t(u) = \frac{1}{u^2 + t^2},$$

and many other (see them in, e.g., Chapter 3 of the book [77], where the corresponding convolutions are studied in the complex case).

3.2.3   Approximation by convolution operators of a paravector variable

We now discuss how the results obtained in the preceding sections can be extended to a more general setting.

For the class of slice monogenic functions one can repeat, with suitable modifications, the preliminary results on slice regular functions mentioned in Chapter 1.

To our goal, we only recall the following:

Definition 3.2.21. Let $B(0;R) = \{x \in \mathbb{R}^{n+1} \mid |x| < R\}$. We say that $f : B(0,R) \to \mathbb{R}_n$ is (right) *W-analytic* in $B(0,R)$ if $f(x) = \sum_{k=0}^{\infty} x^k c_k$, where $c_k \in \mathbb{R}_n$, for all k, for all $x \in B(0,R)$.

Such a function f is slice monogenic in $B(0, R)$.

Given a W-analytic function f on $B(0, R)$, we can define the right convolution operator of a paravector variable by mimicking Definition 3.2.1. Then we can consider the de la Vallée Poussin convolution operator of a paravector variable for a W-analytic function f as above.

One can prove the generalization of Theorem 3.2.4 to this setting by noting that it is based on the validity of the Cauchy formula, on inequalities on norms and (1.17). Similarly, we can state and prove also the Voronovskaya-type theorem, see Theorem 3.2.5, in which, as a consequence of (1.17), the inequality we obtain is

$$\left\| P_{m,r}(f) - f + \frac{e_2 \cdot \partial_s^2 f}{m} + \frac{e_1 \partial_s f}{m} \right\|_d \leq C_n \frac{A_d(f)}{m^2}, \quad m \in \mathbb{N},$$

where C_n is a constant depending on the dimension of $\mathbb{R}_n$. Moreover, we can obtain the analogue of the important Theorem 3.2.9. It is crucial to note that its proof is based on lower estimates. Thus, one has to verify that the fact that the norm in $\mathbb{R}_n$ is not multiplicative has no influence and, a priori, this is not a trivial task.

One may also consider the approximation properties of the convolution obtained by using the Gauss–Weierstrass kernel. With the techniques illustrated above, one can show that the analogues of Theorem 3.2.17 and Theorem 3.2.19 also hold in this setting. Note that, once more, the fact that in the Clifford algebra $\mathbb{R}_n$, $n \geq 3$ one has lesser properties than $\mathbb{H}$ does not hinder the extension of the proofs given in the quaternion case. This is not ensured, in general and, in principle, one has to check that the proofs of the various results can be generalized to this setting.

Chapter 4

Approximation of Slice Regular Functions in Compact Sets

In this chapter, we present the counterparts in the slice regular setting of some classical results in complex analysis, namely approximation results of Runge type, Mergelyan type and Arakelian type. Then, we study approximation by quaternionic Faber polynomials and by quaternionic polynomials in Bergman spaces.

4.1 Runge Type Results

The Runge type results studied in this section are the generalization to the quaternionic setting of the classical results in the complex case; they have been obtained by Colombo, Sabadini and Struppa in [52].

In this section we use the definition of slice regular function given in Section 1.2 (originally introduced in [104]). Given a stem function $F : D \subseteq \mathbb{C} \to \mathbb{H}$, $F(z) = \alpha(z) + i\beta(z)$, we define the function F^c by $F^c(z) = \alpha(z)^c + i\beta(z)^c$ where ^{c} denotes quaternionic conjugation. Let $f = \mathcal{I}(F)$ and denote by $N(f)$ the function $\mathcal{I}(FF^c)$. Note that

$$
\begin{aligned}
F(z)F^c(z) &= (\alpha(z) + i\beta(z))(\alpha(z)^c + i\beta(z)^c) \\
&= \alpha(z)\alpha(z)^c - \beta(z)\beta(z)^c + i(\alpha(z)\beta(z)^c + \beta(z)\alpha(z)^c) \\
&= |\alpha(z)|^2 - |\beta(z)|^2 + i(\alpha(z)\beta(z)^c + (\alpha(z)\beta(z)^c)^c)
\end{aligned}
$$

and

$$
\begin{aligned}
F^c(z)F(z) &= (\alpha(z)^c + i\beta(z)^c)(\alpha(z) + i\beta(z)) \\
&= \alpha(z)^c\alpha(z) - \beta(z)^c\beta(z) + i(\alpha(z)^c\beta(z) + \beta(z)^c\alpha(z)) \\
&= |\alpha(z)|^2 - |\beta(z)|^2 + i(\alpha(z)^c\beta(z) + (\alpha(z)^c\beta(z))^c),
\end{aligned}
$$

so, in general, $FF^c \neq F^cF$. However we have the following result:

© Springer Nature Switzerland AG 2019
S. G. Gal, I. Sabadini, *Quaternionic Approximation*, Frontiers in Mathematics,
https://doi.org/10.1007/978-3-030-10666-9_4

Proposition 4.1.1. *The functions FF^c and F^cF are real slice functions.*

Proof. It is an immediate consequence of the fact that $|\alpha(z)|^2 - |\beta(z)|^2$ is trivially real-valued and $(\alpha(z)^c\beta(z) + (\alpha(z)^c\beta(z))^c)^c = (\alpha(z)^c\beta(z))^c + \alpha(z)^c\beta(z)$ thus $\alpha(z)^c\beta(z) + (\alpha(z)^c\beta(z))^c$ is real-valued.   □

By Z_g we will denote the set of zeros of a function g. By virtue of Proposition 4.1.1, the notion of slice regular inverse of a function is meaningful for all the functions for which FF^c is not identically zero:

Proposition 4.1.2. *Let $F : D \subseteq \mathbb{C} \to \mathbb{H}$ be a stem function such that F^cF is not identically zero on a dense subset of D. Let $f = \mathcal{I}(F)$ and $\Omega_D = \{q = u + Iv \mid z = u + iv \in D, \ I \in \mathbb{S}\}$. Then the slice regular inverse of f is the function*

$$f^{-1} := \mathcal{I}((F^cF)^{-1}F^c)$$

defined on $\Omega_D \setminus Z_{N(FF^c)}$.

Proof. See the proof of Proposition 4.1.1.   □

In particular, we now consider a polynomial in the variable q of the form $a(q) = q^n a_n + q^{n-1} a_{n-1} + \cdots + q a_1 + a_0$. It is a slice regular function obtained as $a = \mathcal{I}(A)$ from the polynomial

$$A(z) = z^n a_n + z^{n-1} a_{n-1} + \cdots + z a_1 + a_0$$

$$= \sum_{k=0}^{n}(x + iy)^k a_k$$

$$= \sum_{k=0}^{n}(u_k + iv_k)a_k,$$

where the real quantities u_k, v_k can be obtained by rewriting $(x + iy)^k$, using the Newton binomial, in the form $u_k + iv_k$.

We have

$$A^cA = \left(\sum_{k=0}^{n}(u_k + iv_k)a_k^c\right)\left(\sum_{k=0}^{n}(u_k + iv_k)a_k\right) = \sum_{k=0}^{2n}(u_k + iv_k)d_k$$

where $d_k = \sum_{j=0}^{k} a_j^c a_{k-j}$ is a real number. Moreover, A^cA is not identically zero if A is not identically zero.

Definition 4.1.3. Given two polynomials $a = \mathcal{I}(A)$, $b = \mathcal{I}(B)$ we call (left) rational function a function of the form $a^{-*} * b := \mathcal{I}((A^cA)^{-1}A^cB)$.

By the arguments used earlier, it is easy to prove the following result:

Proposition 4.1.4. *The singularities of a rational function are isolated 2-spheres.*

Proof. Using the notions introduced above, the singularities of the function $a^{-1}b$ come from the set of zeros of the function $A^c A$ which are isolated points. Thus the singularities are isolated $(n-1)$-spheres, in particular, a real point when a sphere has radius equal to zero. □

Definition 4.1.5. We say that a singularity $q = q_0$ of a function f is a pole if it belongs to an isolated 2-sphere of singularities and it is such that $\lim_{q \to q_0} |f(q)| = +\infty$.

Proposition 4.1.4 has the following immediate consequence:

Corollary 4.1.6. *The singularities of a rational function are poles.*

We are now ready to prove the analogue of Runge's theorem:

Theorem 4.1.7. *Let K be an axially symmetric compact set in $\mathbb{H}$, let Σ be a set having a point in each connected component of $\overline{\mathbb{H}} \setminus K$. For any axially symmetric open set $\Omega \supset K$, for every $f \in \mathcal{R}(\Omega)$, and for any $\varepsilon > 0$, there exists a rational function r whose poles are spheres in Σ such that*

$$|f(q) - r(q)| < \varepsilon$$

for all $q \in K$.

Proof. Let us consider the restriction of the function f to a complex plane $\mathbb{C}_I$. The Splitting Lemma shows that for every $I \in \mathbb{S}$, and every $J \in \mathbb{S}$ that is perpendicular to I, there are two holomorphic functions $F, G : U \cap \mathbb{C}_I \to \mathbb{C}_I$ such that for any $z = u + Iv$, the restriction of f to $\mathbb{C}_I$ can be written as

$$f_I(z) = F(z) + G(z)J.$$

By the classical Runge theorem in the complex case, we can find two rational functions $R(u + Iv)$ and $S(u + Iv)$ with poles in $\Sigma \cap \mathbb{C}_I$ such that

$$|F(u+Iv) - R(u+Iv)| < \frac{\varepsilon}{4}, \quad |G(u+Iv) - S(u+Iv)| < \frac{\varepsilon}{4}, \quad \forall u+Iv \in \Omega \cap \mathbb{C}_I. \quad (4.1)$$

Since $\Omega \cap \mathbb{C}_I$ is symmetric with respect to the real axis, the extension formula allows us to extend the function $r(u + Iv) = R(u + Iv) + S(u + Iv)J$ to the whole Ω as

$$r(q) = r(u + I_q v) = \frac{1}{2}\Big[r(u + Iv) + r(u - Iv) + I_q I[r(u - Iv) - r(u + Iv)] \Big]. \quad (4.2)$$

To compute $|f(q) - r(q)|$ we use (4.2) and the Splitting Lemma on $\mathbb{C}_I$:

$$\begin{aligned}
|f(q) - r(q)| &= \left| \frac{1}{2}\Big[(1 - I_q I)f(u + Iv) + (1 + I_q I)f(u - Iv) \right. \\
&\quad \left. - (1 - I_q I)r(u + Iv) - (1 + I_q I)r(u - Iv)\Big]\right| \\
&= \left| \frac{1}{2}\Big[(1 - I_q I)(F(u + Iv) + G(u + Iv)J - R(u + Iv) - S(u + Iv)J) \right. \\
&\quad \left. + (1 + I_q I)(F(u - Iv) + G(u - Iv)J - R(u - Iv) - S(u - Iv)J)\Big]\right| \\
&\leq |F(u + Iv) - R(u + Iv)| + |G(u + Iv) - S(u + Iv)| \\
&\quad + |F(u - Iv) - R(u - Iv)| + |G(u - Iv) - S(u - Iv)| < \epsilon.
\end{aligned}$$

This concludes the proof. $\qquad\qquad\square$

In particular, we have the following result which allows to approximate a slice regular function with polynomials:

Theorem 4.1.8. *Let K be an axially symmetric compact set such that $\overline{\mathbb{H}} \setminus K$ is connected (and $\overline{\mathbb{C}}_I \setminus (K \cap \mathbb{C}_I)$ is connected for all $I \in \mathbb{S}$) and let $f \in \mathcal{R}(\Omega)$ where $\Omega \supset K$ is an open set. There exists a sequence $\{P_n\}$ of polynomials such that $P_n(q) \to f(q)$ uniformly on K.*

Proof. Our assumptions imply that $\overline{\mathbb{H}} \setminus K$ has only one component. Thus we can apply Theorem 4.1.7 with $\Sigma = \{\infty\}$. $\qquad\qquad\square$

We also have the following version of the Runge theorem which holds for open sets:

Theorem 4.1.9. *Let Ω be an axially symmetric open set in $\overline{\mathbb{H}}$, let Σ be a set having a point in each connected component of $\overline{\mathbb{H}} \setminus \Omega$, and let $f \in \mathcal{R}(\Omega)$. Then f can be approximated by a sequence of rational functions $\{r_n\}$ having their poles in Σ, uniformly on every compact set in Ω. If $\overline{\mathbb{H}} \setminus \Omega$ is a connected set, then we can set $\Sigma = \{\infty\}$ and f can be approximated by polynomials uniformly on every compact set in Ω.*

Proof. Let $\{K_n\}$ be a sequence of axially symmetric compact subsets of Ω such that $K_n \Subset K_{n+1}$, any compact set in Ω is contained in K_n for some n and each component of $\overline{\mathbb{H}} \setminus K_n$ contains a component of $\overline{\mathbb{H}} \setminus \Omega$ for all n. This last assumption implies that each component of $\overline{\mathbb{H}} \setminus K_n$ contains a point in Σ. By Theorem 4.1.7 there exists a function r_n with poles in Σ, such that

$$|f(q) - r_n(q)| < \frac{1}{n}, \quad q \in K_n.$$

If K is an axially symmetric compact subset of Ω, then our assumptions imply that there exists N such that $K \subset K_n$ for all $n \geq N$, thus

$$|f(q) - r_n(q)| < \frac{1}{n}, \quad q \in K, \; n \geq N,$$

and this completes the proof. □

By suitably modifying the proof of the previous result one can prove the following:

Theorem 4.1.10. *Let Ω_1, Ω_2 be axially symmetric open sets in $\mathbb{H}$ such that $\Omega_1 \subset \Omega_2$ and each connected component of $\overline{\mathbb{H}} \setminus \Omega_1$ intersects $\overline{\mathbb{H}} \setminus \Omega_2$. Then every function in $\mathcal{R}(\Omega_1)$ can be approximated by functions in $\mathcal{R}(\Omega_2)$ uniformly on every compact set in Ω_1.*

Finally, we have the following result which is based on the fact that, on axially symmetric s-domains, the two definitions of slice regular functions 1.1.1 and 1.2.5 coincide. The proof can be found, e.g., in Corollary 1.25 in [99].

Proposition 4.1.11. *An axially symmetric s-domain is a domain of slice regularity.*

Remark 4.1.12. The Runge type approximation results hold also for slice monogenic functions with values in a Clifford algebra, see [52].

4.2 Mergelyan Type Results

All the results in this section, except those whose authors are mentioned, were obtained in the first part of the paper of Gal and Sabadini [93].

In complex analysis the Mergelyan's approximation theorem is the ultimate development and generalization of the Weierstrass approximation theorem and Runge's theorem. We reproduce here the statement for the sake of completeness (see [145]):

Theorem 4.2.1. *Let K be a compact subset of the complex plane $\mathbb{C}$ such that $\mathbb{C} \setminus K$ is connected. Then, every continuous function $f : K \to \mathbb{C}$ that is holomorphic in the interior of K can be approximated uniformly on K by polynomials.*

It is interesting to note that all the known proofs of this result are based on the Riemann mapping theorem. So, in order to generalize Mergelyan's result to the case of slice regular functions of quaternionic variable we need first to prove an appropriate analog of the Riemann mapping theorem. However, as we shall see, in the quaternionic setting this result is available only for a particular class of sets that we denote by $\mathfrak{R}(\mathbb{H})$. Its validity is unknown in the general case.

Unfortunately, the validity of the Riemann mapping theorem is not the only obstacle to generalizing Mergelyan's theorem. Another major issue is that, due to noncommutativity, the composition of two slice regular functions is not necessarily a slice regular function. Consequently, Mergelyan's theorem in its full generality seems to fail in the quaternionic setting. However, when the Riemann mapping theorem is true (namely, for sets in $\mathfrak{R}(\mathbb{H})$), also the composition of slice regular functions gives a slice regular function and in fact in this case we can prove a quaternionic Mergelyan theorem. Moreover, an approximation by polynomials in the spirit of Mergelyan's result holds for the class of starlike sets.

4.2.1   Riemann mappings for axially symmetric sets

In this subsection we deal with the generalization to the quaternionic setting of the famous Riemann mapping theorem. Its importance is well known and to our proposes it is enough to recall that it enters in the proof of several approximation results.

Let us begin by recalling the result in the complex plane:

Theorem 4.2.2 (Riemann mapping Theorem). *Let $G \subset \mathbb{C}$ be a simply connected domain, $z_0 \in G$, and let $\mathbb{D} = \{z \in \mathbb{C} : |z| < 1\}$ denote the open unit disk. Then there exists a unique bijective analytic function $f : G \to \mathbb{D}$ such that $f(z_0) = 0$, $f'(z_0) > 0$.*

Remark 4.2.3. It is useful to recall that the theorem holds more generally for simply connected open subsets of the Riemann sphere which both lack at least two points of the sphere.

Our purpose is now to generalize the Riemann mapping theorem to the case of simply connected domains in $\mathbb{H}$. In particular, we need to characterize the open sets which can be mapped bijectively onto the unit ball of $\mathbb{H}$ by a slice regular function f satisfying additional conditions and assigning its value at one point and the value of its derivative at the same point. As we shall see, we have a characterization which holds for the class of axially symmetric s-domains which are simply connected. This class is, in practice, the class of all sets we are interested in.

In [65] the author defines a class of functions called *typically real*. These functions are defined on the open unit disk $\mathbb{D}$, are univalent and take real values just on the real line. When expanded in power series, these functions have real coefficients and so they are (complex) intrinsic. The image of such mappings is symmetric with respect to the real line, see [65], p.55. We have the following result:

Theorem 4.2.4. *Let $G \subset \mathbb{C}$, G nonempty, be a simply connected domain such that $G \cap \mathbb{R} \neq \emptyset$, let $x_0 \in G \cap \mathbb{R}$ be fixed and let $f : G \to \mathbb{D}$ be the unique, bijective, analytic function $f(x_0) = 0$, $f'(x_0) > 0$. Then f is such that f^{-1} is typically real if and only if G is symmetric with respect to the real axis.*

Proof. The map $f : G \to \mathbb{D}$ with the properties $f(x_0) = 0$, $f'(x_0) > 0$ obviously exists by the Riemann mapping theorem. If G is symmetric with respect to the real axis, then by the uniqueness of f we have $\overline{f(z)} = f(\bar{z})$, see, e.g., [3], Exercise 1, p. 232 and so f maps bijectively $G \cap \mathbb{R}$ onto $\mathbb{D} \cap \mathbb{R}$ and so f^{-1} is typically real. Conversely, assume that $f^{-1} : \mathbb{D} \to G$ is typically real. Then G is symmetric with respect to the real line. $\qquad \square$

Corollary 4.2.5. *Let $G \subset \mathbb{C}$, G nonempty, be a simply connected domain such that $G \cap \mathbb{R} \neq \emptyset$, let $x_0 \in G \cap \mathbb{R}$ be fixed, and let $f : G \to \mathbb{D}$ be the unique, bijective, analytic function satisfying $f(x_0) = 0$, $f'(x_0) > 0$. The function $f^{-1} : \mathbb{D} \to G$ is typically real if and only if f is complex intrinsic.*

Proof. If f^{-1} is typically real, then G is symmetric with respect to the real line and $\overline{f^{-1}(w)} = f^{-1}(\overline{w})$. Setting $w = f(z)$, i.e., $z = f^{-1}(w)$, we get

$$f(\bar{z}) = f(\overline{f^{-1}(w)}) = f(f^{-1}(\overline{w})) = \overline{w} = \overline{f(z)},$$

and so f is complex intrinsic. Conversely, let f be complex intrinsic. It means that f is defined on a set G symmetric with respect to the real line and $f(\bar{z}) = \overline{f(z)}$. Then f^{-1} is complex intrinsic by the previous theorem. $\qquad\square$

By Remark 4.2.3, we are immediately lead to the following.

Remark 4.2.6. If $\Omega \subset \overline{\mathbb{C}}$ is a simply connected domain which contains the point ∞, then Corollary 4.2.5 remains valid for $x_0 = \infty$.

Proposition 4.2.7. *Let $G \subset \mathbb{C}$ be symmetric with respect to the real axis and let $f : G \to \mathbb{C}$ be an analytic intrinsic function. Then the function $\mathrm{ext}(f)(q)$ is quaternionic intrinsic.*

Proof. By identifying $\mathbb{C}$ with $\mathbb{C}_J$ for some $J \in \mathbb{S}$ we can assume that f as $f : G \subseteq \mathbb{C}_J \to \mathbb{C}_J$. Since f is intrinsic, i.e., $\overline{f(\bar{z})} = f(z)$, using the extension formula in Corollary 1.3.2, we obtain a function $\mathrm{ext}(f)(q)$, defined on Ω_G where $\Omega_G \subset \mathbb{H}$ denotes the axially symmetric completion of G, which is quaternionic intrinsic: in fact

$$\mathrm{ext}(f)(\bar{q}) = \frac{1 + IJ}{2} f(z) + \frac{1 - IJ}{2} f(\bar{z})$$

$$= \frac{1 + IJ}{2} f(z) + \frac{1 - IJ}{2} \overline{f(z)}$$

$$= \mathrm{Re}f(z) + IJ^2 \mathrm{Im}f(z) = \overline{f(q)}.$$

Thus $\mathrm{ext}(f) \in \mathcal{N}(\Omega_G)$. $\qquad\square$

Note that, for the sake of simplicity, we will write $f(q)$ instead of $\mathrm{ext}(f)(q)$.

Definition 4.2.8. We denote by $\mathfrak{R}(\mathbb{H})$ the class of axially symmetric open sets Ω in $\mathbb{H}$ such that $\Omega \cap \mathbb{C}_I$ is simply connected for one (and hence for all) $I \in \mathbb{S}$.

Note that $\Omega \cap \mathbb{C}_I$ is simply connected for every $I \in \mathbb{S}$ and thus it is connected, so Ω is an s-domain.

In view of Proposition 4.2.7 and of Corollary 4.2.5 we have the following:

Proposition 4.2.9. *Let $\Omega \in \mathfrak{R}(\mathbb{H})$, $\mathbb{B} \subset \mathbb{H}$ be the open unit ball and let $x_0 \in \Omega \cap \mathbb{R}$. Then there exists a unique quaternionic intrinsic slice regular function $f : \Omega \to \mathbb{B}$ which is bijective and such that $f(x_0) = 0$, $f'(x_0) > 0$.*

Proof. Let $\Omega_I = \Omega \cap \mathbb{C}_I$, where $I \in \mathbb{S}$. Then Ω_I is simply connected by hypothesis and symmetric with respect to the real line since Ω is axially symmetric. By Corollary 4.2.5, there exists a bijective, analytic intrinsic map $f_I : \Omega_I \to \mathbb{D}_I$, where $\mathbb{D}_I = \mathbb{B} \cap \mathbb{C}_I$, such that $f(x_0) = 0$, $f'(x_0) > 0$. By Proposition 4.2.7, f_I extends to $f : \Omega \to \mathbb{B}$ and $f \in \mathcal{N}(\Omega)$. Note that for every $J \in \mathbb{S}$ we have $f|_{\mathbb{C}_J} : \Omega_J \to \mathbb{D}_J$, since f takes each complex plane to itself. $\qquad\square$

The class $\mathfrak{R}(\mathbb{H})$ contains all simply connected open sets Ω in $\mathbb{H}$ intersecting the real line for which a map $f : \Omega \to \mathbb{B}$ as in the Riemann mapping theorem is quaternionic intrinsic (and slice regular). In fact, if we assume that the simply connected open set $\Omega \subset \mathbb{H}$ is mapped bijectively onto $\mathbb{B}$ by a map $f \in \mathcal{N}(\Omega)$ such that $f(x_0) = 0$, $f'(x_0) > 0$, $x_0 \in \mathbb{R}$, then $f|_{\mathbb{C}_I} : \Omega \cap \mathbb{C}_I \to \mathbb{B} \cap \mathbb{C}_I = \mathbb{D}_I$. By its uniqueness, $f|_{\mathbb{C}_I}$ is the map prescribed by the Riemann mapping theorem, and since $\Omega \cap \mathbb{C}_I$ is symmetric with respect to the real line, f is complex intrinsic. Thus $\Omega \cap \mathbb{C}_I$ is symmetric with respect to the real axis and simply connected. Since $I \in \mathbb{S}$ is arbitrary, Ω must be also axially symmetric, so it belongs to $\mathfrak{R}(\mathbb{H})$.

4.2.2   Approximation in some particular compact sets

For some particular cases of domains, the approximation by polynomials can easily be proved, as for example in the following result in which we consider starlike regions.

Theorem 4.2.10. *Let $\Sigma \subset \mathbb{H}$ be a bounded region which is starlike with respect to the origin, and such that $\overline{\Sigma}$ is axially symmetric, $\mathbb{H} \setminus \overline{\Sigma}$ is connected and $\overline{\mathbb{C}_I} \setminus (\overline{\Sigma} \cap \mathbb{C}_I)$ is connected for all $I \in \mathbb{S}$. If a function f is slice regular in Σ and continuous in $\overline{\Sigma}$, then the function f can be uniformly approximated in $\overline{\Sigma}$ by polynomials.*

Proof. We can assume that Σ is starlike with respect to the origin. Then the function $\varphi_n(q) = \frac{nq}{n+1} \in \mathcal{N}(\Sigma)$ for all $n \in \mathbb{N}$ and the composition

$$f(\varphi_n(q)) = f\left(\frac{nq}{n+1}\right)$$

is slice regular in $\Sigma_n = \varphi_n^{-1}(\Sigma)$. The function $f\left(\dfrac{nq}{n+1}\right)$ is defined in $\overline{\Sigma}$ and by the uniform continuity of f in $\overline{\Sigma}$, for any $\varepsilon > 0$ and for a suitable n

$$\left| f\left(\frac{nq}{n+1}\right) - f(q) \right| < \varepsilon/2$$

for $q \in \overline{\Sigma}$. Since $f(\frac{nq}{n+1})$ is slice regular in $\overline{\Sigma}$, Runge's result in Theorem 4.1.8 shows that there exists a polynomial $P(q)$ such that

$$\left| f\left(\frac{nq}{n+1}\right) - P(q) \right| < \varepsilon/2$$

for $q \in \overline{\Sigma}$, thus $|f(q) - P(q)| < \varepsilon$ and the statement follows.   □

We now list some examples in the complex case which can be found in, e.g., [80], pp. 82–83, Applications 1.9.8 and 1.9.9, a)–c). We will use them to present some concrete examples of compact sets in $\mathbb{H}$ where the approximation by polynomials is possible.

Example 4.2.11. Let $G \subset \mathbb{C}$ be bounded by the m-cusped hypocycloid H_m, $m = 3, 4, \ldots$, given by the parametric equation

$$z = e^{i\theta} + \frac{1}{m-1} e^{-(m-1)i\theta}, \quad \theta \in [0, 2\pi).$$

The conformal mapping (bijection) $\Psi : \overline{\mathbb{C}} \setminus \overline{\mathbb{D}} \to \overline{\mathbb{C}} \setminus G$ satisfying $\Psi(\infty) = \infty$ and $\Psi'(\infty) > 0$ is known and is given by

$$\Psi(w) = w + \frac{1}{(m-1)w^{m-1}}.$$

Note that the Riemann mapping stated in Theorem 4.2.2 can be expressed with the aid of this mapping Ψ.

Example 4.2.12. Let $G \subset \mathbb{C}$ be the m-leafed symmetric lemniscate, $m = 2, 3, \ldots$, with its boundary given by

$$L_m = \{z \in \mathbb{C}; \ |z^m - 1| = 1\}.$$

Then the corresponding conformal mapping is given by

$$\Psi(w) = w \left(1 + \frac{1}{w^m}\right)^{1/m}.$$

Example 4.2.13. Let $G \subset \mathbb{C}$ be the semidisk

$$SD = \{z \in \mathbb{C}; \ |z| \leq 1 \text{ and } |\arg(z)| \leq \pi/2\}.$$

The corresponding conformal mapping is given by

$$\Psi(w) = \frac{2(w^3 - 1) + 3(w^2 - w) + 2(w^2 + w + 1)^{3/2}}{w(w+1)\sqrt{3}}.$$

Example 4.2.14. Let G be as in one of the above examples. Let

$$\Omega_G = \bigcup_{x+iy \in G, \ I \in \mathbb{S}} (x + Iy)$$

be its axially symmetric completion. Note that $\overline{\Omega}_G$ is a starlike set with respect to the origin. Indeed, for any $q = x + Iy \in \Omega_G$ consider $z = x + iy \in G$. Since $\overline{G}$ is starlike with respect to the origin in the complex plane $\mathbb{C}$, the segment joining z with the origin belongs to $\overline{G}$. Hence, the segment joining q with the origin is contained in $\overline{\Omega}_G$.

Corollary 4.2.15. *Let Σ be a bounded Jordan region which is starlike with respect to a real point, such that $\overline{\Sigma}$ is axially symmetric, $\overline{\mathbb{H}} \setminus \overline{\Sigma}$ is connected, and $\overline{\mathbb{C}_I} \setminus (\overline{\Sigma} \cap \mathbb{C}_I)$ is connected for all $I \in \mathbb{S}$. Let $\alpha \in \overline{\Sigma}$. If a function f is slice regular in Σ and continuous in $\overline{\Sigma}$, then f can be uniformly approximated in $\overline{\Sigma}$ by polynomials which take the value $f(\alpha)$ at the point α.*

Proof. Let $\varepsilon > 0$. By Theorem 4.2.10, there exists a polynomial $P(q)$ such that $|f(q) - P(q)| < \varepsilon/2$ for $q \in \overline{\Sigma}$, so that $|f(\alpha) - P(\alpha)| < \varepsilon/2$. From these two inequalities we deduce that

$$|f(q) - (P(q) + f(\alpha) - P(\alpha))| < \varepsilon$$

for $q \in \overline{\Sigma}$. So the polynomial $P(q) + f(\alpha) - P(\alpha)$ satisfies the conditions in the statement. □

We now prove other approximation results in the more general case of sets in $\mathfrak{R}(\mathbb{H})$. To this end, we need a generalization of Theorem 4, p. 32 in [192] that we prove below.

Theorem 4.2.16. *Let $T \in \mathfrak{R}(\mathbb{H})$ be such that $T \cap \mathbb{C}_I$ is a bounded Jordan region for all $I \in \mathbb{S}$. Let $\{T_n\}$ be a sequence in $\mathfrak{R}(\mathbb{H})$ such that $\overline{T} \subset T_n$, $\overline{T}_{n+1} \subset T_n$, for all $n = 1, 2, \ldots$ and no point exterior to T belongs to T_n for all n. Let us assume that $q = 0$ belongs to T and that the bijective, quaternionic intrinsic functions $\Phi : T \to \mathbb{B}$, $\Phi_n : T_n \to \mathbb{B}$ all map $q = 0$ to $w = 0$ and satisfy $\Phi'(0) > 0$, $\Phi_n'(0) > 0$ for all $n = 1, 2, \ldots$. Then*

$$\lim_{n \to \infty} \Phi_n(q) = \Phi(q) \tag{4.3}$$

uniformly for $q \in \overline{T}$.

Proof. Let $I \in \mathbb{S}$ and consider $T \cap \mathbb{C}_I$, $T_n \cap \mathbb{C}_I$, and the restrictions $\Phi|_{\mathbb{C}_I}$, $\Phi_n|_{\mathbb{C}_I}$ of Φ, Φ_n respectively, to $\mathbb{C}_I$. Since Φ, Φ_n are quaternionic intrinsic, we have

$$\Phi|_{\mathbb{C}_I} : T \cap \mathbb{C}_I \to \mathbb{D}_I,$$

$$\Phi_n|_{\mathbb{C}_I} : T_n \cap \mathbb{C}_I \to \mathbb{D}_I,$$

and these functions satisfy all the hypotheses of Theorem 4 in [192]. Thus we have

$$\lim_{n \to \infty} \Phi_{n|\mathbb{C}_I}(z) = \Phi|_{\mathbb{C}_I}(z) \tag{4.4}$$

uniformly for all $z \in \overline{T} \cap \mathbb{C}_I$. Let K be a compact set in $\overline{T}$. Then for any $q = x + Jy \in K$ we have, using the representation formula (1.1.11)

$$|\Phi_n(q) - \Phi(q)| = \left| \frac{1}{2}(1 - JI)(\Phi_n(x + Iy) - \Phi(x + Iy)) \right.$$

$$\left. + \frac{1}{2}(1 + JI)(\Phi_n(x - Iy) - \Phi(x - Iy)) \right|$$

$$\leq |\Phi_n(x + Iy) - \Phi(x + Iy))| + |\Phi_n(x - Iy) - \Phi(x - Iy)|.$$

Now note that the right-hand side of the last inequality corresponds to the restrictions of Φ_n and Φ to $K \cap \mathbb{C}_I$, which is a compact subset of $\overline{T} \cap \mathbb{C}_I$. So, by (4.4), we have

$$|\Phi_n(x + Iy) - \Phi(x + Iy))| + |\Phi_n(x - Iy) - \Phi(x - Iy)| < \varepsilon,$$

for $n > N(\varepsilon)$, which concludes the proof. □

Corollary 4.2.17. *Under the hypotheses of Theorem 4.2.16, let* $\chi_n : T_n \to T$ *be defined by* $\chi_n(q) = \Phi^{-1}(\Phi_n(q))$. *Then* $\chi_n(0) = 0$, $\chi'_n(0) > 0$ *and*

$$\lim_{n\to\infty} \chi_n(q) = q$$

uniformly in $\overline{T}$.

Proof. The function Φ^{-1} is continuous in $\overline{T}$, so we can apply it to both members of (4.3) and get the statement. $\square$

The following result generalizes further Theorem 4.2.10:

Theorem 4.2.18. *Let* $T \in \mathfrak{R}(\mathbb{H})$ *be bounded and such that* $T \cap \mathbb{C}_I$ *is a Jordan region in the plane* $\mathbb{C}_I$, *for all* $I \in \mathbb{S}$. *If* f *is slice regular in* T *and continuous in* $\overline{T}$, *then in* $\overline{T}$ *the function* $f(q)$ *can be uniformly approximated by polynomials in* q.

Proof. Consider a sequence $\{T_n\}$ and the maps Φ_n, Φ as in the statement of Theorem 4.2.16. Let $\chi_n : T_n \to T$ be defined by $\chi_n(q) = \Phi^{-1}(\Phi_n(q))$. Then $\chi_n \in \mathcal{N}(T_n)$, hence we can consider the composition $f(\chi_n(q))$, which is slice regular in $\overline{T}$. By Corollary 4.2.17 and the continuity of f, for any $\varepsilon > 0$ there exists $N(\varepsilon)$ such that

$$|f(q) - f(\chi_n(q))| < \varepsilon/2$$

for $q \in \overline{T}$. By Theorem 4.1.8, there exists a polynomial $P(q)$ such that

$$|f(\chi_n(q)) - P(q)| < \varepsilon/2$$

for $q \in \overline{T}$. Consequently,

$$|f(q) - P(q)| < \varepsilon$$

for $q \in \overline{T}$, which concludes the proof. $\square$

The proof of the following corollary follows exactly the proof of Corollary 4.2.15, so we omit it.

Corollary 4.2.19. *Let* $T \in \mathfrak{R}(\mathbb{H})$ *be bounded and such that* $T \cap \mathbb{C}_I$ *is a Jordan region in the plane* $\mathbb{C}_I$. *Let* $\alpha \in \overline{T}$. *If a function* f *is slice regular in* T *and continuous in* $\overline{T}$, *then* f *can be uniformly approximated in* $\overline{T}$ *by polynomials which take the value* $f(\alpha)$ *at the point* α.

Remark 4.2.20. Assume that the slice function f is not slice regular, but it is only continuous in a compact subset of $\mathbb{C}$ (say, for example, the closed unit ball). It is well known that, in general, f cannot be uniformly approximated by polynomials. However, in this case, we can prove that f can be uniformly approximated by polynomials in z and z^{-1}. This can be done by virtue of the Weierstrass theorem on approximation by trigonometric polynomials, which asserts that a real-valued function $g(\theta)$ which is continuous for all θ and periodic of period 2π can be uniformly approximated for all θ by a trigonometric polynomial of the form $\sum_{n=0}^{N}(\cos(n\theta)a_n + \sin(n\theta)b_n)$. It is immediate that if the function g is complex-

(resp. quaternionic-) valued the theorem holds with $a_n, b_n \in \mathbb{C}$ (resp. $\mathbb{H}$). Let us now consider a continuous slice function $f(q) = f(x + Iy) = \alpha(x, y) + I\beta(x, y)$. Take $q \in \mathbb{H}$ such that $|q| = 1$ and write

$$q^n = \cos(n\theta) + I\sin(n\theta), \quad q^{-n} = \cos(n\theta) - I\sin(n\theta).$$

Then

$$\cos(n\theta) = \frac{1}{2}(q^n + q^{-n}), \quad \sin(n\theta) = \frac{1}{2}I(q^{-n} - q^n).$$

Moreover, $f(q) = f(x + Iy) = f(\cos\theta + I\sin\theta) = \alpha(\theta) + I\beta(\theta)$, where α, β are $\mathbb{H}$-valued.

This discussion allows us to state the following generalization of Theorem 7, §2.5 in [192]:

Theorem 4.2.21. *Let Σ be the boundary of an open set belonging to $\mathfrak{R}(\mathbb{H})$ and containing the origin, and assume that $\Sigma \cap \mathbb{C}_I$ is a Jordan curve for any $I \in \mathbb{S}$. Then any continuous slice function f on Σ can be uniformly approximated on Σ by polynomials in q and q^{-1}.*

Proof. By hypothesis, the interior Ω of Σ can be mapped onto the unit ball $\mathbb{B}$ of $\mathbb{H}$ by a function $\Phi \in \mathcal{N}(\Omega)$. Let us denote by Ψ the inverse of Φ, i.e., $q = \Psi(w)$, $w \in \mathbb{B}$. By the Weierstrass's theorem on approximation by trigonometric polynomials mentioned in Remark 4.2.20, for any $\varepsilon > 0$ there exists a polynomial $P(w, 1/w)$ such that

$$|f(\Psi(w)) - P(w, 1/w)| < \varepsilon/2, \quad w \in \partial\mathbb{B} \cap \mathbb{C}_J;$$

in fact, $z = \Psi(w) = \cos(\theta) + J\sin(\theta)$ if and only if $w \in \mathbb{C}_J$. The inequality can be written as

$$|f(z) - P(\Phi(z), \Phi(z)^{-1})| < \varepsilon/2, \quad z \in \Sigma \cap \mathbb{C}_J.$$

Let us now extend the polynomial P from $\mathbb{C}_J$ to $\mathbb{H}$ using the extension formula for slice continuous functions, see [105] (and recall that $\overline{\Phi(z)} = \Phi(\bar{z})$):

$$P(\Phi(q), \Phi(q)^{-1}) = \frac{1 - IJ}{2} P(\Phi(z), \Phi(z)^{-1}) + \frac{1 + IJ}{2} P(\Phi(\bar{z}), \Phi(\bar{z})^{-1})$$

where $q = x + Iy$. Thus, for any $q \in \Sigma$, we have

$$\begin{aligned}
|f(q) - P(\Phi(q), \Phi(q)^{-1})| &= \left| \frac{1 - IJ}{2} f(z) + \frac{1 + IJ}{2} f(\bar{z}) \right. \\
&\quad \left. - \frac{1 + IJ}{2} P(\Phi(\bar{z}), \Phi(\bar{z})^{-1}) - \frac{1 - IJ}{2} P(\Phi(z), \Phi(z)^{-1}) \right| \\
&\leq \left| \frac{1 - IJ}{2} \right| \cdot |f(z) - P(\Phi(z), \Phi(z)^{-1})| \\
&\quad + \left| \frac{1 + IJ}{2} \right| \cdot |f(\bar{z}) - P(\Phi(\bar{z}), \Phi(\bar{z})^{-1})| < \varepsilon.
\end{aligned}$$

The function $P(\Phi(q), \Phi(q)^{-1})$ is slice regular in the region R bounded by Σ and by the boundary of a ball with center at the origin. By Theorem 4.1.7, there exists a rational function Q with poles at zero and at infinity, i.e., a polynomial $Q(q, q^{-1})$ in q, q^{-1}, such that

$$|P(\Phi(q), \Phi(q)^{-1}) - Q(q, q^{-1})| < \varepsilon/2$$

for $q \in \overline{R}$. Taking $q \in \Sigma$ we obtain, in particular, that

$$|f(q) - Q(q, q^{-1})| < \varepsilon,$$

which finishes the proof. $\square$

4.3 Arakelian Type Results

Except for the results whose authors are mentioned, all the other results in this section were obtained in the paper of Gal and Sabadini [89].

Arakelian's theorem concerns uniform approximation by entire functions on closed (possibly unbounded) subsets in $\mathbb{C}$, see [10] or [71] pp. 9–34, [171]. Specifically, we give the following:

Definition 4.3.1. We say that $E \subset \mathbb{C}$ is an Arakelian set, if it is closed (possibly unbounded), without holes, and such that for every closed disk $D \subset \mathbb{C}$, the union of all holes of $E \bigcup D$ is a bounded set.

Note that here a hole of E is any bounded component of $\mathbb{C} \setminus E$.
The Runge-type version of the Arakelian's theorem can be stated as follows:

Theorem 4.3.2. ([10], [171]) *If $E \subset \mathbb{C}$ is an Arakelian set and f is a complex-valued analytic function in an open neighborhood of E, then for any $\varepsilon > 0$, there exists an entire function h on $\mathbb{C}$, such that $|f(z) - h(z)| < \varepsilon$, for all $z \in E$.*

Remark 4.3.3. Theorem 4.3.2 (originally proved in [10]) was obtained by Rosay and Rudin in [171] in its most general form as a consequence of the Mergelyan's approximation theorem. In fact, in [171], the hypothesis of analyticity in a neighborhood of E is replaced by the analyticity on the interior of E and continuity on E. But, because the extension of the Mergelyan's theorem in its full generality is not available in the quaternionic setting, we can deal only with Arakelian's theorem of Runge type.

Consider the power series $\sum_{k=0}^{\infty} q^k a_k$, $a_k \in \mathbb{H}$, and denote its partial sums by $S_n(q) = \sum_{k=0}^{n} q^k a_k$, $n \in \mathbb{N}$. Note that $S_n(q)$ is a polynomial of degree $\leq n$.
Following the classical nomenclature, we give the following definition:

Definition 4.3.4. We say that a set is a *hole of a set* $\Omega \subset \mathbb{H}$ if it is one of the bounded components of the complement of Ω in $\mathbb{H}$.

The following consequence of the Runge-type approximation result 4.1.8 will be useful in what follows.

Theorem 4.3.5. *Let K be an axially symmetric compact set, such that $\overline{\mathbb{C}_I}\backslash(K\cap\mathbb{C}_I)$ is connected for some (and thus for every) $I \in \mathbb{S}$, and let Ω be an axially symmetric open set with $K \subset \Omega$. For any $f \in \mathcal{R}(\Omega)$, there exists a sequence of right polynomials $P_n(q)$, $n \in \mathbb{N}$, such that $P_n \to f$ uniformly on K, as $n \to \infty$.*

The following results are of independent interest, but the proof of the second will be useful in the sequel.

Lemma 4.3.6. *Let $\phi : \mathbb{H} \to \mathbb{H}$ be a slice function, infinitely differentiable and with compact support. Then for every $q \in \mathbb{H}$ we have*

$$\phi(q) = -\frac{1}{2\pi}\int\int_{\mathbb{C}_I} S^{-1}(s,q)\frac{\partial}{\partial\bar{s}}\phi(s)ds_I \wedge d\bar{s},$$

where

$$S^{-1}(s,q) = -(q^2 - 2\mathrm{Re}(s)q + |s|^2)^{-1}(q - \bar{s}),$$

and $s = u + Iv$, $ds = du + Idv$, $ds_I = -Ids$.

Proof. Consider the restriction $\phi|_{\mathbb{C}_I}$ of ϕ to a complex plane $\mathbb{C}_I$, $I \in \mathbb{S}$ and consider a fixed $J \in \mathbb{S}$ orthogonal to I. Then $\phi|_{\mathbb{C}_I}(z) = \phi_1(z) + \phi_2(z)J$ with $\mathbb{C}_I$-valued functions ϕ_1, ϕ_2. Theorem 1 p. 103 in [153] states that

$$\phi_j(z) = -\frac{1}{2\pi}\int\int_{\mathbb{C}_I} (s-z)^{-1}\frac{\partial}{\partial\bar{s}}\phi_j(s)ds_I \wedge d\bar{s}, \quad j = 1,2$$

and so, since $ds_I \wedge d\bar{s}J = Jds_I \wedge d\bar{s}$ we deduce that

$$\phi(z) = \phi|_{\mathbb{C}_I}(z) = \phi_1(z) + \phi_2(z)J$$
$$= -\frac{1}{2\pi}\int\int_{\mathbb{C}_I} (s-z)^{-1}\frac{\partial}{\partial\bar{s}}\phi(s)ds_I \wedge d\bar{s}. \tag{4.5}$$

Since ϕ is a slice function, it is completely determined by the values of any restriction, by the representation formula in Theorem 1.1.11, thus if $z = x + Iy$, $q = x + Ly$, $L \in \mathbb{S}$, we have

$$\phi(q) = \frac{1}{2}(\phi(z) + \phi(\bar{z})) + \frac{1}{2}LI(\phi(\bar{z}) - \phi(z)). \tag{4.6}$$

Substituting in (4.6) the corresponding integrals as in (4.5), and recalling that

$$S^{-1}(s,q) = \frac{1}{2}((s-z)^{-1} + (s-\bar{z})^{-1} + \frac{1}{2}LI((s-\bar{z})^{-1} - (s-z)^{-1})), \tag{4.7}$$

and that

$$-(q^2 - 2\mathrm{Re}(s)q + |s|^2)^{-1}(q-\bar{s}) = (s-\bar{q})(s^2 - 2\mathrm{Re}(q)s + |q|^2)^{-1}$$

(see Proposition 4.4.6 in [51]), we complete the proof. □

Lemma 4.3.7. *Let* $\phi : \mathbb{H} \to \mathbb{H}$ *be a slice function, infinitely differentiable and with compact support. Let*

$$f(q) = -\frac{1}{2\pi} \int \int_{\mathbb{C}_I} S^{-1}(s,q)\,\phi(s)\,ds_I \wedge d\bar{s}.$$

Then $f \in C^\infty(\mathbb{H})$ *and* $\left(\frac{\partial}{\partial x} + L\frac{\partial}{\partial y}\right) f(x + Ly) = \phi(x + Ly)$ *for every* $L \in \mathbb{S}$.

Proof. Consider the restriction $\phi|_{\mathbb{C}_I}$ of ψ to a complex plane $\mathbb{C}_I$, $I \in \mathbb{S}$ and fix an arbitrary $J \in \mathbb{S}$ orthogonal to I. Then $\phi|_{\mathbb{C}_I}(z) = \phi_1(z) + \phi_2(z)J$, where ϕ_1, ϕ_2 are $\mathbb{C}_I$-valued. By Theorem 2 p. 104 in [153], we have that the functions f_1, f_2 defined by

$$f_j(z) = -\frac{1}{2\pi} \int \int_{\mathbb{C}_I} (s - z)^{-1}\phi_j(s)ds_I \wedge d\bar{s}, \quad j = 1, 2,$$

where $z \in \mathbb{C}_I$, are infinitely differentiable and $\frac{\partial}{\partial \bar{z}} f_j(z) = \phi_j(z)$, $j = 1, 2$, hence

$$\frac{\partial}{\partial \bar{z}} f(z) = \phi(z). \tag{4.8}$$

Set $f(z) = f_1(z) + f_2(z)J$, $z \in \mathbb{C}_I$, and let $q = x + Ly$ be any point in $\mathbb{H}$. Define

$$f(q) = f(x + Ly) = \frac{1}{2}(f(z) + f(\bar{z})) + \frac{1}{2}LI(f(\bar{z}) - f(z)).$$

Formula (4.7) allows to rewrite f as

$$f(q) = -\frac{1}{2\pi} \int \int_{\mathbb{C}_I} S^{-1}(s,q)\phi(s)ds_I \wedge d\bar{s}.$$

It is immediate that f is infinitely differentiable. Moreover,

$$\begin{aligned}
\frac{\partial}{\partial \bar{q}} f(q) &= \frac{1}{2}\left(\frac{\partial}{\partial x} + L\frac{\partial}{\partial y}\right)\left[\frac{1}{2}(f(z) + f(\bar{z})) + \frac{1}{2}LI(f(\bar{z}) - f(z))\right] \\
&= \frac{1}{4}\left[\frac{\partial}{\partial x} f(z) + \frac{\partial}{\partial x} f(\bar{z}) + LI\frac{\partial}{\partial x} f(\bar{z}) - LI\frac{\partial}{\partial x} f(z) \right. \\
&\qquad \left. + L\frac{\partial}{\partial y} f(z) + L\frac{\partial}{\partial y} f(\bar{z}) - I\frac{\partial}{\partial y} f(\bar{z}) + I\frac{\partial}{\partial y} f(z)\right] \\
&= \frac{1}{4}\left[\frac{\partial}{\partial \bar{z}} f(z) + \frac{\partial}{\partial z} f(\bar{z}) + LI\frac{\partial}{\partial z} f(\bar{z}) - LI\frac{\partial}{\partial \bar{z}} f(z)\right] \\
&= \frac{1}{2}[\phi(z) + \phi(\bar{z}) + LI(\phi(\bar{z}) - \phi(z))] = \phi(q).
\end{aligned}$$

where we used (4.8). The assertion follows. $\qquad\square$

The main result of this section is the following.

Theorem 4.3.8. *Let $E \subset \mathbb{H}$ be a (not necessarily bounded) closed, axially symmetric set with no holes and such that for any closed ball B centered at a real number, the union of all holes of the set $E \bigcup B$ is a bounded set. If f is a slice regular function in a neighborhood of E, then for any $\varepsilon > 0$, there exists an entire slice regular function h such that $|f(q) - h(q)| < \varepsilon$, for all $q \in E$.*

Proof. Our hypotheses on E imply that there are closed balls B_i, $i = 1, 2, 3, \ldots$, centered at the origin, whose union is $\mathbb{H}$, such that the interior of B_{i+1} contains the compact set $B_i \bigcup \overline{H_i}$, where H_i is the union of the holes of $E \bigcup B_i$. Define

$$E_0 := E, \quad E_i := E \bigcup B_i \bigcup \overline{H_i}, \quad i \in \mathbb{N}.$$

Clearly, E_i has no holes. By induction we will construct a sequence (h_n); $n = 0, 1, 2, \ldots$ of slice regular functions whose limit will be the entire slice regular function h in the statement.

Let $\varepsilon > 0$ be arbitrary but fixed. Set $h_0 = f$, and note that f is slice regular in a neighborhood of $E_0 = E$. Then, for $i \geq 1$, assume that we have constructed a function h_{i-1} that is slice regular in a neighborhood of E_{i-1}. We will construct h_i, a function slice regular in a neighborhood of E_i and with the property

$$|h_i(q) - h_{i-1}(q)| < 2^{-i}\varepsilon, \quad \text{for all } q \in E_{i-1}.$$

From the construction of B_i, it follows that there exists an open ball Δ centered at the origin that contains $B_i \bigcup \overline{H_i}$ and whose closure still lies in the interior of B_{i+1}.

We now construct a slice continuously differentiable function ψ defined on $\mathbb{H}$, with the properties $0 \leq \psi(q) \leq 1$ for all $q \in \mathbb{H}$, $\psi(q) = 1$ for all $q \in \Delta$ and $\psi(q) = 0$ for all $q \in \mathbb{H} \setminus B_{i+1}$.

Let us fix $I \in \mathbb{S}$ and choose a continuously differentiable function ψ defined on $\mathbb{C}_I$ identified with $\mathbb{R}^2$, with the properties $0 \leq \psi(x, y) \leq 1$ for all $(x, y) \in \mathbb{R}^2$, $\psi(x, y) = 1$ for all (x, y) such that $x + Iy \in \Delta \cap \mathbb{C}_I$, and $\psi(x, y) = 0$ for all (x, y) such that $x + Iy \in \mathbb{C}_I \setminus (B_{i+1} \cap \mathbb{C}_I)$. Such a function exists by the Urysohn's theorem (see, e.g., [151], p. 281, Theorem 4.4.3, or, e.g., [153], p. 101, Theorem 2).

We can also assume that $\psi(x, -y) = \psi(x, y)$. In fact, if this is not the case, we set $\tilde{\psi}(x, y) = \frac{1}{2}(\psi(x, y) + \psi(x, -y))$ and use $\tilde{\psi}$ instead of ψ. Now set $\psi(x + Jy) = \psi(x, y)$ for $x + Jy \in \mathbb{H}$. Note that, by its definition, ψ is a slice function where $\alpha(x, y) = \psi(x, y)$ and $\beta(x, y) = 0$.

Since E_{i-1} has no holes, the same is valid for $E_{i-1} \bigcap B_{i+1}$. Moreover, by their definition, $E_{i-1} \bigcap B_{i+1}$ is a compact, axially symmetric set and $\overline{\mathbb{H}} \setminus (E_{i-1} \bigcap B_{i+1})$ is connected. Indeed, it is immediate that $E_{i-1} \bigcap B_{i+1}$ is compact and $\overline{\mathbb{H}} \setminus (E_{i-1} \bigcap B_{i+1})$ is connected. Moreover, E_{i-1} is axially symmetric, in fact $\overline{H_i}$ is the union of the holes of $E \bigcup B_i$, but a hole is a bounded component of the complement $(E \bigcup B_i)^c$ of $E \bigcup B_i$. Since $(E \bigcup B_i)^c = E^c \bigcap B_i^c$ is an intersection of axially symmetric sets, it is axially symmetric itself, and so are all its bounded components. Finally, since a union of axially symmetric sets is axially symmetric, the assertion follows.

Now fix $I \in \mathbb{S}$ and set

$$\mathcal{I} := \frac{1}{2\pi} \int\!\!\int_{E_{i-1} \cap \mathbb{C}_I} |S^{-1}(w,q)| \left| \frac{\partial \psi}{\partial \overline{w}}(w)(P(w) - h_{i-1}(w)) \right| dw_I \wedge d\overline{w},$$

$$E_{i-1} \cap B_{i+1} = (E_{i-1} \cap \Delta) \cup (E_{i-1} \cap B_{i+1} \setminus \Delta),$$

and

$$(E_{i-1} \cap \Delta) \cap (E_{i-1} \cap D_{i+1} \setminus \Delta) = \emptyset.$$

Since ψ vanishes outside B_{i+1} we can rewrite the integral $\mathcal{I}$ as

$$\begin{aligned}
\mathcal{I} &= \frac{1}{2\pi} \int\!\!\int_{E_{i-1} \cap B_{i+1} \cap \mathbb{C}_I} |S^{-1}(w,q)| \left| \frac{\partial \psi}{\partial \overline{w}}(w)(P(w) - h_{i-1}(w)) \right| dw_I \wedge d\overline{w} \\
&= \frac{1}{2\pi} \int\!\!\int_{E_{i-1} \cap \Delta \cap \mathbb{C}_I} |S^{-1}(w,q)| \left| \frac{\partial \psi}{\partial \overline{w}}(w)(P(w) - h_{i-1}(w)) \right| dw_I \wedge d\overline{w} \\
&\quad + \frac{1}{2\pi} \int\!\!\int_{E_{i-1} \cap (B_{i+1} \setminus \Delta) \cap \mathbb{C}_I} |S^{-1}(w,q)| \left| \frac{\partial \psi}{\partial \overline{w}}(w)(P(w) - h_{i-1}(w)) \right| dw_I \wedge d\overline{w} \\
&= \frac{1}{2\pi} \int\!\!\int_{E_{i-1} \cap (B_{i+1} \setminus \Delta) \cap \mathbb{C}_I} |S^{-1}(w,q)| \left| \frac{\partial \psi}{\partial \overline{w}}(w)(P(w) - h_{i-1}(w)) \right| dw_I \wedge d\overline{w},
\end{aligned}$$

where we used the fact that ψ is constant on Δ.

Let $M_{i+1} < +\infty$ be the maximum of $|\partial\psi/\partial\overline{w}|$ on B_{i+1}; note that the maximum exists since ψ is continuously differentiable on B_{i+1}. Denote

$$R_{i+1} = \int\!\!\int_{B_{i+1} \cap \mathbb{C}_I} |S^{-1}(w,q)| dw_I \wedge dw,$$

where the integral exists, i.e., is finite (see, e.g., the reasoning on the pages 127–128, in [96]). By its definition, R_{i+1} does not depend on $I \in \mathbb{S}$. In fact, the value of the integral in the formula for R_{i+1} is the same, for every $I \in \mathbb{S}$.

By Runge's approximation theorem, there exists a polynomial P such that

$$|P - h_{i-1}| < \min\left\{ 2^{-i-1}\varepsilon,\ 2^{-i-1}\varepsilon \cdot \frac{2\pi}{M_{i+1} \cdot R_{i+1}} \right\}, \qquad \text{on } E_{i-1} \cap B_{i+1}. \quad (4.9)$$

Let us pick a polynomial P satisfying (4.9) and compute $\mathcal{I}$. We have:

$$\mathcal{I} = \frac{1}{2\pi} \int \int_{E_{i-1} \cap \mathbb{C}_I} |S^{-1}(w,q)| \left| \frac{\partial \psi}{\partial \bar{w}}(w)(P(w) - h_{i-1}(w)) \right| dw_I \wedge d\bar{w}$$

$$= \frac{1}{2\pi} \int \int_{E_{i-1} \cap (B_{i+1} \backslash \Delta) \cap \mathbb{C}_I} |S^{-1}(w,q)| \left| \frac{\partial \psi}{\partial \bar{w}}(w)(P(w) - h_{i-1}(w)) \right| dw_I \wedge d\bar{w}$$

$$\leq 2^{-i-1}\varepsilon \cdot \frac{2\pi}{M_{i+1} \cdot R_{i+1}} \cdot M_{i+1} \cdot \frac{1}{2\pi} \int \int_{E_{i-1} \cap (B_{i+1} \backslash \Delta) \cap \mathbb{C}_I} |S^{-1}(w,q)| \, dw_I \wedge d\bar{w}$$

$$\leq 2^{-i-1}\varepsilon \cdot \frac{1}{R_{i+1}} \cdot \int_{B_{i+1}} |S^{-1}(w,q)| \, dw_I \wedge d\bar{w} \leq 2^{-i-1}\varepsilon. \qquad (4.10)$$

Now select a neighborhood V of E_{i-1} such that (4.10) holds with V in place of E_{i-1}. This is possible because for any function F we can write

$$\left| \int \int_{V \cap \mathbb{C}_I} F(w) \, dw_I \wedge d\bar{w} \right|$$

$$= \left| \int \int_{E_{i-1} \cap \mathbb{C}_I} F(w) \, dw_I \wedge d\bar{w} + \int \int_{V \backslash E_{i-1} \cap \mathbb{C}_I} F(w) \, dw_I \wedge d\bar{w} \right|$$

$$\leq \left| \int \int_{E_{i-1} \cap \mathbb{C}_I} F(w) \, dw_I \wedge d\bar{w} \right| + \int \int_{V \backslash E_{i+1} \cap \mathbb{C}_I} |F(w)| \, dw_I \wedge d\bar{w}.$$

By the absolute continuity of the Lebesgue integral, if we choose V sufficiently close to E_{i-1}, then the integral

$$\int \int_{V \backslash E_{i+1} \cap \mathbb{C}_I} |F(w)| \, dw_I \wedge d\bar{w}$$

can be made sufficiently small.

Next, consider

$$\frac{1}{2\pi} \int \int_{V \cap \mathbb{C}_I} S^{-1}(w,q) \left(\frac{\partial}{\partial \bar{w}} \psi(w)(P(w) - h_{i-1}(w)) \right) dw_I \wedge d\bar{w}. \qquad (4.11)$$

Since $\psi(x + Ly) = \alpha(x,y)$ is a slice function, α is even in its second coordinate. Thus $\frac{\partial}{\partial x}\alpha(x,y) + L\frac{\partial}{\partial y}\alpha(x,y)$ is a slice function and then so is the function

$$\left(\frac{\partial}{\partial x}\alpha(x,y) + L\frac{\partial}{\partial y} \right) \psi(x + Ly)(P(x + Ly) - h_{i-1}(x + Ly)).$$

Note that, in general, one has to write

$$\frac{\partial \psi}{\partial \bar{w}}(w) * (P(w) - h_{i-1}(w))$$

to guarantee that the product is still a slice function. However, since ψ is real-valued the $*$-product coincides with the pointwise product.

We now set

$$r(x + Ly) = \frac{1}{2\pi} \int \int_{V \cap \mathbb{C}_I} S^{-1}(w, x + Ly) \left(\frac{\partial}{\partial \bar{w}} \psi(w)(P(w) - h_{i-1}(w)) \right) dw_I \wedge d\bar{w},$$

and

$$h_i(q) := \psi(q)P(q) + (1 - \psi(q))h_{i-1}(q) + r(q), \quad q \in (\Delta \cup V) \cap \mathbb{C}_L.$$

Reasoning as in the proof of Lemma 4.3.7, for all $x + Ly \in V \cap \mathbb{C}_L$ we obtain

$$\left(\frac{\partial}{\partial x} + L \frac{\partial}{\partial y} \right) r(x + Ly) = - \left(\frac{\partial}{\partial x} + L \frac{\partial}{\partial y} \right) \psi(x + Ly)(P(x + Ly) - h_{i-1}(x + Ly)).$$

Moreover, a simple computation shows that

$$\left(\frac{\partial}{\partial x} + L \frac{\partial}{\partial y} \right) h_i(x + Ly) = \left(\frac{\partial}{\partial x} + L \frac{\partial}{\partial y} \right) \psi(x + Ly)(P(x + Ly) - h_{i-1}(x + Ly))$$
$$+ \left(\frac{\partial}{\partial x} + L \frac{\partial}{\partial y} \right) r(x + Ly) = 0, \quad x + Ly \in V \cap \mathbb{C}_L.$$

Since $L \in \mathbb{S}$ is arbitrary, the slice function h_i is slice regular in V. The integral defining r extends only over $V \setminus \Delta$, and r is in the kernel of $\left(\frac{\partial}{\partial x} + L \frac{\partial}{\partial y} \right)$ in Δ since $\frac{\partial}{\partial \bar{w}} \psi(w) = 0$. By the arbitrariness of $L \in \mathbb{S}$, the slice function $h_i = P + r$ is slice regular in Δ. In conclusion, we have constructed a function h_i that is slice regular in the neighborhood $V \cup \Delta$ of E_i.

Thus, iterating the procedure, we produce a sequence $(h_i)_{i \in \mathbb{N}}$, with each h_i slice regular in a neighborhood of E_i, such that $|h_i(q) - h_{i-1}(q)| < 2^{-i}\varepsilon$ on E_i. Indeed, for all $q \in E_i$ we have

$$|h_i(q) - h_{i-1}(q)| = |\psi(P - h_{i-1})(q) + r(q)|$$
$$\leq |P(q) - h_{i-1}(q)| + |r(q)|$$
$$\leq 2(2^{-i-1}\varepsilon + 2^{-i-1}\varepsilon) = 2^{-i}\varepsilon.$$

This easily implies that the sequence $(h_j)_{j \in \mathbb{N}}$, and the fact that $(E_i)_{i \in \mathbb{N}}$ covers $\mathbb{H}$, implies that $h = \lim_{j \to \infty} h_j$ is the required entire function. $\square$

Remark 4.3.9. If E is an Arakelian set in $\mathbb{C}$, then its axially completion in $\mathbb{H}$ satisfies the conditions in Theorem 4.3.8.

We now describe how to adapt the result in the case of Clifford algebra-valued slice monogenic functions. We have the following result:

Theorem 4.3.10. *Let $E \subset \mathbb{R}^{n+1}$ be a (not necessarily bounded) closed, axially symmetric set with no holes, and such that for any closed ball B centered at a real number, the union of all holes of the set $E \bigcup B$ is a bounded set. If f is a slice monogenic function in a neighborhood of E, then for any $\varepsilon > 0$, there exists an entire slice monogenic function h on $\mathbb{R}^{n+1}$, such that $|f(x) - h(x)| < \varepsilon$, for all $x \in E$.*

Proof. Since, as we observed, the Runge theorem is valid also for this class of functions and the proof of Lemma 4.3.7 applies also to the case of Clifford algebra-valued functions, the proof of Theorem 4.3.8 carries out in this setting by simply making the suitable changes. □

4.4   Approximation by Faber Type Polynomials

Except for the results for which we indicate the authors, all the other results in this section were obtained in the Gal and Sabadini paper [91].

From a historical point of view, after their introduction by Faber in [67], the Faber polynomials became a very useful tool of study in the theory of holomorphic functions. The Faber series represent a natural analogue, on compact sets in $\mathbb{C}$, of the Taylor series in compact disks. For details we refer the reader to, e.g., [183].

We briefly recall the definition and properties of expansion in series of Faber polynomials in the complex plane $\mathbb{C}$. Let $E \subset \mathbb{C}$ be a compact set such that $\overline{\mathbb{C}} \setminus E$ is simply connected. Note that if E is a continuum in $\mathbb{C}$, i.e., a compact connected set, then $\overline{\mathbb{C}} \setminus E$ is simply connected, where $\overline{\mathbb{C}}$ denotes the extended complex plane.

As customary, let $\mathbb{D}$ be the open unit disk. By the well-known Riemann mapping theorem, there exists a unique bijective biholomorphic (meromorphic) function

$$\Phi : \overline{\mathbb{C}} \setminus E \to \overline{\mathbb{C}} \setminus \overline{\mathbb{D}}, \quad \Phi(z) = dz + d_0 + \sum_{j=1}^{+\infty} \frac{d_j}{z^j},$$

satisfying the conditions

$$\Phi(\infty) = \infty, \quad \Phi'(\infty) = d > 0.$$

Let us denote by $\Psi : \overline{\mathbb{C}} \setminus \overline{\mathbb{D}} \to \overline{\mathbb{C}} \setminus E$ the inverse of Φ; then

$$\Psi(w) = cw + c_0 + \sum_{j=-\infty}^{-1} w^j c_j, \quad \Psi(\infty) = \infty$$

and

$$\Psi'(\infty) = c = \frac{1}{d} > 0.$$

Let $n \in \mathbb{N} \cup \{0\}$. One possible way of defining Faber polynomials involves Laurent series. Specifically, we can consider

$$[\Phi(z)]^n = \left(dz + d_0 + \frac{d_1}{z} + \frac{d_2}{z^2} + \cdots \right)^n = d^n z^n + \sum_{k=-\infty}^{n-1} d_{n,k} z^k, \qquad (4.12)$$

which is valid for large $|z|$.

Definition 4.4.1. The polynomial part of (4.12) is called the *Faber polynomial of exact degree* n and is denoted by $F_n(z)$.

An alternative way to introduce the Faber polynomials is with the aid of the function Ψ and the relation (see, e.g., [183], p. 35):

$$\frac{\Psi'(w)}{\Psi(w) - z} = \sum_{n=0}^{\infty} \frac{F_n(z)}{w^{n+1}}, \quad z \in E_R, \ |w| > R,$$

where $\Gamma_R = \{\Psi(w); |w| = R\}$ and $E_R = \text{int}(\Gamma_R)$.

Remark 4.4.2. An important problem in complex analysis is to find sufficient conditions under which the conformal mapping Φ can be analytically and bijectively continued on the boundary Γ of the domain E or even across its boundary Γ. This is equivalent to asking whether $\Psi = \Phi^{-1}$ can be analytically and bijectively continued to $\overline{\mathbb{D}}$ or even to $\overline{\mathbb{C}} \setminus \overline{B(0;r)}$, where $0 < r < 1$ and $B(0;r) = \{z \in \mathbb{C}; |z| < r\}$. In the case when Γ is a Jordan curve, then by the Carathéodory theorem [29], we are in the first case above, while when Γ is an analytic regular curve, we are in the second case above (see, e.g., [183], p. 51, proof of Theorem 1).

Let C be a Jordan curve in $\mathbb{C}$, let $\Omega = \text{int}(C)$ and set

$$A(\overline{\Omega}) = \{ f : \overline{\Omega} \to \mathbb{C} \ ; \ f \text{ is analytic in } \Omega \text{ and continuous in } \overline{\Omega} \}.$$

Definition 4.4.3. The *Faber coefficients* of $f \in A(\overline{\Omega})$ are defined as

$$a_n(f) = \frac{1}{2\pi i} \int_{|w|=1} \frac{f(\Psi(w))}{w^{n+1}} dw, \quad n = 0, 1, 2, \ldots. \qquad (4.13)$$

The series $\sum_{n=0}^{\infty} a_n(f) \cdot F_n(z)$ is called the *formal Faber series attached to* f.

One of the main problems is to find under which conditions

$$f(z) = \sum_{n=0}^{\infty} a_n(f) \cdot F_n(z), \quad \text{for all } z \in \Omega.$$

Remark 4.4.4. In the above formula for $a_n(f)$ we used in fact Remark 4.4.2. However, there are other cases in which, without having any information on the boundary, we can still define the Faber coefficients and the Faber series. To provide an example, we consider the case when $\overline{\Omega} := K$ is a continuum, so that $\overline{\mathbb{C}} \setminus K$ is simply connected. For any $f \in A(K)$ we can define the Faber coefficients of f as

$$a_n(f) = \frac{1}{2\pi i} \int_{|w|=\beta} \frac{f(\Psi(w))}{w^{n+1}} dw, \quad n = 0, 1, 2, \ldots$$

and the Faber series as $\sum_{n=0}^{\infty} a_n(f) \cdot F_n(z)$, where $\beta > 1$ is arbitrary, but fixed. The Cauchy theorem implies that the coefficients $a_n(f)$ do not depend on the choice of β.

The following important results can be found, e.g., in [183], pp. 51–52.

Theorem 4.4.5. (i) *If the boundary C of a bounded simply connected domain Ω is a regular analytic curve and $f \in A(\overline{\Omega})$, then $f(z) = \sum_{n=0}^{\infty} a_n(f) \cdot F_n(z)$ for all $z \in \Omega$ and the Faber series converges uniformly inside Ω.*

(ii) *Let $K \subset \mathbb{C}$ be a continuum (that is, a connected compact set). If $f : K \to \mathbb{C}$ is analytic on K, then $f(z) = \sum_{n=0}^{\infty} a_n(f) \cdot F_n(z)$ uniformly in K.*

Concerning estimates of Faber polynomials, the following result holds (see, e.g., [183], p. 43, inequality (8)):

Theorem 4.4.6. *Let $K \subset \mathbb{C}$ be a continuum and suppose that f is analytic in K, that is, there exists $R > 1$ such that f is analytic in Ω_R, where Ω_R denotes the interior of the closed level curve $\Gamma_R = \{\Psi(w); |w| = R\}$. Then, for any $1 < r < R$ we have $|F_n(z)| \leq C(r) \cdot r^n$, for all $n = 0, 1, \ldots$, and all $z \in \overline{\Omega_r}$, where $C(r) > 0$ is independent of n.*

The literature is very rich on properties of the Faber polynomials and Faber expansions and we refer the interested reader to, e.g., the books [74], [183].

Our main goal is to extend, in the quaternionic setting $\mathbb{H}$, the concept of Faber polynomials, the estimates in Theorem 4.4.6 and the above Theorem 4.4.5, within the framework of slice regular functions. The extension in its full generality runs into two main unavoidable obstacles: as we have already discussed, a Riemann mapping theorem in the quaternionic setting is at this moment available only for the particular class of axially symmetric sets. However, it is not reductive to consider axially symmetric sets, since they are the natural domains of definition of slice regular functions. Another obstacle is that, in general, the composition of two slice regular functions is not necessarily slice regular.

As a consequence, it seems that the extension of the concept of Faber polynomials in its full generality is not possible in the quaternionic setting. However, when the Riemann mapping theorem holds, namely when Ψ is a quaternionic intrinsic function, we are precisely in the case in which the composition of two slice regular functions $f \circ \Psi$ is slice regular. This will be the case we will consider in the sequel, and it will be also crucial in the definition of Faber coefficients, see $f \circ \Psi$ in (4.13), under the integral sign.

4.4.1 Biholomorphic bijections between complements of compact axially symmetric sets

In this subsection we will use the Riemann mapping Theorem 4.2.2 to construct and study biholomorphic bijections between complements of compact sets in $\mathbb{C}$ and $\mathbb{H}$. This topic is of great importance, since such bijections enter in the definitions and the properties of Faber polynomials.

The following result is well known in the complex case: we provide it with its proof for the reader's convenience and also because the reasoning will be useful in the sequel.

Theorem 4.4.7. *If $K \subset \mathbb{C}$ is a compact set such that $\overline{\mathbb{C}} \setminus K$ is simply connected, then there exist two unique bijective biholomorphic (meromorphic) functions $\Phi :$ $\overline{\mathbb{C}} \setminus K \to \overline{\mathbb{C}} \setminus \overline{\mathbb{D}}$, $\Psi : \overline{\mathbb{C}} \setminus \overline{\mathbb{D}} \to \overline{\mathbb{C}} \setminus K$, $\Psi(z) = \Phi^{-1}(z)$, satisfying $\Phi(\infty) = \Psi(\infty) = \infty$, $\Phi'(\infty) = \frac{1}{\Psi'(\infty)} > 0$.*

Proof. Since $\overline{\mathbb{C}} \setminus K$ is simply connected, by the Riemann mapping Theorem 4.2.2 and Remark 4.2.3, there exists a biholomorphic bijection $f : \overline{\mathbb{C}} \setminus K \to \mathbb{D}$, with $f(\infty) = 0$ and $f'(\infty) > 0$. This means that the function $\Phi(z) = \frac{1}{f(z)}$ has $z = \infty$ as a simple pole, that is $\Phi(1/z)$ has $z = 0$ as a simple pole. From the Laurent expansion around $z = 0$, it follows that we can write

$$\Phi(1/z) = \frac{d}{z} + \sum_{j=0}^{+\infty} d_j z^j,$$

which implies that Φ is of the form

$$\Phi(z) = dz + d_0 + \sum_{j=1}^{+\infty} \frac{d_j}{z^j}.$$

Choosing the unique Φ with $\Phi(\infty) = \infty$ and $\Phi'(\infty) = d > 0$, we take $\Psi : \overline{\mathbb{C}} \setminus \overline{\mathbb{D}} \to \overline{\mathbb{C}} \setminus K$ as the inverse of Φ. This immediately gives that Ψ is of the form $\Psi(w) = cw + c_0 + \sum_{j=-\infty}^{-1} w^j c_j$, with $c = 1/d$. $\square$

Our purpose is to generalize this theorem to some classes of subsets in $\mathbb{H}$, more precisely we establish results which work for the class of axially symmetric compact sets whose complements in $\overline{\mathbb{H}}$ are simply connected.

In the case of complex intrinsic functions we have:

Corollary 4.4.8. *Let $K \subset \mathbb{C}$ be a compact set such that $\overline{\mathbb{C}} \setminus K$ is simply connected. There exist two unique bijective, biholomorphic (meromorphic), intrinsic functions $\Phi : \overline{\mathbb{C}} \setminus K \to \overline{\mathbb{C}} \setminus \overline{\mathbb{D}}$, $\Psi : \overline{\mathbb{C}} \setminus \overline{\mathbb{D}} \to \overline{\mathbb{C}} \setminus K$, $\Psi(z) = \Phi^{-1}(z)$, satisfying $\Phi(\infty) = \Psi(\infty) = \infty$, $\Phi'(\infty) = \frac{1}{\Psi'(\infty)} > 0$, if and only if K is symmetric with respect to the real axis.*

Proof. Assume that K is symmetric with respect to the real axis. Then $\overline{\mathbb{C}} \setminus K$ is also symmetric with respect to the real axis, which by the above Remark 4.2.6 implies that the bijective biholomorphic function $f : \overline{\mathbb{C}} \setminus K \to \mathbb{D}$ which satisfies $f(\infty) = 0$ is an intrinsic function. Then, arguing as in the proof of Theorem 4.4.7, we easily conclude that the functions $\Phi(z)$ and $\Psi(z)$ are intrinsic functions.

Conversely, suppose that the bijective holomorphic function Ψ in the statement is intrinsic. Then it is immediate that $\overline{\mathbb{C}} \setminus K$ is symmetric with respect to the real line, which clearly implies that K is symmetric with respect to the real line.

$\square$

Remark 4.4.9. Let $K \subset \mathbb{C}$ be compact, symmetric with respect to the real axis, such that $\overline{\mathbb{C}} \setminus K$ is simply connected and the Riemann mapping $f : \overline{\mathbb{C}} \setminus K \to \mathbb{D}$ in the proof of Corollary 4.4.8 is intrinsic, i.e., $\overline{f(\bar{z})} = f(z)$.

We can identify $\mathbb{C}$ with $\mathbb{C}_J$ for some $J \in \mathbb{S}$, and using the extension formula in Corollary 1.3.2, we obtain a function $\text{ext}(f)(q)$ which is quaternionic intrinsic, see Proposition 4.2.7. Thus $\text{ext}(f) \in \mathcal{N}(\Omega_{\overline{\mathbb{C}} \setminus K})$, where $\Omega_{\overline{\mathbb{C}} \setminus K} \subset \mathbb{H}$ denotes the axially symmetric completion of $\overline{\mathbb{C}_I} \setminus K$. For the sake of simplicity, we will write $f(q)$ instead of $\text{ext}(f)(q)$.

Similarly, it follows that for the functions $\Phi(z)$ and $\Psi(z)$ in the statement of Corollary 4.4.8 we have $\text{ext}(\Phi) \in \mathcal{N}(\Omega_{\overline{\mathbb{C}} \setminus K})$, $\text{ext}(\Psi) \in \mathcal{N}(\Omega_{\overline{\mathbb{C}} \setminus \overline{\mathbb{D}}})$. For the sake of simplicity again we will simply write $\Phi(q)$ and $\Psi(q)$ instead of $\text{ext}(\Phi)(q)$ and $\text{ext}(\Phi)(q)$.

Definition 4.4.10. We will denote by $\mathfrak{R}_c(\overline{\mathbb{H}})$ the class of axially symmetric compact sets K in $\mathbb{H}$ such that $(\overline{\mathbb{H}} \setminus K) \cap \mathbb{C}_I$ is simply connected for one (and hence for all) $I \in \mathbb{S}$.

Note that $(\overline{\mathbb{H}} \setminus K) \cap \mathbb{C}_I$ is simply connected for every $I \in \mathbb{S}$ and thus it is connected, so $\overline{\mathbb{H}} \setminus K$ is an s-domain.

A function is said to be hypermeromorphic or slice regular meromorphic if it is of the form $g^{-*} * f$ where f, g are slice regular in Ω and f is not identically zero. In the sequel, we will consider intrinsic slice regular meromorphic functions which are obtained when both f, g are quaternionic intrinsic. In this case, a slice regular meromorphic function is of the simpler form $g^{-1}f$.

In view of Remark 4.4.9 we get the following quaternionic analogue of Corollary 4.4.8:

Corollary 4.4.11. *Let $K \in \mathfrak{R}_c(\overline{\mathbb{H}})$ and $\mathbb{B} \subset \mathbb{H}$ the open unit ball. Then there exist two unique quaternionic intrinsic slice hypermeromorphic functions*

$$\Phi : \overline{\mathbb{H}} \setminus K \to \overline{\mathbb{H}} \setminus \mathbb{B}, \quad \Psi : \overline{\mathbb{H}} \setminus \mathbb{B} \to \overline{\mathbb{H}} \setminus K, \quad \Psi = \Phi^{-1},$$

which are bijective and such that

$$\Phi(\infty) = \Psi(\infty) = \infty, \quad \Phi'(\infty) = \frac{1}{\Psi'(\infty)} > 0.$$

Proof. Let us consider $(\overline{\mathbb{H}} \backslash K)_I := (\overline{\mathbb{H}} \backslash K) \cap \mathbb{C}_I$. Then $(\overline{\mathbb{H}} \backslash K)_I$ is simply connected by hypothesis and symmetric with respect to the real line, since K is axially symmetric. As a consequence, $\overline{\mathbb{H}} \backslash K$ is axially symmetric. Let us set $\mathbb{D}_I = \mathbb{B} \cap \mathbb{C}_I$. By Corollary 4.4.8 there exist two bijective, bihyperholomorphic intrinsic maps

$$\Phi_I : (\overline{\mathbb{H}} \backslash K)_I \to \overline{\mathbb{C}_I} \backslash \overline{\mathbb{D}_I}, \quad \Psi_I : \overline{\mathbb{C}_I} \backslash \overline{\mathbb{D}_I} \to (\overline{\mathbb{H}} \backslash K)_I,$$

such that

$$\Phi_I(\infty) = \Psi_I(\infty) = \infty, \quad \Phi'_I(\infty) = \frac{1}{\Psi'_I(\infty)} > 0.$$

By Remark 4.4.9, Φ_I and Ψ_I extend to

$$\Phi : \overline{\mathbb{H}} \backslash K \to \overline{\mathbb{H}} \backslash \overline{\mathbb{B}}, \quad \Psi : \overline{\mathbb{H}} \backslash \overline{\mathbb{B}} \to \overline{\mathbb{H}} \backslash K$$

and $\Phi \in \mathcal{N}(\overline{\mathbb{H}} \backslash K)$, $\Psi \in \mathcal{N}(\overline{\mathbb{H}} \backslash \overline{\mathbb{B}})$.

Note that for every $J \in \mathbb{S}$ we have $\Phi|_{\mathbb{C}_J} : (\overline{\mathbb{H}} \backslash K)_J \to \overline{\mathbb{C}_J} \backslash \overline{\mathbb{D}_J}$ and $\Psi|_{\mathbb{C}_J} : \overline{\mathbb{C}_J} \backslash \overline{\mathbb{D}_J} \to (\overline{\mathbb{H}} \backslash K)_J$, since Φ and Ψ map each complex plane to itself. $\qquad\square$

Proposition 4.4.12. *The class $\mathfrak{R}_c(\overline{\mathbb{H}})$ contains all the possible compact sets K in $\mathbb{H}$ with $\overline{\mathbb{H}} \backslash K$ simply connected, for which the maps Φ and Ψ as in Corollary 4.4.11 are quaternionic intrinsic map.*

Proof. If we assume that $\overline{\mathbb{H}} \backslash K$ is mapped onto $\overline{\mathbb{H}} \backslash \overline{\mathbb{B}}$ by a quaternionic intrinsic map Φ such that $\Phi(\infty) = \infty$, $\Phi'(\infty) > 0$, then

$$\Phi|_{\mathbb{C}_I} : (\overline{\mathbb{H}} \backslash K) \cap \mathbb{C}_I \to (\overline{\mathbb{H}} \backslash \overline{\mathbb{B}}) \cap \mathbb{C}_I.$$

By its uniqueness, $\Phi|_{\mathbb{C}_I}$ is the map provided by the Riemann mapping theorem and it is complex intrinsic. Thus $(\overline{\mathbb{H}} \backslash K) \cap \mathbb{C}_I$ is symmetric with respect to the real axis and simply connected. Since $I \in \mathbb{S}$ is arbitrary, $\overline{\mathbb{H}} \backslash K$ must be also axially symmetric, which implies that K is axially symmetric and therefore $K \in \mathfrak{R}_c(\overline{\mathbb{H}})$. $\qquad\square$

4.4.2 Quaternionic Faber polynomials and Faber series

We now introduce and study properties of the Faber quaternionic polynomials attached to a set $K \in \mathfrak{R}_c(\overline{\mathbb{H}})$.

Let Φ and Ψ be the intrinsic slice hypermeromorphic functions given by Corollary 4.4.11, i.e.,

$$\Phi : \overline{\mathbb{H}} \backslash K \to \overline{\mathbb{H}} \backslash \overline{\mathbb{B}}, \quad \Psi = \Phi^{-1},$$

where

$$\Phi(q) = dq + d_0 + \sum_{i=-\infty}^{-1} q^i d_i, \quad d, d_i \in \mathbb{R}$$

and

$$\Psi(q) = qc + c_0 + \sum_{i=-\infty}^{-1} q^i c_i, \quad |q| > 1, \ c, c_i \in \mathbb{R}.$$

Reasoning as in the complex case, we can associate with each $K \in \mathfrak{R}_c(\overline{\mathbb{H}})$ the Faber polynomials as follows.

Let $n \in \mathbb{N} \bigcup \{0\}$ and consider $[\Phi(q)]^n$. Since $\Phi(q)$ is slice regular and intrinsic, $[\Phi(q)]^n$ is slice regular in $\overline{\mathbb{H}} \setminus K$ and we have:

$$[\Phi(q)]^n = \left(qd + d_0 + \frac{d_1}{q} + \frac{d_2}{q^2} + \cdots \right)^n$$

$$= d^n q^n + \sum_{k=-\infty}^{n-1} q^k d_{n,k}$$

$$= q^n d^n + \sum_{k=-\infty}^{-1} q^k d_{n,k} + \sum_{k=0}^{n-1} q^k d_{n,k} \qquad (4.14)$$

$$:= F_n(q) + \sum_{k=-\infty}^{-1} q^k d_{n,k}$$

$$:= F_n(q) + H_n(q),$$

where it is clear that $F_n(q)$ is an intrinsic slice regular polynomial and $H_n(q)$ is an intrinsic function that is slice regular outside K. It is immediate that H_n satisfies $|H_n(q)| \le C \cdot \frac{1}{|q|}$, for $|q|$ sufficiently large.

Definition 4.4.13. The polynomial part $F_n(q)$ of the above expression (4.14) is called the *Faber quaternionic polynomial of exact degree n* attached to the set K.

The construction shows that all $F_n(q)$ are intrinsic polynomials.

Remark 4.4.14. Replacing in the above computations $q = \Psi(w)$ for all $|w| > 1$, we easily obtain

$$F_n(\Psi(w)) = w^n - H_n(\Psi(w)) = w^n - \sum_{k=1}^{\infty} nw^{-k} b_{n,k}, \qquad (4.15)$$

for suitable coefficients $b_{n,k}$.

An alternative way to introduce the Faber quaternionic polynomials is the following (recall that $S^{-1}(p,q)$ denotes the slice regular Cauchy kernel):

Proposition 4.4.15. *Let $\Gamma_R = \{z \in \mathbb{H}; \ z = \Psi(w), |w| = R, \ R > 1\}$ and let Ω_R be its interior. Then the Faber polynomials are the Laurent coefficients of the function $S^{-1}(\Psi(w), q)\Psi'(w)$, i.e.,*

$$S^{-1}(\Psi(w), q)\Psi'(w) = \sum_{n=0}^{\infty} F_n(q) w^{-n-1}, \quad q \in \Omega_R, \ |w| > R. \qquad (4.16)$$

Proof. The functions $[\Phi(q)]^n$ and $F_n(q)$ are defined in $\overline{\mathbb{H}} \setminus K$ and hence so is $H_n(q)$. Since $H_n(q) = \sum_{k=-\infty}^{-1} q^k d_{n,k}$, it is clear that H_n vanishes at infinity. Then, the Cauchy formula for an unbounded domain, see [51], p. 38, implies

$$\frac{1}{2\pi} \int_{\Gamma_R \cap \mathbb{C}_I} S^{-1}(s,q)_I H_n(s) = 0, \quad q \in \Omega_R. \tag{4.17}$$

Using (4.14) we deduce that

$$F_n(q) = \frac{1}{2\pi} \int_{\Gamma_R \cap \mathbb{C}_I} S^{-1}(s,q) \, ds_I \, [\Phi(s)]^n, \quad q \in \Omega_R. \tag{4.18}$$

Let us set $s = \Psi(w)$ and consider $s \in \mathbb{C}_I$, so that also $w \in \mathbb{C}_I$. Then $ds_I = \Psi'(w) dw_I$, and substituting in (4.18) we obtain

$$F_n(q) = \frac{1}{2\pi} \int_{\{|w|=R\} \cap \mathbb{C}_I} S^{-1}(\Psi(w),q) \, \Psi'(w) dw_I \, w^n, \quad q \in \Omega_R. \tag{4.19}$$

This proves that $F_n(q)$ is the coefficient of w^{-n-1} in the Laurent expansion of $S^{-1}(\Psi(w),q)\Psi'(w)$, (see [51], p. 41, where the case of left slice regular functions is treated; here the function $S^{-1}(\Psi(w),q)\Psi'(w)$ is right slice regular in w). Hence (4.16) holds. □

For $K \in \mathfrak{R}_c(\overline{\mathbb{H}})$, consider the set

$$\mathcal{R}(K) = \{f : K \to \mathbb{H}; \ f \text{ is slice regular in int}(K) \text{ and continuous on } K\}.$$

Inspired by Remark 4.4.2 and recalling that the Cauchy integral formula (1.13) is independent of the choice of the complex plane on which we integrate, we give the following definition:

Definition 4.4.16. The Faber coefficients of a function $f \in \mathcal{R}(K)$ are defined as

$$a_n(f) = \frac{1}{2\pi} \int_{\partial \mathbb{B} \cap \mathbb{C}_I} w^{-n-1} dw_I f(\Psi(w)), \quad n = 1, 2, \ldots, \quad dw_I = -I \, dw,$$

assuming that $\Psi : \overline{\mathbb{H}} \setminus \mathbb{B} \cap \mathbb{C}_I \to \overline{\mathbb{H}} \setminus K$ can be extended slice regularly and bijectively on $\partial \mathbb{B} \cap \mathbb{C}_I$.

Definition 4.4.17. If $f(q) = \sum_{n=0}^{\infty} F_n(q) a_n(f)$ for all $q \in K$, then we say that f can be expanded in a *Faber series* on K.

When we consider a continuum then, by analogy with Remark 4.4.4, we can still give a definition of Faber coefficients, even though we have no information of the boundary of the set.

Definition 4.4.18. Let $K \in \mathfrak{R}_c(\overline{\mathbb{H}})$ be such that for all $I \in \mathbb{S}$, the set $K \cap \mathbb{C}_I$ is a continuum. The *Faber coefficients* attached to f and K are

$$a_n(f) = \frac{1}{2\pi} \int_{\partial \mathbb{B}_R \cap \mathbb{C}_I} w^{-n-1} \, dw_I \, f(\Psi(w)), \quad n = 1, 2, \ldots, \quad dw_I = -I dw,$$

for an arbitrary fixed $R > 1$.

We now extend to the quaternionic setting Theorem 4.4.5 and Theorem 4.4.6 presented in Introduction. We first prove the results in the case when we deal with a continuum.

Theorem 4.4.19. *Let $K \in \mathfrak{R}_c(\mathbb{H})$ be such that for all $I \in \mathbb{S}$, $K \cap \mathbb{C}_I$ is a continuum. Suppose that f is slice regular in K, and there exists $R > 1$ such that f is analytic in Ω_R, where Ω_R denotes the interior of the closed level hypersurface $\Gamma_R = \{\Psi(w); |w| = R\}$. Then, for any $1 < r < R$ we have $|F_n(q)| \leq C(r)r^n$, for all $n = 0, 1, \ldots$, and all $q \in \overline{\Omega_r}$, where $C(r) > 0$ is independent of n.*

Proof. First of all, we observe that last line of (4.14) defines two functions $F_n(q)$, $H_n(q)$ which are slice regular in the interior and in the exterior of K, respectively. Thus, by the Cauchy formula,

$$H_n(q) = -\frac{1}{2\pi} \int_{\partial \mathbb{B}_R \cap \mathbb{C}_I} S^{-1}(s, q) \, ds_I \, (F_n(s) - [\Phi(s)]^n)$$

$$= \frac{1}{2\pi} \int_{\partial \mathbb{B}_R \cap \mathbb{C}_I} S^{-1}(s, q) \, ds_I \, [\Phi(s)]^n, \qquad q \in \overline{\Omega_R}^c. \qquad (4.20)$$

Consider the function $F_n(\Psi(w)) = \tilde{F}_n(w)$, and let R_1 be such that $1 < R_1 < R$. From (4.15) and (4.20) we have

$$\tilde{F}_n(w) = w^n - \frac{1}{2\pi} \int_{\partial \mathbb{B}_{R_1} \cap \mathbb{C}_I} S^{-1}(\Psi(v), \Psi(w)) \, \Psi'(v) \, dv_I \, v^n \qquad (4.21)$$

Setting $|w| = r$ and $R_1 < r < R$ in (4.21), we obtain

$$|\tilde{F}_n(w)| \leq r^n + R_1^{n+1} \max_{|v|=R_1, |w|=r} \left(|S^{-1}(\Psi(v), \Psi(w)) \, \Psi'(v)| \right)$$

$$\leq r^n + r^{n+1} M(r, R_1),$$

$M = M(r, R_1)$, which gives

$$|\tilde{F}_n(w)| \leq r^n M_1,$$

with $M_1 = M_1(r, R_1)$. Now the result follows by setting $C(r) = M_1$. $\qquad \square$

Theorem 4.4.20. *Let $K \in \mathfrak{R}_c(\mathbb{H})$ be such that for every $I \in \mathbb{S}$, $K \cap \mathbb{C}_I$ is a continuum. If $f \in \mathcal{R}(K)$, then $f(q) = \sum_{n=0}^{\infty} F_n(q) a_n(f)$, uniformly on K. The rate of convergence of the series is geometric.*

Proof. To say that $f \in \mathcal{R}(K)$ means that $f \in \mathcal{R}(\Omega_R)$, where $R > 1$. Let ρ be such that $1 < \rho < R$. Then for any $q \in K$, $f(q)$ can be expressed as

$$f(q) = \frac{1}{2\pi} \int_{\Gamma_\rho \cap \mathbb{C}_I} S^{-1}(s, q) ds_I f(s)$$

$$= \frac{1}{2\pi} \int_{|w|=\rho \cap \mathbb{C}_I} S^{-1}(\Psi(w), q) \, dw_I \, \Psi'(w) f(\Psi(w)). \qquad (4.22)$$

The series (4.16) for $q \in K$, $|w| = \rho$ converges uniformly and substituting this series expansion in (4.22) one obtains $f(q) = \sum_{n=0}^{\infty} F_n(q)a_n(f)$, where a_n are given by Definition 4.4.18 with ρ instead of R, or by formula

$$a_n = \frac{1}{2\pi} \int_{\partial \mathbb{B}_\rho \cap \mathbb{C}_I} w^{-n-1} dw_I f(\Psi(w)) \tag{4.23}$$

with $\rho > 1$.

To find the rate of convergence, let us choose $1 < r < \rho < R$. For $w \in \partial \mathbb{B}_\rho \cap \mathbb{C}_I$ we have $|w| = \rho$ and $f \circ \Psi$ is obviously continuous, and thus bounded, on $w \in \partial \mathbb{B}_\rho \cap \mathbb{C}_I$. By (4.23) we immediately get $|a_n(f)| \leq \frac{M}{\rho^n}$ for all $n \in \mathbb{N}$. Combining this with the estimate in Theorem 4.4.19, we conclude that

$$|F_n(q)a_n(f)| \leq c_1 \cdot \left(\frac{r}{\rho}\right)^n, \quad \text{for all } n \in \mathbb{N}.$$

This obviously implies a geometric rate of convergence for the series $\sum_{n=0}^{\infty} F_n(q)a_n(f)$. □

To prove next result we require suitable assumptions on the boundary of the set we consider.

Theorem 4.4.21. *Let $\Omega \subset \mathbb{H}$ be a bounded, simply connected, axially symmetric domain such that $\Omega \in \mathfrak{R}_c(\overline{\mathbb{H}})$. Assume that the boundary $\partial \Omega$ is such that $\partial \Omega \cap \mathbb{C}_I$ is an analytic Jordan curve γ_I in $\mathbb{C}_I$ with nonzero derivative no interior point of which is a cluster point of the set of boundary points not belonging to the curve γ_I. Then every function $f \in R(\Omega)$ can be expanded in series of the Faber polynomials attached to the continuum $\overline{\Omega}$ and the series converges uniformly inside Ω.*

Proof. Let $\Phi : \overline{\mathbb{H}} \backslash \Omega \to \overline{\mathbb{H}} \backslash \mathbb{B}$ as in Corollary 4.4.11. We claim that since $\partial \Omega \cap \mathbb{C}_I$ is an analytic curve in $\mathbb{C}_I$ with nonzero derivative, the function Φ can be continued slice regularly and univalently across the boundary of Ω. To prove this fact, we follow the argument in [110] p. 44.

Let z_I be the variable in $\mathbb{C}_I$ and let γ_I be given by $z_I = z_I(t)$, $t \in [a, b]$. Let $t_0 \in (a, b)$ and $z_0 = z_I(t_0)$. Since γ_I is analytic, the function $z_I(t)$ can be expanded in power series in a neighborhood of $t = t_0$. By taking complex values of t belonging to a suitable complex neighborhood U_{t_0} of t_0, we obtain a function $z_I(t)$ analytic. Since $z_I'(t_0) \neq 0$, by taking U_{t_0} sufficiently small, we can assume that U_{t_0} is mapped univalently by $z_I(t)$ into $V_I \ni z_0$, $V_I \subset \mathbb{C}_I$. Thus, we can assume that $U_{t_0} \cap \mathbb{C}_I^+$ (or $U_{t_0} \cap \mathbb{C}_I^-$) is mapped in $V \cap \Omega$ and has a common boundary arc with γ_I. The function

$$\Phi \circ z_I : U_{t_0} \cap \mathbb{C}_I^+ \to V \cap \Omega$$

is holomorphic. In fact $\Phi : \Omega \cap \mathbb{C}_I \to \mathbb{C}_I$ is intrinsic and slice regular, so it is a holomorphic map for all $I \in \mathbb{S}$, and maps univalently $U_{t_0} \cap \mathbb{C}_I^+$ into the neighborhood of $w_0 = \Phi(z_0)$ lying in $\mathbb{B} \cap \mathbb{C}_I$ and adjacent to the boundary. Standard arguments show that $\Phi(z_I(t))$ is continuous in $U_{t_0} \cap \mathbb{C}_I^+$ up to the real axis and thus, by the symmetry principle, it extends across the real axis. In particular, $\Phi(z_I(t))$ is holomorphic at t_0 and with nonzero derivative. Also the inverse function $t = t(z_I)$ is regular at z_0 and $t'(z_0) \neq 0$. Thus $\Phi(z_I(t(z_I))) = \Phi(z_I)$ is holomorphic at z_0 and with nonzero derivative.

So, since the restriction of the function Φ to any complex plane $\mathbb{C}_I$ can be analytically and univalently continued on $\partial\Omega \cap \mathbb{C}_I$, the map $w = \Phi(q)$ can be continued slice regularly and univalently across the boundary of Ω and it is then univalent into $\mathbb{B}_{\rho_0} = \{w \in \mathbb{H}; \; |w| < \rho_0\}$. Thus $q = \Psi(w)$ is slice regular (and quaternionic intrinsic) in $|w| > \rho_0$ except at infinity where it has a simple pole.

Let $q \in \Omega$ then $q \in \Omega_\rho$ for a suitable ρ such that $\rho_0 < \rho < 1$. The Cauchy formula gives

$$f(q) = \frac{1}{2\pi} \int_{\Gamma_\rho \cap \mathbb{C}_I} S^{-1}(s, q) \, ds_I f(s)$$

$$= \frac{1}{2\pi} \int_{\partial\mathbb{B}_\rho \cap \mathbb{C}_I} S^{-1}(\Psi(w), q) \, \Psi'(w) \, dw_I f(\Psi(w)). \qquad (4.24)$$

Formula (4.16) gives the series

$$S^{-1}(\Psi(w), q)\Psi'(w) = \sum_{n=0}^{\infty} F_n(q) w^{-n-1}, \quad q \in \Omega_\rho, \; |w| \geq \rho,$$

which converges uniformly for $|w| \geq \rho$ and q in any fixed compact set K in Ω_ρ. In fact, one can show that the inequality stated in Theorem 4.4.19 holds also in this case, thus establishing that the series above has the geometric series as a majorant.

Finally, substituting this series in (4.24) we obtain

$$f(q) = \frac{1}{2\pi} \int_{\partial\mathbb{B}_\rho \cap \mathbb{C}_I} \left(\sum_{n=0}^{\infty} F_n(q) w^{-n-1} \right) dw_I f(\Psi(w))$$

$$= \sum_{n=0}^{\infty} F_n(q) \frac{1}{2\pi} \int_{\partial\mathbb{B}_\rho \cap \mathbb{C}_I} w^{-n-1} dw_I f(\Psi(w))$$

$$= \sum_{n=0}^{\infty} F_n(q) a_n,$$

where a_n is given by (4.23), and the statement follows.   □

Remark 4.4.22. The technical hypotheses in the statement of Theorem 4.4.21 are imposed by the fact that the possibility of approximation in a domain depends strongly on the smoothness and geometrical properties of its boundary.

The results obtained in this section open the possibility of extending other results on Faber polynomials and Faber expansions from the complex to the quaternionic setting. As examples, we refer to asymptotic properties of Faber polynomials, conditions for convergence and uniqueness for Faber series, Faber polynomials and the theory of univalent functions, Faber operators and their properties, and other properties. For details in the complex case, see, e.g., the books [183] and [74].

Example 4.4.23. To obtain some concrete examples of compact sets in $\mathbb{H}$ where the expansion by Faber polynomials is possible we can consider the set G as in Example 4.2.11, or 4.2.12, or 4.2.13 and its axially symmetric completion $\Omega_G = \bigcup_{x+iy \in G, \ I \in \mathbb{S}} (x + Iy)$. It is worth recalling that $\overline{\Omega}_G$ is a starlike set with respect to the origin.

4.5 Approximation by Polynomials in Bergman Spaces

It was shown in [41] that in the framework of slice regular functions two kinds of Bergman spaces can be considered. Before to introduce them, we first recall some known facts about Bergman spaces in the complex case (for other details see, e.g., [120], Chapter 1, Section 1.1 or [66], pp. 30–32).

Let $\mathbb{D}$ be the open unit disk. For $0 < p < +\infty$, $-1 < \alpha < +\infty$ and $\rho_\alpha(z) = (\alpha + 1)(1 - |z|^2)^\alpha$, the weighted Bergman space, denoted by $\mathcal{A}^p_\alpha(\mathbb{D})$, is the space of all analytic functions in $\mathbb{D}$ such that

$$\left(\int_{\mathbb{D}} |f(z)|^p dA_\alpha(z) \right)^{1/p} < +\infty,$$

where $dA_\alpha(z) = \rho_\alpha(z)dA(z)$, with $dA(z) = \frac{1}{\pi}dxdy = \frac{1}{\pi}rdrd\theta$, $z = x + iy = re^{i\theta}$, the normalized area measure on $\mathbb{D}$.

The norm

$$\|f\|_{p,\alpha} = \left(\int_{\mathbb{D}} |f(z)|^p dA_\alpha(z) \right)^{1/p},$$

for $1 \leq p < +\infty$, makes $(\mathcal{A}^p_\alpha(\mathbb{D}), \|\cdot\|_{p,\alpha})$ a Banach space, while for $0 < p < 1$, $\mathcal{A}^p_\alpha$ is a complete metric space with the metric $d(f,g) = \|f - g\|^p_{p,\alpha}$.

It is worth noting that, according to Proposition 1.3, p. 4 in [120] (see also Theorem 3, p. 30 and Theorem 4, p. 31 in [66]), for any $0 < p < +\infty$, the set of polynomials is dense in $\mathcal{A}^p_\alpha(\mathbb{D})$ in the norm $\|\cdot\|_{p,\alpha}$ (for $1 \leq p < +\infty$) and in the quasi-norm $\|\cdot\|_{p,\alpha}$ (for $0 < p < 1$). The proof of this result is not constructive: one shows that for any $f \in \mathcal{A}^p_\alpha(\mathbb{D})$, there exists a sequence of polynomials $(p_n)_{n \in \mathbb{N}}$ such that $\|p_n - f\|_{p,\alpha} \to 0$ as $n \to \infty$.

A classical constructive proof for the above denseness result in Bergman spaces on the unit disk, resorts to the partial sum of the Taylor series of f at 0, this being valid only for $1 < p < +\infty$. Also, we can mention Theorem 3.5 in the paper [167], Theorem 1.3 in [166], Theorem 1.2 in [34], where convolution polynomials were used to obtain estimates in terms of the modulus of continuity. In addition, based also on the convolution polynomials, constructive proofs for the denseness of polynomials with much better quantitative estimates (in terms

of higher order L^p-moduli of smoothness and in terms of L^p-best approximation)
were obtained in the recent paper [81].

We now consider Bergman spaces in the quaternionic setting. We begin with
the following definition, which has not been previously considered in the literature
in this generality.

Below we denote the unit ball $B(0;1)$ by $\mathbb{B}_1$.

Definition 4.5.1. Let $0 < p < +\infty$ and $-1 < \alpha < +\infty$. For

$$\rho_\alpha(q) = (\alpha + 1)(1 - |q|^2)^\alpha,$$

the *weighted Bergman space of the first kind* on the open unit ball $\mathbb{B}_1$ is denoted
by $\mathcal{A}_\alpha^p(\mathbb{B}_1)$ and is defined as the space of all functions $f \in \mathcal{R}(\mathbb{B}_1)$, such that

$$\|f\|_{p,\alpha} := \left[\int_{\mathbb{B}_1} |f(q)|^p dA_\alpha(q) \right]^{1/p} < +\infty,$$

where $dA_\alpha(q) = \rho_\alpha(q) dm(q)$, with $dm(q)$ representing the Lebesgue volume element in $\mathbb{R}^4$.

Remark 4.5.2. With the same techniques used in the complex case, one can verify
that for $1 \le p < +\infty$, $\| \cdot \|_{p,\alpha}$ has the properties of a norm, while for $0 < p < 1$,
$\|f - g\|_{p,\alpha}^p$ has the properties of a quasi-norm.

To introduce the weighted Bergman spaces of the second kind, we need the
following definition:

Definition 4.5.3. Let $I \in \mathbb{S}$, $-1 < \alpha < +\infty$ and $0 < p < +\infty$, let us set $\mathbb{B}_{1,I} = \mathbb{B}_1 \cap \mathbb{C}_I$ and

$$\|f\|_{p,\alpha,I} = \left(\int_{\mathbb{B}_{1,I}} |f(q)|^p dA_{\alpha,I}(q) \right)^{1/p},$$

where $dA_{\alpha,I}(q) = (\alpha + 1)(1 - |q|^2)^\alpha dA_I(q)$, and $dA_I(q)$ is the normalized area
measure on $\mathbb{B}_{1,I}$.

The space of all f with the property that $\|f\|_{p,\alpha,I} < +\infty$ will be denoted by
$\mathcal{A}_{\alpha,I}^p(\mathbb{B}_{1,I})$.

We are now in position to introduce:

Definition 4.5.4. The *weighted Bergman space of second kind* is denoted by
$\mathcal{A}_\alpha^{(2),p}(\mathbb{B}_1)$ and is defined as the space of all $f \in \mathcal{R}(\mathbb{B}_1)$ with the property that
for some $I \in \mathbb{S}$ we have $f \in \mathcal{A}_{\alpha,I}^p(\mathbb{B}_{1,I})$.

Remark 4.5.5. By Proposition 3.4 in [31], which is basically based on the repre-
sentation formula, if $f \in \mathcal{A}_{\alpha,I}^p(\mathbb{B}_{1,I})$ for some $I \in \mathbb{S}$, then we have $f \in \mathcal{A}_{\alpha,J}^p(\mathbb{B}_{1,J})$
for all $J \in \mathbb{S}$.

For various properties of the Bergman spaces of the first and second kind, see, e.g., [31] and [42].

The goal of the present section is to extend the above mentioned results on the denseness of polynomials in the complex case, to quaternionic weighted Bergman spaces of the first and second kind. Firstly, we study approximation in weighted Bergman spaces of the first kind. The result obtained is valid under the hypothesis that $\alpha \geq 0$. Then, we study approximation in weighted Bergman spaces of the second kind, obtaining a result of general validity. We also obtain quantitative estimates in terms of higher order moduli of smoothness and of best approximation quantity.

Except for the results whose authors are mentioned explicitly, all the other results in this section were obtained in the paper of Gal and Sabadini [84].

4.5.1 Approximation in Bergman spaces of first kind

The Bergman spaces of first and second kind defined above have been introduced in [41]. Their weighted versions were studied in [31], but only for the second kind, while the weighted Bergman spaces of the first kind are new. Using the above notations we can state and prove the main result of the polynomial approximation in this setting.

Theorem 4.5.6. *Let* $\alpha \geq 0$. *For every* $f \in A_\alpha^p(\mathbb{B}_1)$, *there exists a sequence of polynomials* $(p_n)_{n \in \mathbb{N}}$ *such that* $\|p_n - f\|_{p,\alpha} \to 0$ *as* $n \to +\infty$.

Proof. We divide the proof in two steps.

Step 1. Let $0 < r < 1$, $f \in A_\alpha^p(\mathbb{B}_1)$ and define $f_r(q) = f(rq)$.

First, we will prove that for any $\varepsilon > 0$, there exists $r \in (0,1)$ (depending on c and sufficiently close to 1), such that

$$\|f_r - f\|_{p,\alpha} = \int_{\mathbb{B}_1} |f_r(q) - f(q)|^p dA_\alpha(q) < \varepsilon. \tag{4.25}$$

It is evident that $\overline{\mathbb{B}_1}$ is strictly included in $\mathbb{B}_{1/r}$ and that $f_r \in \mathcal{R}(\mathbb{B}_{1/r})$. For any $\delta \in (0,1)$ and $r \in (0,1)$, we have the obvious inequality

$$\int_{\mathbb{B}_1} |f_r(q) - f(q)|^p dA_\alpha(q)$$
$$\leq \int_{\mathbb{B}_\delta} |f_r(q) - f(q)|^p dA_\alpha(q) + \int_{\delta < |q| < 1} |f_r(q) - f(q)|^p dA_\alpha(q)$$
$$:= I_1(\delta, r) + I_2(\delta, r). \tag{4.26}$$

Since for $a, b \geq 0$, we have $(a+b)^p \leq 2^{p-1}(a^p + b^p)$ if $1 \leq p < +\infty$ and $(a+b)^p \leq a^p + b^p$ if $0 < p < 1$, we can write for all $0 < p < +\infty$:

$$(a+b)^p \leq C_p(a^p + b^p), \quad \text{where } C_p = \max\{2^{p-1}, 1\}. \tag{4.27}$$

Therefore

$$I_2(\delta, r) \leq C_p \int_{\delta < |q| < 1} |f_r(q)|^p dA_\alpha(q) + C_p \int_{\delta < |q| < 1} |f(q)|^p dA_\alpha(q). \qquad (4.28)$$

Now, let $\varepsilon > 0$ be fixed. Since for any $\delta > 0$ we have

$$\int_{\mathbb{B}_1} |f(q)|^p dA_\alpha(q) = \int_{|q| \leq \delta} |f(q)|^p dA_\alpha(q) + \int_{\delta < |q| < 1} |f(q)|^p dA_\alpha(q),$$

and since it is evident that

$$\int_{\mathbb{B}_1} |f(q)|^p dA_\alpha(q) = \lim_{\delta \nearrow 1} \int_{|q| \leq \delta} |f(q)|^p dA_\alpha(q),$$

it follows that

$$\lim_{\delta \nearrow 1} \int_{\delta < |q| < 1} |f(q)|^p dA_\alpha(z) = 0.$$

Therefore, there exists $0 < \delta_1 < 1$ (depending on ε and C_p, but independent of r), sufficiently close to 1, such that for any $\delta \in (\delta_1, 1)$ we have

$$\int_{\delta < |q| < 1} |f(q)|^p dA_\alpha(q) < \varepsilon/(4C_p). \qquad (4.29)$$

In what follows, we observe that because at this step we are in fact interested only in r sufficiently close to 1, in all the previous and next relations, we can assume that $0 < r_0 < r < 1$, where r_0 is an absolutely independent constant.

Let us denote the first Lebesgue integral (in $\mathbb{R}^4$) which appears in (4.28) by

$$I_3 := \int_{\delta < |q| < 1} |f(rq)|^p dA_\alpha(q).$$

Denoting $q = x + iy + jz + ku$, the spherical coordinates $\rho \in [0, 1], \varphi_1 \in [0, \pi], \varphi_2 \in [0, \pi], \varphi_3 \in [0, 2\pi]$ in $\mathbb{R}^4$ are given by (see, e.g., [19])

$$\rho = \sqrt{x^2 + y^2 + z^2 + u^2}, \quad x = \rho \cos(\varphi_1), \quad y = \rho \sin(\varphi_1) \cos(\varphi_2),$$

$$z = \rho \sin(\varphi_1) \sin(\varphi_2) \cos(\varphi_3), \quad u = \rho \sin(\varphi_1) \sin(\varphi_2) \sin(\varphi_3)$$

and

$$dm(q) = \rho^3 \sin^2(\varphi_1) \sin(\varphi_2) dr d\varphi_1 d\varphi_2 d\varphi_3,$$

where

$$\varphi_1 = \arccos\left(\frac{x}{\sqrt{x^2 + y^2 + z^2 + u^2}}\right), \quad \varphi_2 = \arccos\left(\frac{y}{\sqrt{y^2 + z^2 + u^2}}\right),$$

$$\varphi_3 = \arccos\left(\frac{z}{\sqrt{z^2 + u^2}}\right), \quad \text{if } u \geq 0,$$

$$\varphi_3 = 2\pi - \arccos\left(\frac{z}{\sqrt{z^2 + u^2}}\right), \text{ if } u < 0.$$

Now, making the substitution $w = rq$ in I_3, i.e., $q = \frac{w}{r}$, and using the above formulas, we easily obtain that

$$I_3 = \int_{r\delta < |w| < r} |f(w)|^p (\alpha + 1) \left(1 - \left|\frac{w}{r}\right|^2\right)^\alpha dm(w/r)$$

$$= \frac{1}{r^4} \cdot \int_{r\delta < |w| < r} |f(w)|^p (\alpha + 1) \left(1 - \left|\frac{w}{r}\right|^2\right)^\alpha dm(w)$$

$$\leq \frac{1}{r_0^4} \cdot \int_{r\delta < |w| < r} |f(w)|^p (\alpha + 1) \left(1 - \left|\frac{w}{r}\right|^2\right)^\alpha dm(w),$$

for all $r \in [r_0, 1)$ and all $\delta \in (\delta_1, 1)$.

Let $\alpha \geq 0$ be fixed. Since $r < 1$,

$$\left(1 - \left|\frac{w}{r}\right|^2\right)^\alpha \leq \left(1 - |w|^2\right)^\alpha$$

and therefore by using (4.29), for all $r \in [r_0, 1)$ and all $\delta \in (\delta_1, 1)$ we obtain

$$I_3 \leq \frac{1}{r_0^4} \cdot \int_{r\delta < |w| < r} |f(w)|^p (\alpha + 1) \left(1 - |w|^2\right)^\alpha dm(w)$$

$$\leq \frac{1}{r_0^4} \cdot \int_{r\delta < |w| < 1} |f(w)|^p (\alpha + 1) \left(1 - |w|^2\right)^\alpha dm(w)$$

$$\leq \frac{1}{r_0^4} \cdot \int_{r\delta < |w| < \delta} |f(w)|^p (\alpha + 1) \left(1 - |w|^2\right)^\alpha dm(w)$$

$$+ \frac{1}{r_0^4} \cdot \int_{\delta < |w| < 1} |f(w)|^p (\alpha + 1) \left(1 - |w|^2\right)^\alpha dm(w)$$

$$\leq \frac{1}{r_0^4} \cdot \int_{r\delta < |w| < \delta} |f(w)|^p (\alpha + 1) \left(1 - |w|^2\right)^\alpha dm(w) + \frac{\varepsilon}{4r_0^4 C_p}.$$

Since $f \in A_\alpha^p(\mathbb{B}_1)$, the Lebesgue integral $\int |f(w)|^p (\alpha + 1) \left(1 - |w|^2\right)^\alpha dm(w)$ is absolutely continuous as a set function on the measurable subsets in $\mathbb{B}_1$. Therefore, for the given $\varepsilon > 0$, there exists $\eta \in (0, 1]$ (depending on ε only), such that for all Lebesgue measurable $E \subset \mathbb{B}_1$ with $m(E) \leq \eta$, we have

$$\int_E |f(w)|^p (\alpha + 1) \left(1 - |w|^2\right)^\alpha dm(w) < \varepsilon.$$

Now, denoting by $\text{Vol}(\mathbb{B}_R)$ the volume of the ball $\mathbb{B}_R$ in $\mathbb{R}^n$, it is known that (see, e.g., [136], or [56], p. 125) $\text{Vol}(\mathbb{B}_R) = \frac{2\pi^{n/2}}{n\Gamma(n/2)} \cdot R^n$. Taking $n = 4$, we readily see that

$$m(\{r\delta < |w| < \delta\}) = m(\{|w| < \delta\}) - m(\{|w| \le r\delta\})$$

$$= \text{Vol}(\mathbb{B}_\delta) - \text{Vol}(\mathbb{B}_{r\delta}) = \frac{\pi^2}{2}\delta^4 - \frac{\pi^2}{2}r^4\delta^4$$

$$= \frac{\pi^2}{2}\delta(1-r)[\delta^3(1+r+r^2+r^3)]$$

$$\le 2\pi^2\delta(1-r).$$

Choosing $r_0' = 1 - \frac{\eta}{2\pi^2} \in (0,1)$ (r_0' depends on ε only), we get

$$2\pi^2\delta(1-r) \le 2\pi^2(1-r) \le 2\pi^2(1-r_0') = \eta, \text{ for all } r \ge \max\{r_0, r_0'\},$$

which by the above mentioned absolute continuity of the integral implies

$$I_3 \le \frac{1}{r_0^4} \cdot \int_{r\delta < |w| < \delta} |f(w)|^p (\alpha+1)\left(1-|w|^2\right)^\alpha dm(w) + \frac{\varepsilon}{4r_0^4 C_p}$$

$$\le \frac{\varepsilon}{r_0^4} + \frac{\varepsilon}{4r_0^4 C_p},$$
(4.30)

for all $\delta \in (\delta_1, 1)$ and $r \ge \max\{r_0, r_0'\}$.

Combining now (4.29) with (4.30), from (4.28) we obtain that

$$I_2(\delta, r) < \varepsilon/2$$
(4.31)

for any $\delta \in (\delta_1, 1)$ and $r \ge \max\{r_0, r_0'\}$. Now, let $\delta \in (\delta_1, 1)$ be fixed. Since both f and f_r are slice regular in $\overline{\mathbb{B}_\delta}$, there exists $M_\delta > 0$ (the maximum of the slice derivative of f on $\overline{\mathbb{B}_\delta}$), such that by the mean value theorem on each slice $\mathbb{C}_I$ (in fact, by the mean value theorem in the complex case), we get

$$|f(rq) - f(q)| \le M_\delta \cdot |q| \cdot |r-1|.$$

Since M_δ is independent of I, it follows that

$$|f(rq) - f(q)| \le M_\delta \cdot |q| \cdot |r-1| \le M_\delta|r-1|,$$

for all $q \in \overline{\mathbb{B}_\delta}$. Therefore, given $\varepsilon > 0$, it is clear that there exists $r > \max\{r_0, r_0'\}$ (sufficiently close to 1), such that

$$|r-1|^p < \frac{\varepsilon}{2M_\delta^p K},$$

with

$$K = \int_{\mathbb{B}_1} dA_\alpha(q) < +\infty$$

(see the next Step 2). Consequently,

$$I_1(\delta, r) = \int_{\mathbb{B}_\delta} |f_r(q) - f(q)|^p dA_\alpha(q) < \varepsilon/2$$

which by (4.26) and (4.31) immediately implies (4.25).

Step 2. Let $r \in (0, 1)$ satisfy (4.25). Denoting by $T_n(f)(q)$ the partial sum of order n of the Taylor expansion of f in the open unit ball $\mathbb{B}_1$, it is clear that $T_n(f)(rq)$ is a polynomial of degree n and that $p_n(q) = T_n(f)(rq)$, $n \in \mathbb{N}$, is a sequence of polynomials uniformly convergent to f_r on the closed unit ball $\overline{\mathbb{B}_1}$.

Denoting by $\| \cdot \|$ the uniform norm on $\overline{\mathbb{B}_1}$, for each $\varepsilon > 0$ there exists $n_0 \in \mathbb{N}$, such that $\|f_r - p_n\|^p < \varepsilon$, for all $n \geq n_0$. Choosing $n \geq n_0$, it easily follows that

$$\|f_r - p_n\|_{p,\alpha} \leq \|f_r - p_n\|^p \cdot \int_{\mathbb{B}_1} (\alpha + 1)(1 - |q|^2)^\alpha dm(q)$$
$$< (\alpha + 1) \cdot \varepsilon \cdot \int_{\mathbb{B}_1} (1 - |q|^2)^\alpha dm(q), \tag{4.32}$$

where $dm(q)$ is the Lebesgue volume element in $\mathbb{R}^4$.

Moreover, one can prove that

$$\int_{\mathbb{B}_1} (1 - |q|^2)^\alpha dm(q) < +\infty, \quad \text{for } -1 < \alpha < 0.$$

We note that this fact is obvious for $\alpha \geq 0$.

Indeed, let us introduce the spherical coordinates $r \in [0, 1], \varphi_1 \in [0, \pi], \varphi_2 \in [0, \pi], \varphi_3 \in [0, 2\pi]$ in $\mathbb{R}^4$ as in Step 1. They are related with the Cartesian coordinates x, y, z, u by

$$r = \sqrt{x^2 + y^2 + z^2 + u^2}, \quad x = r\cos(\varphi_1), \quad y = r\sin(\varphi_1)\cos(\varphi_2),$$

$$z = r\sin(\varphi_1)\sin(\varphi_2)\cos(\varphi_3), \quad u = r\sin(\varphi_1)\sin(\varphi_2)\sin(\varphi_3),$$

and

$$dm(q) = r^3 \sin^2(\varphi_1)\sin(\varphi_2)drd\varphi_1 d\varphi_2 d\varphi_3.$$

Consequently, we obtain

$$\int_{\mathbb{B}_1} (1 - |q|^2)^\alpha dm(q)$$
$$= \int_0^{2\pi} \int_0^\pi \int_0^\pi \int_0^1 \frac{1}{(1 - r^2)^{-\alpha}} \cdot r^3 \sin^2(\varphi_1)\sin(\varphi_2)drd\varphi_1 d\varphi_2 d\varphi_3 < +\infty,$$

since for $\alpha \in (-1, 0)$ we have

$$\int_0^1 \frac{r^3}{(1 - r^2)^{-\alpha}} dr \leq \int_0^1 \frac{r}{(1 - r^2)^{-\alpha}} dr = \frac{1}{2} \int_0^1 \frac{ds}{s^{-\alpha}} = \frac{1}{2(-\alpha + 1)}.$$

In conclusion, the inequality

$$\|f - p_n\|_{p,\alpha} \leq C_p(\|f - f_r\|_{p,\alpha} + \|f_r - p_n\|_{p,\alpha}),$$

where $C_p = 1$ if $p \geq 1$, and $C_p > 1$, if $0 < p < 1$, and relations (4.25) and (4.32) yield the desired conclusion. $\qquad\square$

Remark 4.5.7. In the case when $\alpha \in (-1, 0)$, although the method of proof of Theorem 4.5.6 works at Step 2, it does not work at Step 1. Indeed, in this case, for I_3 we get the relation

$$I_3 = \frac{1}{r^4} \cdot \int_{r\delta < |w| < r} |f(w)|^p (\alpha + 1) \left(1 - \left|\frac{w}{r}\right|^2\right)^\alpha dm(w)$$

$$= \frac{r^{-2\alpha}}{r^4} \cdot \int_{r\delta < |w| < r} |f(w)|^p (\alpha + 1) \left(r^2 - |w|^2\right)^\alpha dm(w),$$

but now $(r^2 - |w|^2)^\alpha > (1 - |w|^2)^\alpha$ for all $|w| < r$ (and not $(r^2 - |w|^2)^\alpha \leq (1 - |w|^2)^\alpha$ as we would need for our proof). For this reason, the case $-1 < \alpha < 0$ in Theorem 4.5.6 remains as an open problem.

4.5.2 Approximation in Bergman spaces of second kind

In this subsection we prove the denseness of polynomials in weighted Bergman spaces of the second kind, including a result with quantitative estimates in terms of higher-order moduli of smoothness and in terms of the best approximation quantity.

Before we state our main result, we need a technical fact.

Proposition 4.5.8. *Let $p \geq 1$ (resp. $0 < p < 1$) and let $\|\cdot\|_{p,\alpha,I}$ be the norm (resp. quasi-norm) in $\mathcal{A}_{\alpha,I}^p(\mathbb{B}_{1,I})$. Then $\|\cdot\|_{p,\alpha,I}$ and $\|\cdot\|_{p,\alpha,J}$ are equivalent for any $I, J \in \mathbb{S}$.*

Proof. The proof follows the same reasoning as in the case of Proposition 3.1 in [42]. In fact it depends just on the analogous inequalities in the complex case and the representation formula. More precisely, we take $|\cdot|^p$ in the representation formula and then use the inequalities $(a+b)^p \leq 2^{p-1}(a^p + b^p)$, if $1 \leq p < +\infty$, and $(a+b)^p \leq a^p + b^p$, if $0 < p < 1$, for all $a, b \geq 0$. We note here that the concept of equivalence for two quasi-norms (corresponding to the case $0 < p < 1$) coincides with the concept of equivalence of two norms. $\square$

The first main result of this subsection is the following.

Theorem 4.5.9. *Let $0 < p < +\infty$, $-1 < \alpha < +\infty$ and $f \in \mathcal{A}_\alpha^{(2),p}(\mathbb{B}_1)$. There exists a sequence of polynomials $(P_n)_{n \in \mathbb{N}}$ such that for any $I \in \mathbb{S}$ we have $\|P_n - f\|_{p,\alpha,I} \to 0$ as $n \to +\infty$.*

Proof. We proceed in two steps.

Step 1. Fix $I_0 \in \mathbb{S}$. For $0 < r < 1$, define $f_r(q) = f(rq)$, $q \in \mathbb{B}_1$. By hypothesis, we know that $f \in \mathcal{A}_{\alpha,I_0}^p(\mathbb{B}_{1,I_0})$.

We first prove that for any $\varepsilon > 0$, there exists $r_0 := r_{I_0} \in (0,1)$ (sufficiently close to 1 and depending on I_0), such that

$$\|f_{r_0} - f\|_{p,\alpha,I_0} = \int_{\mathbb{B}_{1,I_0}} |f_{r_0}(q) - f(q)|^p dA_{\alpha,I_0}(q) < \varepsilon. \tag{4.33}$$

Obviously, $f_{r_0} \in \mathcal{R}(\mathbb{B}_{1/r_0})$ and $\overline{\mathbb{B}_{1,I_0}}$ is strictly included in $\mathbb{B}_{1/r_0, I_0}$.

For any $\delta \in (0,1)$ and $r \in (0,1)$, it is clear that we have the inequality

$$\int_{\mathbb{B}_{1,I_0}} |f_r(q) - f(q)|^p dA_{\alpha, I_0}(q) \le \int_{\overline{\mathbb{B}_{\delta, I_0}}} |f_r(q) - f(q)|^p dA_{\alpha, I_0}(q)$$

$$+ \int_{q \in \mathbb{C}_{I_0}, \delta < |q| < 1} |f_r(q) - f(q)|^p dA_{\alpha, I_0}(q) := I_1(\delta, r, I_0) + I_2(\delta, r, I_0).$$

Using (4.27) we obtain

$$I_2(\delta, r, I_0)$$

$$\le C_p \int_{q \in \mathbb{C}_{I_0}, \delta < |q| < 1} |f_r(q)|^p dA_{\alpha, I_0}(q) + C_p \int_{q \in \mathbb{C}_{I_0}, \delta < |q| < 1} |f(q)|^p dA_{\alpha, I_0}(q).$$

$$(4.34)$$

Now, let $\varepsilon > 0$ be fixed. Since for any $\delta > 0$ we have

$$\int_{\mathbb{B}_{1,I_0}} |f(q)|^p dA_{\alpha, I_0}(q)$$

$$= \int_{q \in \mathbb{C}_{I_0}, |q| \le \delta} |f(q)|^p dA_{\alpha, I_0}(q) + \int_{q \in \mathbb{C}_{I_0}, \delta < |q| < 1} |f(q)|^p dA_{\alpha, I_0}(q)$$

and since obviously

$$\int_{\mathbb{B}_{1,I_0}} |f(q)|^p dA_{\alpha, I_0}(q) = \lim_{\delta \nearrow 1} \int_{q \in \mathbb{C}_{I_0}, |q| \le \delta} |f(q)|^p dA_{\alpha, I_0}(q),$$

we see that $\lim_{\delta \nearrow 1} \int_{q \in \mathbb{C}_{I_0}, \delta < |q| < 1} |f(q)|^p dA_{\alpha, I_0}(q) - 0$. Therefore, there exists $0 < \delta_1 < 1$ (depending on ε, I_0 and C_p), sufficiently close to 1, such that for any $\delta \in (\delta_1, 1)$ we have

$$\int_{q \in \mathbb{C}_{I_0}, \delta < |q| < 1} |f(q)|^p dA_{\alpha, I_0}(q) < \varepsilon/(4C_p). \qquad (4.35)$$

In what follows, we will prove that there exists $0 < \delta_2 < 1$ (depending on ε, I_0 and p, but independent of r), sufficiently close to 1, such that for any $\delta \in (\delta_2, 1)$ we have

$$\int_{q \in \mathbb{C}_{I_0}, \delta < |q| < 1} |f_r(q)|^p dA_{\alpha, I_0}(q) < \varepsilon/(4C_p), \quad \text{for all } r \in (0,1). \qquad (4.36)$$

We will show that in fact

$$\int_{q \in \mathbb{C}_{I_0}, \delta < |q| < 1} |f_r(q)|^p dA_{\alpha, I_0}(q) \le \int_{q \in \mathbb{C}_{I_0}, \delta < |q| < 1} |f(q)|^p dA_{\alpha, I_0}(q),$$

which combined with (4.35) will yield (4.36).

Indeed, we can argue as in the complex case. Let $q \in \mathbb{C}_{I_0}$, $q = \rho e^{I_0 \theta}$. We can write

$$\int_{q \in \mathbb{C}_{I_0}, \delta < |q| < 1} |f_r(q)|^p dA_{\alpha, I_0}(q)$$

$$= \int_\delta^1 \int_0^{2\pi} |f(r\rho e^{I_0 \theta})|^p (\alpha + 1)(1 - \rho^2)^\alpha \frac{1}{\pi} \rho d\rho d\theta$$

$$= \int_\delta^1 \left[\int_0^{2\pi} |f(r\rho e^{I_0 \theta})|^p d\theta \right] (\alpha + 1)(1 - \rho^2)^\alpha \frac{1}{\pi} \rho d\rho$$

$$\leq \int_\delta^1 \left[\int_0^{2\pi} |f(r e^{I_0 \theta})|^p d\theta \right] (\alpha + 1)(1 - \rho^2)^\alpha \frac{1}{\pi} \rho d\rho$$

$$= \int_{q \in \mathbb{C}_{I_0}, \delta < |q| < 1} |f(q)|^p dA_{\alpha, I_0}(q).$$

For the above last inequality we used the property in the complex plane that $F(r) = \int_0^{2\pi} |f(r e^{I_0 \theta})|^p d\theta$ is increasing as function of r and that $r\rho \leq r$, see, e.g., [66], p. 26 or p. 30. Therefore, as the above integrals are taken in the same slice $\mathbb{C}_{I_0}$, it follows that the above property remains valid (for a complete proof of this inequality in the quaternionic setting, see [68], p. 6, Proposition 3.1).

Now put $\delta_0 = \max\{\delta_1, \delta_2\}$ and fix a $\delta \in (\delta_0, 1)$. Since both f and f_r are slice regular in $\overline{B(0; \delta)}$, there exists $M_{\delta, r} > 0$ (the maximum of the slice derivative of f on $\overline{B(0; \delta)}$), such that, by the mean value theorem,

$$|f(rq) - f(q)| \leq M_{\delta, r} \cdot |q| \cdot |r - 1|.$$

Indeed, for $q \in \overline{B(0; \delta)}$ on a fixed slice, since rq remains on the same slice, the previous inequality follows from the corresponding result valid in the complex setting.

Therefore, given $\varepsilon > 0$, it is clear that there exists r' (sufficiently close to 1 and depending on δ), such that for all $r \in (r', 1)$, we have $I_1(\delta, r, I_0) < \varepsilon/2$. Taking here an $r_0 > r'$, we obtain

$$I_1(\delta, r_0, I_0) < \varepsilon/2.$$

Combining (4.35), (4.36), and (4.34) written for $r = r_0$, we also obtain

$$I_2(\delta, r_0, I_0) < \varepsilon/2, \tag{4.37}$$

which immediately yields (4.33).

Step 2. Let $r_0 \in (0, 1)$ satisfy (4.33). Denoting by $T_n(f)(q)$ the partial sum of order n of the Taylor expansion of f in the open unit ball $\mathbb{B}_1$, it is clear that $T_n(f)(r_0 q)$ is a polynomial of degree n and that $p_n(q) = T_n(f)(r_0 q)$, $n \in \mathbb{N}$, is a sequence of polynomials that converges uniformly to f_{r_0} in the closed unit ball $\overline{\mathbb{B}_1}$.

Denoting by $\|\cdot\|$ the uniform norm on $\overline{\mathbb{B}_1}$, we see that for $\varepsilon > 0$ there exists $n_1 \in \mathbb{N}$, such that

$$\|f_{r_0} - p_n\|^p < \varepsilon, \quad \text{for all } n \geq n_1.$$

Choosing an $n_0 > n_1$, it easily follows that

$$\|f_{r_0} - p_{n_0}\|_{p,\alpha,I_0} \leq \|f_{r_0} - p_{n_0}\|^p \cdot \int_{\mathbb{B}_{1,I_0}} (\alpha+1)(1 - |q|^2)^\alpha dA_{\alpha,I_0}(q)$$

$$< (\alpha+1) \cdot \varepsilon \cdot \int_{\mathbb{B}_{1,I_0}} (1 - |q|^2)^\alpha dA_{I_0}(q). \tag{4.38}$$

However,

$$\int_{\mathbb{B}_{1,I_0}} (1 - |q|^2)^\alpha dA_{I_0}(q) < +\infty.$$

Indeed, this is obvious for $\alpha \geq 0$. If $-1 < \alpha < 0$ then we have

$$E := \int_{\mathbb{B}_{1,I_0}} (1 - |q|^2)^\alpha dA_{I_0}(q) < +\infty.$$

Indeed, passing to polar coordinates in $\mathbb{B}_{1,I_0}$, $q = \rho e^{I_0 \theta}$, we easily obtain

$$E = \int_0^{2\pi} \left[\int_0^1 (1 - \rho^2)^\alpha \cdot \frac{1}{\pi} \rho d\rho \right] d\theta = \int_0^1 v^\alpha dv = \frac{1}{1 - \alpha}.$$

Hence, using the inequality

$$\|f - p_{n_0}\|_{p,\alpha,I_0} \leq C_p(\|f - f_r\|_{p,\alpha,I_0} + \|f_r - p_{n_0}\|_{p,\alpha,I_0}),$$

where $C_p = 1$ if $p \geq 1$, and $C_p > 1$, if $0 < p < 1$, and relations (4.33) and (4.38), we conclude that for any $\varepsilon > 0$, there exists a polynomial p_{n_0}, such that $\|f - p_{n_0}\|_{p,\alpha,I_0} < \varepsilon$.

Choosing now $\varepsilon = \frac{1}{n}$, the above inequality shows that there exists a sequence of polynomials $(P_n(q))_n$, such that P_n converges (as $n \to \infty$) to f in the norm (quasi-norm) $\|\cdot\|_{p,\alpha,I_0}$. By Proposition 4.5.8, any other norm (or quasi-norm), depending on p) $\|\cdot\|_{p,\alpha,I}$ with $I \in \mathbb{S}$, is equivalent to the norm (quasi-norm) $\|\cdot\|_{p,\alpha,I_0}$. It follows that the sequence of polynomials $(P_n)_n$ converges to f in any norm (quasi-norm) $\|\cdot\|_{p,\alpha,I}$, which proves the theorem. $\qquad\square$

For $1 \leq p < +\infty$ we now present a constructive proof for the denseness result in Theorem 4.5.9, with quantitative estimates in terms of higher order moduli of smoothness and in terms of the best approximation quantity. To this end, we need more notations and a definition.

Definition 4.5.10. Let $0 < p < +\infty$, $I \in \mathbb{S}$ and $f \in \mathcal{A}_{\alpha,I}^p(\mathbb{B}_{1,I})$. The L^p-modulus of smoothness of k-th order on $\mathbb{C}_I$ is defined by

$$\omega_k(f; \delta)_{\mathcal{A}_{\alpha,I}^p(\mathbb{B}_{1,I})} = \sup_{0 \leq |h| \leq \delta} \left\{ \int_{\mathbb{B}_{1,I}} |\Delta_h^k f(q)|^p dA_{\alpha,I}(q) \right\}^{1/p},$$

where $k \in \mathbb{N}$ and

$$\Delta_h^k f(q) = \sum_{s=0}^{k} (-1)^{k+s} \binom{k}{s} f(qe^{Ish}).$$

The *best approximation quantity* is defined by

$$E_n(f; \mathbb{B}_{1,I})_{p,\alpha} = \inf\{\|f - P\|_{p,\alpha,I}; P \in \mathcal{P}_n\},$$

where $\mathcal{P}_n$ denotes the set of all quaternionic polynomials of degree $\leq n$.

Note that exactly as in the case of the L^p-moduli of smoothness for functions of real variable (see, e.g., [63], pp. 44-45), it can be proved that

$$\lim_{\delta \to 0} \omega_k(f; \delta)_{\mathcal{A}_{\alpha,I}^p(\mathbb{B}_{1,I})} = 0,$$

$$\omega_k(f; \lambda \cdot \delta)_{\mathcal{A}_{\alpha,I}^p(\mathbb{B}_{1,I})} \leq (\lambda + 1)^k \cdot \omega_k(f; \delta)_{\mathcal{A}_{\alpha,I}^p(\mathbb{B}_{1,I})}, \qquad (4.39)$$

if $1 \leq p < +\infty$ and

$$[\omega_k(f; \lambda \cdot \delta)_{\mathcal{A}_{\alpha,I}^p(\mathbb{B}_{1,I})}]^p \leq (\lambda + 1)^k \cdot [\omega_k(f; \delta)_{\mathcal{A}_{\alpha,I}^p(\mathbb{B}_{1,I})}]^p, \qquad (4.40)$$

if $0 < p < 1$.

Indeed, this is immediate from the fact that denoting (for fixed q) $g(x) = f(qe^{Ix})$, we get $\Delta_h^k f(q) = \overline{\Delta}_h^k g(0)$, where

$$\overline{\Delta}_h^k g(x_0) = \sum_{s=0}^{k} (-1)^{s+k} \binom{k}{s} g(x_0 + sh).$$

Now, for any $1 \leq p < +\infty$ and $f \in \mathcal{R}(\mathbb{B}_1)$, let us define the convolution operators

$$L_n(f)(q) = \int_{-\pi}^{\pi} f(qe^{I_q t}) \cdot K_n(t) dt, \ q \in \mathbb{H}.$$

Here $K_n(t)$ is a positive and even trigonometric polynomial with the property that $\int_{-\pi}^{\pi} K_n(t) dt = 1$.

In particular, we can consider the Fejér kernel

$$K_n(t) = \frac{1}{2\pi n} \cdot \left(\frac{\sin(nt/2)}{\sin(t/2)} \right)^2,$$

in which case we will denote $L_n(f)(q)$ by $F_n(f)(q)$.

For

$$K_{n,r}(t) = \frac{1}{\lambda_{n,r}} \cdot \left(\frac{\sin(nt/2)}{\sin(t/2)} \right)^{2r},$$

where r will be chosen as the smallest integer with $r \geq \frac{p(m+1)+2}{2}$, $m \in \mathbb{N}$, and the constants $\lambda_{n,r}$ are chosen so that $\int_{-\pi}^{\pi} K_{n,r}(t)dt = 1$, we define

$$I_{n,m,r}(f)(q) = -\int_{-\pi}^{\pi} K_{n,r}(t) \sum_{k=1}^{m+1} (-1)^k \binom{m+1}{k} f(qe^{I_q kt})dt, \quad q \in \mathbb{H}.$$

Further, let

$$V_n(f)(q) = 2F_{2n}(f)(q) - F_n(f)(q), \quad q \in \mathbb{H}.$$

Following the reasoning in [85] and [93], for fixed $I \in \mathbb{S}$, if $q \in \mathbb{C}_I$ then $L_n(f)(q)$, $I_{n,m,r}(f)(q)$ and $V_n(f)(q)$ are polynomials in q on $\mathbb{C}_I$, with the coefficients independent of I and depending only on the coefficients in the series expansion of f. Therefore, as functions of q, $L_n(f)(q)$, $I_{n,m,r}(f)(q)$ and $V_n(f)(q)$ are polynomials on the whole $\mathbb{H}$.

The second main result of this section is the following.

Theorem 4.5.11. *Let $1 \leq p < +\infty$, $-1 < \alpha < +\infty$, $m \in \mathbb{N}\bigcup\{0\}$, and let $f \in A_\alpha^{(2),p}(\mathbb{B}_1)$ be arbitrary but fixed.*

(i) *$I_{n,m,r}(f)(q)$ is a quaternionic polynomial of degree $\leq r(n-1)$, which for any $I \in \mathbb{S}$ satisfies the estimate*

$$\|I_{n,m,r}(f) - f\|_{p,\alpha,I} \leq C_{p,m,r} \cdot \omega_{m+1}\left(f; \frac{1}{n}\right)_{A_{\alpha,I}^p(\mathbb{B}_{1,I})}, \quad n \in \mathbb{N},$$

where $m \in \mathbb{N}$, r is the smallest integer such that $r \geq \frac{p(m+1)+2}{2}$ and $C(p,m,r) > 0$ is a constant independent of f, n and I.

(ii) *$V_n(f)(z)$ is a quaternionic polynomials of degree $\leq 2n-1$, satisfying for any $I \in \mathbb{S}$ the estimate*

$$\|V_n(f) - f\|_{p,\alpha,I} \leq [2^{(p-1)/p} \cdot (2^p + 1)^{1/p} + 1] \cdot E_n(f; \mathbb{B}_{1,I})_{p,\alpha}, \quad n \in \mathbb{N}.$$

Proof. For the fact that the convolution operators $I_{n,m,r}(f)(q)$ and $V_n(f)(q)$ are quaternionic polynomials of the corresponding degrees, see Chapter 3.

(i) In the sequel we will apply the following well-known Jensen type inequality for integrals: if $\int_{-\pi}^{+\pi} G(u)du = 1$, $G(u) \geq 0$ for all $u \in [-\pi, \pi]$ and $\varphi(t)$ is a convex function over the range of the measurable function of real variable F, then

$$\varphi\left(\int_{-\pi}^{+\pi} F(u)G(u)du\right) \leq \int_{-\pi}^{+\pi} \varphi(F(u))G(u)du.$$

Let $m \in \mathbb{N}$ and r be the smallest integer such that $r \geq \frac{p(m+1)+2}{2}$.

We now argue as in the complex case, following [81]. Thus, choosing $\varphi(t) = t^p$, $1 \leq p < \infty$, we get

$$|f(q) - I_{n,m,r}(f)(q)|^p = \left| \int_{-\pi}^{\pi} \Delta_t^{m+1} f(q) K_{n,r}(t) dt \right|^p$$

$$\leq \left[\int_{-\pi}^{\pi} |\Delta_t^{m+1} f(q)| K_{n,r}(t) dt \right]^p$$

$$\leq \int_{-\pi}^{\pi} |\Delta_t^{m+1} f(q)|^p K_{n,r}(t) dt.$$

Multiplying both members by $\rho_\alpha = (\alpha + 1)(1 - |q|^2)^\alpha$, integrating on $\mathbb{B}_{1,I}$ with respect to $dA_I(q)$, applying Fubini's theorem, we obtain

$$\int_{\mathbb{B}_{1,I}} |I_{n,m,r}(f)(q) - f(q)|^p dA_{\alpha,I}(q)$$

$$\leq \int_{-\pi}^{\pi} \left[\int_{\mathbb{B}_{1,I}} |\Delta_t^{m+1} f(q)|^p dA_{\alpha,I}(q) \right] K_{n,r}(t) dt$$

$$\leq \int_{-\pi}^{\pi} \omega_{m+1}(f; |t|)_{\mathcal{A}_{\alpha,I}^p(\mathbb{B}_{1,I})}^p \cdot K_{n,r}(t) dt$$

$$\leq \int_{-\pi}^{\pi} \omega_{m+1}(f; 1/n)_{\mathcal{A}_{\alpha,I}^p(\mathbb{B}_{1,I})}^p (n|t| + 1)^{(m+1)p} \cdot K_{n,r}(t) dt.$$

But by [135], p. 57, relation (5), for $r \in \mathbb{N}$ with $r \geq \frac{p(m+1)+2}{2}$ we get

$$\int_{-\pi}^{\pi} (n|t| + 1)^{(m+1)p} \cdot K_{n,r}(t) dt \leq C_{p,m,r} < +\infty, \qquad (4.41)$$

which proves the estimate in (i).

(ii) Now, let $f, g \in \mathcal{A}_\alpha^{(2),p}(\mathbb{B}_1)$ and $1 \leq p < +\infty$. By the convexity of $\varphi(t) = t^p$ we get the obvious inequality $(a+b)^p \leq 2^{p-1}(a^p + b^p)$, valid for all $a, b \geq 0$, which for all $q \in \mathbb{B}_{1,I} \subset \mathbb{C}_I$ implies

$$|V_n(f)(q) - V_n(g)(q)| \leq 2|F_{2n}(f)(q) - F_{2n}(g)(q)| + |F_n(f)(q) - F_n(g)(q)|$$

$$\leq 2 \int_{-\pi}^{\pi} |f(qe^{It}) - g(qe^{It})| \cdot K_{2n}(t) dt + \int_{-\pi}^{\pi} |f(qe^{It}) - g(qe^{It})| \cdot K_n(t) dt$$

and

$$|V_n(f)(q) - V_n(g)(q)|^p \le 2^{p-1}\left[\left(2\int_{-\pi}^{\pi} |f(qe^{It}) - g(qe^{It})| \cdot K_{2n}(t)dt\right)^p\right.$$

$$+ \left.\left(\int_{-\pi}^{\pi} |f(qe^{It}) - g(qe^{It})| \cdot K_n(t)dt\right)^p\right]$$

$$\le 2^{p-1}\left[2^p\int_{-\pi}^{\pi} |f(qe^{It}) - g(qe^{It})|^p \cdot K_{2n}(t)dt\right.$$

$$+ \left.\int_{-\pi}^{\pi} |f(qe^{It}) - g(qe^{It})|^p \cdot K_n(t)dt\right].$$

Integrating this inequality on $\mathbb{B}_{1,I}$ with respect to $dA_{\alpha,I}(q)$ and reasoning above in (i), we obtain

$$\|V_n(f) - V_n(g)\|^p_{p,\alpha,I}$$

$$\le 2^{p-1}\left[2^p\int_{-\pi}^{\pi}\left(\int_{\mathbb{B}_{1,I}} |f(qe^{It}) - g(qe^{It})|^p dA_{\alpha,I}(q)\right) K_{2n}(t)dt\right.$$

$$+ \left.\int_{-\pi}^{\pi}\left(\int_{\mathbb{B}_{1,I}} |f(qe^{It}) - g(qe^{It})|^p dA_{\alpha,I}(q)\right) K_n(t)dt\right].$$

Denoting $F(q) = |f(q) - g(q)|^p$, $q \in \mathbb{C}_I$, writing $q = r\cos(\theta) + Ir\sin(\theta)$ in polar coordinates, and taking into account that

$$dA_{\alpha,I}(q) = (\alpha + 1)(1 - |q|^2)^\alpha dA_I(q) = (\alpha + 1)(1 - |q|^2)^\alpha \frac{1}{\pi}rdrd\theta,$$

simple calculations (made exactly as in the complex case) lead to the equality

$$\int_{\mathbb{B}_{1,I}} |F(qe^{It})|^p dA_{\alpha,I}(q) = \int_{\mathbb{B}_{1,I}} |F(q)|^p dA_{\alpha,I}(q), \quad \text{for all } t,$$

which in conjunction with the above inequality immediately yields

$$\|V_n(f) - V_n(g)\|^p_{p,\alpha,I} \le 2^{p-1}[2^p\|f - g\|^p_{p,\alpha,I} + \|f - g\|^p_{p,\alpha,I}]$$
$$= 2^{p-1}(2^p + 1)\|f - g\|^p_{p,\alpha,I},$$

that is,

$$\|V_n(f) - V_n(g)\|_{p,\alpha,I} \le 2^{(p-1)/p} \cdot (2^p + 1)^{1/p}\|f - g\|_{p,\alpha,I}.$$

Now, let us denote by $P^*_{n,I}$ a polynomial of best approximation by elements in the set $\mathcal{P}_n$ of polynomials of degree $\le n$, in the norm in $\|\cdot\|_{p,\alpha,I}$, that is,

$$E_n(f; \mathbb{B}_{1,I})_{p,\alpha} = \inf\{\|f - P\|_{p,\alpha,I}; P \in \mathcal{P}_n\} = \|f - P^*_{n,I}\|_{p,\alpha,I}.$$

Note that since $\mathcal{P}_n$ is finite-dimensional (for fixed n), the polynomial $P_{n,I}^*$ exists.

Reasonings similarly to the complex case in [75], p. 425, we get $V_n(P_{n,I}^*) = P_{n,I}^*$, for all $q \in \mathbb{B}_{1,I}$ (in fact for all $q \in \mathbb{C}_I$), and consequently,

$$
\begin{aligned}
\|f - V_n(f)\|_{p,\alpha,I} &\leq \|f - P_{n,I}^*\|_{p,\alpha,I} + \|V_n(P_{n,I}^*) - V_n(f)\|_{p,\alpha,I} \\
&\leq E_n(f; \mathbb{B}_{1,I})_{p,\alpha} + 2^{(p-1)/p} \cdot (2^p + 1)^{1/p} \|P_{n,I}^* - f\|_{p,\alpha,I} \\
&= [2^{(p-1)/p} \cdot (2^p + 1)^{1/p} + 1] \cdot E_n(f; \mathbb{B}_{1,I})_{p,\alpha},
\end{aligned}
$$

which proves (ii) and the theorem.   □

Remark 4.5.12. The approximation results obtained in the case of the Bergman space can also be proved for Bloch and Besov spaces with similar techniques, see [95].

Chapter 5

Overconvergence, Equiconvergence and Universality Properties

In this chapter we study in the quaternionic setting the overconvergence of Chebyshev and Legendre polynomials, the Walsh equiconvergence result and the universality properties of power series.

5.1 Overconvergence of Chebyshev and Legendre Polynomials

In this section we show that, in the quaternionic case, the Chebyshev and Legendre expansions converge uniformly to a slice regular function in all compact subsets lying in the interior of a quaternionic ellipsoid. The results in this section were originally obtained by the authors in [92].

The classical Chebyshev and Legendre orthogonal polynomials are well known and serve as an important tool in approximation theory. For the sake of completeness, we recall below the basics.

Definition 5.1.1. The Chebyshev polynomials are defined by

$$
\begin{aligned}
&T_0(x) - 1, \\
&T_k(x) = \cos(k \arccos x) \\
&\qquad = \frac{k}{2} \sum_{j=0}^{[k/2]} \frac{(-1)^j (k-j-1)!}{j!(k-2j)!} \cdot (2x)^{k-2j}, \quad k = 1, 2, \dots .
\end{aligned}
\tag{5.1}
$$

© Springer Nature Switzerland AG 2019
S. G. Gal, I. Sabadini, *Quaternionic Approximation*, Frontiers in Mathematics,
https://doi.org/10.1007/978-3-030-10666-9_5

Let

$$P_0(x) = \frac{1}{\sqrt{\pi}}T_0(x) = \frac{1}{\sqrt{\pi}},$$

$$P_k(x) = \frac{\sqrt{2}}{\sqrt{\pi}}T_k(x), \qquad \text{for} \quad k \geq 1.$$

Then $(P_k)_k$ is an orthonormal system of polynomials on $[-1, 1]$ with the weight function $\rho(x) = \frac{1}{\sqrt{1-x^2}}$, that is,

$$\int_{-1}^{+1} \rho(x)P_k(x)P_j(x)dx = 0, \text{ if } k \neq j, m,$$

$$\int_{-1}^{+1} \rho(x)P_k(x)P_j(x)dx = 1, \text{ if } k = j.$$

Let $f : [-1, 1] \to \mathbb{R}$ be a function and let $a_k(f)$ be the coefficients of f with respect to the system $(P_k)_k$,

$$a_k(f) = \int_{-1}^{+1} \rho(x)f(x)P_k(x)dx.$$

Under suitable hypotheses on $f : [-1, 1] \to \mathbb{R}$, for example the Dini–Lipschitz condition, see, e.g., [36] p. 129, the partial sums

$$S_n(f)(x) = \sum_{j=0}^{n} P_j(x)a_j(f) \tag{5.2}$$

converge uniformly to f on $[-1, +1]$.

These results can be extended to analytic functions, see, e.g., [125], pp. 177–178. An important result in this framework is the following:

Theorem 5.1.2. *If $f : E_r \to \mathbb{C}$ is an analytic function in the interior of the (closed) ellipse E_r with foci at -1 and $+1$ and with semiaxes $a = \frac{r+r^{-1}}{2}$, $b = \frac{r-r^{-1}}{2}$, $r > 1$, then $S_n(f)(z)$ converge uniformly to f, in any compact subset of E_r, with the rate of convergence of a geometric series.*

Definition 5.1.3. The *Legendre polynomials* are defined by

$$L_k(x) = \frac{1}{2^k k!} \cdot \frac{d^k}{dx^k}[(x^2 - 1)^k], \quad k = 0, 1, \dots .$$

Set

$$Q_k(x) = \frac{\sqrt{2k+1}}{\sqrt{2}}L_k(x).$$

It is well known that $(Q_k)_k$ is an orthonormal system of polynomials on $[-1, +1]$ with the weight function $\rho(x) = 1$, that is,

$$\int_{-1}^{+1} Q_k(x)Q_j(x)dx = \begin{cases} 0, & \text{if } k \neq j, \\ 1, & \text{if } k = j. \end{cases}$$

Let us denote the coefficients of f with respect to the system $(Q_k)_k$ by

$$b_k(f) = \int_{-1}^{+1} f(x)Q_k(x)dx.$$

Under suitable hypotheses on $f : [-1, 1] \to \mathbb{R}$, the partial sums

$$R_n(f)(x) = \sum_{j=0}^{n} Q_j(x)b_j(f) \tag{5.3}$$

converge uniformly to f on $[-1, +1]$. This result was extended to the complex case as follows (see, e.g., [57], p. 312, Theorem 12.4.7).

Theorem 5.1.4. *If $f : E_r \to \mathbb{C}$ is an analytic function in the interior of the (closed) ellipse E_r with foci at -1 and $+1$ and with semiaxes $a = \frac{r+r^{-1}}{2}$, $b = \frac{r-r^{-1}}{2}$, $r > 1$, then $R_n(f)(z)$ converge uniformly to f, in any compact proper subset of E_r, with the rate of convergence of a geometric series.*

Theorems 5.1.2 and 5.1.4 can be extended to the quaternionic setting, in fact, we will prove that for any function f that is slice regular in a quaternionic ellipsoid containing their real interval of orthogonality, the Chebyshev and Legendre expansions converge uniformly to f in all compact subsets in the interior in the ellipsoid. We will also show that the rate of convergence is geometric. To start with, we introduce the notion of "quaternionic ellipsoid".

Definition 5.1.5. Let $\rho > 1$. We call *quaternionic ellipsoid* $\mathbb{E}_\rho \subset \mathbb{H}$ of semiaxes $a = \frac{\rho+\rho^{-1}}{2}$ and $b = \frac{\rho-\rho^{-1}}{2}$, the image of the closed ball $\overline{B(0; \rho)}$, under the map $F(q) = \frac{q+q^{-1}}{2}$.

Our first result is about Chebyshev expansions and can be stated as follows.

Theorem 5.1.6. *If $f : \mathbb{E}_\rho \to \mathbb{H}$ is a slice regular function in the interior of the quaternionic ellipsoid $\mathbb{E}_\rho$ with semiaxes $a = \frac{\rho+\rho^{-1}}{2}$, $b = \frac{\rho-\rho^{-1}}{2}$, $\rho > 1$. Then the partial sums*

$$S_n(f)(q) = \sum_{j=0}^{n} P_j(q)a_j(f)$$

converge uniformly to f for $n \to \infty$, in any compact proper subset of $\mathbb{E}_\rho$, with the rate of convergence of a geometric series.

Proof. The proof follows the pattern of the one given in the complex case [125], p. 178. We first compute the coefficients $a_n(f)$ of f. To this end, we set $x = \cos t$ and $q = e^{It}$ for $I \in \mathbb{S}$. Then we have

$$a_k(f) = \frac{1}{\pi} \int_{-\pi}^{\pi} f(\cos t) \cos(kt)\, dt$$

$$= \frac{1}{\pi} \int_{\partial \mathbb{B} \cap \mathbb{C}_I} \frac{q + q^{-1}}{2}\, dq_I\, f\left(\frac{q + q^{-1}}{2}\right),$$

where the last integral is evidently independent of the complex plane $\mathbb{C}_I$. Let r be such that $1 < r < \rho$. Then, by Remark 1.3.8, the function

$$g(q) = f\left(\frac{q + q^{-1}}{2}\right)$$

is slice regular in the closed annular domain D defined by $r^{-1} \le |q| \le r$ since the function $w = \frac{q + q^{-1}}{2}$ is slice regular and intrinsic. Moreover, if $q \in \partial D$, then $w = \frac{q + q^{-1}}{2} \in \partial \mathbb{E}_r$. Thus we have

$$a_k(f) = \frac{1}{2\pi} \int_{\partial B_{1/r} \cap \mathbb{C}_I} q^{k-1}\, dq_I\, g(q) + \frac{1}{2\pi} \int_{\partial B_r \cap \mathbb{C}_I} q^{-k-1}\, dq_I\, g(q),$$

which yields

$$|a_k(f)| \le 2r^{-k} \max_{q \in D} |g(q)| = 2r^{-k} M.$$

Let $r' > 1$ be such that $r' < r$, let $q \in \mathbb{E}_{r'}$ and $q = (w + w^{-1})/2$. Then

$$\frac{1}{2}|w^k + w^{-k}| \le \frac{1}{2}(|w|^k + |w|^{-k})$$

$$\le \frac{1}{2}((r')^k + (r')^{-k}) < (r')^k, \quad k = 0, 1, \ldots,$$

and so the series

$$\frac{1}{2}a_0(f) + \sum_{k=1}^{\infty} \frac{(w^k + w^{-k})}{2} a_k(f)$$

has the series $\sum_{k=0}^{\infty}(r'/r)^k$ as a majorant, and this last series is convergent. Thus the sequence $S_n(f)(q)$ converges uniformly to f, in any compact proper subset of $\mathbb{E}_{r'}$, and since r' is arbitrary and close to ρ, the assertion is proved.   □

The case of Legendre expansions can be treated by using the following lemma.

Lemma 5.1.7. *Let $f_n, f : \Omega \subseteq \mathbb{H} \to \mathbb{H}$ and suppose that*

$$\lim_{n \to \infty} |f_n(q)|^{1/n} = |f(q)| \tag{5.4}$$

on Ω and uniformly on a subset $\Omega' \subseteq \Omega$. Let (c_n) be a sequence of quaternions such that

$$\lim_{n \to \infty} |c_n|^{1/n} = \frac{1}{r}, \quad r > 0. \tag{5.5}$$

Then the series $\sum_{n=0}^{\infty} f_n(q) c_n$ converges uniformly at all points of $\Omega' = \Omega'_s$ for which the inequality $|f(q)| \leq s$ is satisfied, where $0 < s < r$.

Proof. Let

$$\Omega_r := \Omega \cap \{q \in \mathbb{H}; \ |f(q)| < r\}$$

and let $q \in \Omega_r$. From (5.4) and (5.5) it follows that for n sufficiently large and $0 < r' < r'' < r$

$$|f_n(q)|^{1/n} \leq r' \quad \text{and} \quad |c_n|^{1/n} \leq \frac{1}{r''}$$

and so

$$|f_n(q) c_n| \leq \left(\frac{r'}{r''}\right)^n.$$

Thus the series $\sum_{n=0}^{\infty} f_n(q) c_n$ is convergent since it has a geometric series as a majorant.

Let

$$\Omega'_s := \Omega' \cap \{q \in \mathbb{H}; \ |f(q)| \leq s\}$$

and let $q \in \Omega'_s$, with $s < r$. The estimate

$$|f_n(q)|^{1/n} \leq r', \quad s < r' < r$$

holds uniformly in Ω'_s and so, reasoning as before, we conclude that the convergence is uniform. $\qquad\square$

We can now prove our result on the Legendre expansion:

Theorem 5.1.8. *If $f : \mathbb{E}_\rho \to \mathbb{H}$ is a slice regular function in the interior of the quaternionic ellipsoid $\mathbb{E}_\rho$ with semiaxes $a = \frac{\rho + \rho^{-1}}{2}$, $b = \frac{\rho - \rho^{-1}}{2}$, $\rho > 1$, then the partial sums $R_n(f)(q)$ converge uniformly to f, in any compact interior of $\mathbb{E}_\rho$, with the rate of convergence of a geometric series.*

Proof. We follow the lines of the proof of Theorem 12.4.7, pp. 313–314, in [57]. We know that f is slice regular in $\mathbb{E}_\rho$, $\rho > 1$ and, by definition,

$$b_n(f) = \frac{2n+1}{2} \int_{-1}^{+1} Q_n(t) f(t) dt. \tag{5.6}$$

If r is such that $1 < r < \rho$, then f is obviously slice regular in $\mathbb{E}_r$ up to its boundary. Thus, by the Cauchy formula,

$$f(t) = \frac{1}{2\pi} \int_{\partial \mathbb{E}_r \cap \mathbb{C}_I} (q - t)^{-1} dq_I f(q) \tag{5.7}$$

for $-1 \leq t \leq 1$, where I is any element in the sphere $\mathbb{S}$. Using (5.7) and (5.6), we deduce that

$$
\begin{aligned}
b_n(f) &= \frac{2n+1}{4\pi} \int_{-1}^{+1} Q_n(t) \left(\int_{\partial \mathbb{E}_r \cap \mathbb{C}_I} (q-t)^{-1} dq_I f(q) \right) dt \\
&= \frac{2n+1}{4\pi} \int_{\partial \mathbb{E}_r \cap \mathbb{C}_I} \left(\int_{-1}^{+1} (q-t)^{-1} Q_n(t) dt \right) dq_I f(q) \\
&= \frac{2n+1}{2\pi} \int_{\partial \mathbb{E}_r \cap \mathbb{C}_I} \widetilde{Q}_n(q) dq_I f(q),
\end{aligned}
$$

where

$$
\widetilde{Q}_n(q) = \frac{1}{2} \int_{-1}^{+1} (q-t)^{-1} Q_n(t) dt.
$$

So we have

$$
|b_n(f)| \leq \frac{2n+1}{2\pi} C_{I,r} \cdot \max_{q \in \mathbb{E}_r \cap \mathbb{C}_I} |\widetilde{Q}_n(q)| \cdot \max_{q \in \mathbb{E}_r \cap \mathbb{C}_I} |f(q)|,
$$

where $C_{I,r}$ is the length of $\partial \mathbb{E}_r \cap \mathbb{C}_I$. It is immediate that if $q \in \mathbb{C}_I$ and if we set $w = \frac{q+q^{-1}}{2}$, then $w \in \mathbb{C}_I$. Thus we have (see, e.g., [57], Lemma 12.4.6)

$$
\widetilde{Q}_n(w) = \sum_{k=n+1}^{\infty} \frac{\sigma_{nk}}{q^k}, \tag{5.8}
$$

where σ_{nk} are suitable real constants such that

$$
|\sigma_{nk}| \leq \pi, \quad n = 0, 1, 2, \ldots, \ k \geq n+1.
$$

When q varies in $\mathbb{H}$, namely when I varies in $\mathbb{S}$, the result obviously still holds. When $q \in \partial \mathbb{E}_r \cap \mathbb{C}_I$, then w belongs to the circle obtained by intersecting $|w| = r$ with $\mathbb{C}_I$, so

$$
\max_{q \in \mathbb{E}_r \cap \mathbb{C}_I} |\widetilde{Q}_n(q)| \leq \sum_{k=n+1}^{\infty} \frac{\pi}{r^k} = \frac{\pi}{r^n(r-1)}.
$$

This gives

$$
|b_n(f)| \leq \frac{2n+1}{2\pi} \frac{C}{r^n}
$$

for a suitable constant C which does not depend on n. Since $r < \rho$ is arbitrary, we deduce that

$$
\varlimsup_{n \to \infty} |b_n(f)|^{1/n} \leq \frac{1}{\rho}.
$$

Lemma 5.1.7 finally gives that $R_n(f)(q)$ converges uniformly to f on any compact subset of $\mathbb{E}_\rho$, with a geometric rate of convergence. $\qquad \square$

5.2 Walsh Equiconvergence Type Results

In this section we address the problem of generalizing the Walsh equiconvergence result in the quaternionic setting (the source is the paper by Gal and Sabadini [87] for the quaternionic case, as well as other works explicitly mentioned below for the complex case). We start the section by discussing the complex analysis counterpart of Walsh equiconvergence result.

Let $\rho > 1$ and $\mathbb{D}_\rho = \{z \in \mathbb{C}; |z| < \rho\}$. We consider the set

$$A_\rho(\mathbb{C}) = \{f : \mathbb{D}_\rho \to \mathbb{C}; \ f \text{ is analytic in } \mathbb{D}_\rho\}.$$

For any $f \in A_\rho(\mathbb{C})$ we can write $f(z) = \sum_{k=0}^\infty z^k a_k$ (with $a_k \in \mathbb{C}$), and consider the partial sum

$$s_{n-1}(f)(z) = \sum_{k=0}^{n-1} z^k a_k.$$

Let $L_{n-1}(f)(z)$ be the Lagrange interpolation polynomial of degree at most $n - 1$, interpolating f at the n zeros of the equation $z^n - 1 = 0$. It is known that both $s_{n-1}(f)(z)$ and $L_{n-1}(f)(z)$ converge uniformly in any closed disk of radius $R < \rho$. Walsh proved the beautiful and surprising result that the difference $s_{n-1}(f)(z) - L_{n-1}(f)(z)$ converges uniformly to zero in a larger set, namely in any closed disk in $\mathbb{D}_{\rho^2}$. This phenomenon is called Walsh equiconvergence and can be stated as follows (see, e.g., [125], Chapter 1, p.8, Theorem 6, or Walsh [16], pp. 153–154).

Theorem 5.2.1. *If $f \in A_\rho(\mathbb{C})$, then $L_{n-1}(f)(z)$ converges geometrically to $f(z)$ in any closed disk $\overline{\mathbb{D}}_R$ with $1 < R < \rho$ and, in addition*

$$\lim_{n\to\infty} [s_{n-1}(f)(z) - L_{n-1}(f)(z)] = 0, \text{ for all } |z| < \rho^2;$$

the convergence is again uniform and with geometric rate of convergence in any $\overline{\mathbb{D}}_R$ with $1 < R < \rho^2$. The result is the best possible, in the sense that in the case of the function

$$g = \frac{1}{z - \rho} \in A_\rho(\mathbb{C}),$$

for $z = \rho^2$ we have

$$\lim_{n\to\infty} [s_{n-1}(g)(z) - L_{n-1}(g)(z)] \neq 0.$$

In the sequel, we will prove an analog of the Walsh equiconvergence theorem (Theorem 5.2.10), its converse (Theorem 5.2.14), as well as an extension of Walsh's result, see Theorem 5.2.11. As in the complex case, we show that our results are sharp. This generalization is far from trivial: in the complex case, given distinct points $z_1, \ldots, z_n$ and the values $w_1, \ldots, w_n$, the polynomial interpolating these values can be constructed according to the so-called Lagrange interpolation formula (see, e.g., [193], p.50):

$$p(z) = \sum_{k=1}^{n} w_k \frac{w(z)}{(z - z_k)w'(z)},$$ (5.9)

where

$$w(z) = (z - z_1) \cdots (z - z_n).$$

This formula is based on the fact that $w(z)/(z - z_k)$ vanishes at all points z_i different from z_k. In the quaternionic framework this procedure cannot work, as the following simple example shows.

Example 5.2.2. Consider two distinct points q_1, q_2 and the polynomial

$$w(q) = (q - q_1) * (q - q_2).$$

This polynomial vanishes at q_1 and at

$$\tilde{q}_2 = (q_2 - q_1)^{-1} q_2 (q_2 - q_1),$$

see Section 1.4, so it interpolates the points $(q_1, 0)$ and $(\tilde{q}_2, 0)$. To divide by a polynomial of the form $(q - \alpha)$ means to multiply by $(q - \alpha)^{-*}$, the $*$-reciprocal of $(q - \alpha)$, on the left or on the right. Suppose we divide by multiplying on the left by $(q - q_1)^{-*}$. It turns out that the polynomial $(q - q_1)^{-*} * w(q) = (q - q_2)$ does not vanish at $\tilde{q}_2$.

Despite the fact that there are several difficulties, it is possible to solve the problem of interpolating n points over the quaternions. We note that the problem has been addressed and solved in the general framework of division rings by Lam, see [129], who studied the Vandermonde matrix, its invertibility and interpolation. Other papers, in which similar results are obtained in the specific case of the algebra of quaternions, are [123] in which the quaternionic Vandermonde matrix and so the so-called double determinant are studied, and [34].

We will show that it is possible to construct in integral form the Lagrange interpolation polynomial $L_{n-1}(f)(q)$ of degree $n - 1$ interpolating the values of a function slice regular in a ball $B(0, R)$ at the roots of the unit, i.e., the roots of $q^n - 1 = 0$.

5.2.1   The interpolation problem

We consider now the problem of interpolating the values of a slice regular function f at the roots of the unity $q^n - 1 = 0$ over the quaternions. Proposition 1.4.4 shows that $q^n - 1$ has an infinite number of zeros so, in principle, we have an infinite number of points where to interpolate. However, the zeros belong to a finite number of spheres and this is enough to guarantee that we can interpolate at a finite number of points, see Proposition 1.4.3. More precisely, Proposition 1.4.3 implies that it is enough to assign the n values of the function f at n points corresponding to:

- the real root $+1$ of the equation $q^n - 1 = 0$ and k pairs of points (not necessarily conjugated) belonging to a sphere of roots $[\alpha_j]$, $j = 1, \ldots, k$, if $n = 2k + 1$;

- the real roots ± 1 of the equation $q^n - 1 = 0$ and $k - 1$ pairs of points (not necessarily conjugated) belonging to a sphere of roots $[\alpha_j]$, $j = 1, \ldots, k - 1$, if $n = 2k$.

As in the complex case, the search for polynomials $p(q) = \sum_{j=0}^{n-1} q^j a_j$ of degree $n - 1$ interpolating (q_i, w_i), $i = 1, \ldots, n$ leads to the system of equations

$$a_0 + q_1 a_1 + \cdots + q_1^{n-1} a_{n-1} = w_1,$$

$$\ldots$$

$$a_0 + q_n a_1 + \cdots + q_n^{n-1} a_{n-1} = w_n,$$

which can be written in matrix form as

$$V(q_1, \ldots, q_n)\underline{a} = \underline{w}, \qquad (5.10)$$

where $\underline{a} = [a_0, \ldots, a_{n-1}]^T$, $\underline{w} = [w_0, \ldots, w_n]^T$, where the matrix of coefficients is the Vandermonde matrix

$$V(q_1, \ldots, q_n) = \left[1 \ q_i \ \cdots \ q_i^{n-1} \right]_{i=1}^n .$$

The Vandermonde matrix in division rings has been studied in [128] and [130] in which the authors provide a sufficient condition for the invertibility of $V(q_1, \ldots, q_n)$ and also for having a unique solution to the interpolation problem $f(q_i) = w_i$, $i = 1, \ldots, n$, see [128, Corollary 24]. The condition, in our case, can be expressed as follows:

Theorem 5.2.3. *Let $q_1, \ldots, q_n \in \mathbb{H}$ be such that no three of them belong to the same equivalence class. Then the Vandermonde matrix $V(q_1, \ldots, q_n)$ is invertible and for any $w_1, \ldots, w_n$ there exists a unique solution to the interpolation problem* (5.10).

The case of quaternions was studied in the paper [123] where the authors prove the result already given in [128, Corollary 24] with a different method. Using the language of [123], the so-called double determinant of the Vandermonde matrix $V(q_1, \ldots, q_n)$ is nonzero if and only if no three of the q_j's are in the same equivalence class, see [123, Theorem 4.1]. This condition is necessary and sufficient in order to have a unique solution of (5.10), by virtue of the Cramer rule proved in [34, Theorem 2.3]. The Vandermonde matrix and the interpolation problem were considered more recently also in [155].

In the specific case of the roots of unity, the interpolation problem admits a unique solution if we assign the values at the real root(s) of $q^n - 1 = 0$ and at two elements in each equivalence class which is a root of the equation. One should note that the computations do not depend on the two selected elements in each equivalence class, since, by Proposition 1.4.3, the obtained interpolation polynomial has uniquely determined values at all points belonging to the same equivalence class.

5.2.2   Walsh equiconvergence theorem

We begin by giving a definition which, as we shall see, corresponds to the analog of the Lagrange interpolation polynomial at the roots of the unity:

Definition 5.2.4. Let $f : B(0; \rho) \to \mathbb{H}$ be (left) slice regular and let

$$\Gamma_{R,I} = \{u \in \mathbb{H}, \ u = x + Iy; \ |u| = R\},$$

$1 < R < \rho$, in other words, $\Gamma_{R,I}$ is the intersection of the ball with center at the origin and radius R with the complex plane $\mathbb{C}_I$. We define

$$L_{n-1}(f)(q) = \frac{1}{2\pi} \int_{\Gamma_{R,I}} (\xi - q)^{-*} * (\xi^n - q^n)(\xi^n - 1)^{-1} d\xi_I f(\xi), \quad q \in B(0; R), \quad (5.11)$$

where $d\xi_I = d\xi/I$ and the $*$-product is computed with respect to the variable q.

Formula (5.11) seems to depend on the complex plane on which the integral is computed. However, this is not the case:

Proposition 5.2.5. *The definition of $L_{n-1}(f)(q)$ does not depend on the choice of $I \in \mathbb{S}$.*

Proof. Let $I \in \mathbb{S}$ be arbitrary and let $J \in \mathbb{S}$ be such that $I \perp J$. By the splitting lemma, the restriction of the slice regular function f to $\mathbb{C}_I$ can be written as $f|_{\mathbb{C}_I}(\xi) = F(\xi) + G(\xi)J$, where $F, G : B(0, \rho) \cap \mathbb{C}_I \to \mathbb{C}_I$ are holomorphic functions of the complex variable ξ. The Lagrange polynomials of the holomorphic functions F and G can be obtained as

$$L_{n-1}(F)(z) = \frac{1}{2\pi} \int_{\Gamma_{R,I}} (\xi - z)^{-1}(\xi^n - z^n)(\xi^n - 1)^{-1} d\xi_I F(\xi),$$

$$L_{n-1}(G)(z) = \frac{1}{2\pi} \int_{\Gamma_{R,I}} (\xi - z)^{-1}(\xi^n - z^n)(\xi^n - 1)^{-1} d\xi_I G(\xi).$$

Thus, by the additivity of L_{n-1},

$$L_{n-1}(F + GJ)(z) = \frac{1}{2\pi} \int_{\Gamma_{R,I}} (\xi - z)^{-1}(\xi^n - z^n)(\xi^n - 1)^{-1} d\xi_I (F(\xi) + G(\xi)J),$$

i.e.,

$$L_{n-1}(f|_{\mathbb{C}_I})(z) = \frac{1}{2\pi} \int_{\Gamma_{R,I}} (\xi - z)^{-1}(\xi^n - z^n)(\xi^n - 1)^{-1} d\xi_I(f|_{\mathbb{C}_I}(\xi)),$$

where $z \in B(0; R) \cap \mathbb{C}_I$, $d\xi_I = d\xi/I$. By the representation formula, the value of the polynomial $L_{n-1}(f|_{\mathbb{C}_I})(z)$ at any other point of $q = x + Jy \in B(0; R)$ belonging to $[x + Iy]$ can be computed as

$$L_{n-1}(f)(q) = \frac{1}{2}\left[L_{n-1}((1+JI)f|_{\mathbb{C}_I})(z) + (1-JI)L_{n-1}(f|_{\mathbb{C}_I})(z)\right].$$

Since

$$\frac{1}{2}\left[(1+JI)(\xi - z)^{-1}(\xi^n - z^n) + (1-JI)(\xi - z)^{-1}(\xi^n - z^n)\right]$$
$$= (\xi - q)^{-*} * (\xi^n - q^n),$$

we obtain formula (5.11).   □

Remark 5.2.6. Alternatively, to prove that the integral (5.11) does not depend on $I \in \mathbb{S}$ we can use the fact that $L_{n-1}(f)(q)$ is a polynomial of degree $n-1$ and it is unique, see Theorem 5.2.3. This fact implies that the definition does not depend on the choice of the complex plane.

As we discussed above, a polynomial interpolating n roots of the unity, not three of them in the same equivalence class, does in fact interpolate f at all the possible solutions to the equation $q^n - 1 = 0$.

We use this fact to prove that $L_{n-1}(f)$ is a polynomial of degree $n - 1$ interpolating f at the roots of $q^n - 1 = 0$. To this end, we denote the roots of $q^n - 1 = 0$ as

$$q_{k,n} = \cos(2k\pi/n) + J_{n,k}\sin(2k\pi/n), \quad k = 0, 1, \ldots, n-1, \quad J_{n,k} \in \mathbb{S}.$$

Proposition 5.2.7. *Given the slice regular function $f : B(0; \rho) \to \mathbb{H}$, the polynomial $L_{n-1}(f)$ in the variable q defined in (5.11) has degree $n - 1$ and*

$$L_{n-1}(f)(q_{k,n}) = f(q_{k,n}), \text{ for all } k = 0, 1, \ldots, n-1,$$

where $q_{k,n} = \cos(2k\pi/n) + J_{n,k}\sin(2k\pi/n)$, for all $J_{n,k} \in \mathbb{S}$. In other words, $L_{n-1}(f)$ interpolates f at the roots of the unity.

Proof. It is immediate that the polynomial $\xi^n - q^n$ vanishes when $q = \xi$, thus $\xi - q$ is a left factor of this polynomial, and we can write

$$(\xi^n - q^n) = (\xi - q) * P_{n-1}(\xi, q)$$

where $P_{n-1}(\xi, q)$ has degree $n-1$ in q. Consequently,

$$L_{n-1}(f)(q) = \frac{1}{2\pi} \int_{\Gamma_{R,I}} (\xi - q)^{-*} * (\xi - q) * P_{n-1}(\xi, q)(\xi^n - 1)^{-1} d\xi_I f(\xi)$$

$$= \frac{1}{2\pi} \int_{\Gamma_{R,I}} P_{n-1}(\xi, q)(\xi^n - 1)^{-1} d\xi_I f(\xi)$$

and $L_{n-1}(f)(q)$ is a polynomial of degree $n-1$ in q because so is $P_{n-1}(\xi, q)$. Let us consider the root of the unity

$$q_{k,n} = \cos(2k\pi/n) + J_{n,k} \sin(2k\pi/n)$$

and let us compute $L_{n-1}(f)(q_{k,n})$. Recalling that the evaluation is not a homomorphism with respect to the $*$-product, we cannot substitute $q_{n,k}$ in place of q in the expression $(\xi - q)^{-*} * (\xi^n - q^n)$. However, formula (1.8) gives

$$(\xi - q)^{-*} * (\xi^n - q^n) = (\xi - q)^{-*}(\xi^n - \tilde{q}^n),$$

where $\tilde{q} = f(q)^{-1} q f(q)$ with $f(q) = (\xi - q)^{-*}$. Hence, $\tilde{q}^n = f(q)^{-1} q^n f(q)$ and so

$$(\tilde{q}^n)|_{q=q_{k,n}} = f(q_{k,n})^{-1} q_{k,n}^n f(q_{k,n}) = f(q_{k,n})^{-1} f(q_{k,n}) = 1,$$

yielding

$$((\xi - q)^{-*} * (\xi^n - q^n)) |_{q=q_{k,n}} = (\xi - q_{k,n})^{-*}(\xi^n - 1).$$

It is worthwhile mentioning that, in this case, the result also follows by direct computation. In fact we have:

$$(\xi - q)^{-*} * (\xi^n - q^n) = (q^2 - 2\operatorname{Re}(\xi)q + |\xi|^2)^{-1}(\bar{\xi} - q) * (\xi^n - q^n)$$

$$= (q^2 - 2\operatorname{Re}(\xi)q + |\xi|^2)^{-1}(\bar{\xi}\xi^n - q\xi^n - q^n\bar{\xi} + q^{n+1}).$$

This yields

$$((\xi - q)^{-*} * (\xi^n - q^n)) |_{q=q_{k,n}}$$
$$= (q_{k,n}^2 - 2\operatorname{Re}(\xi)q_{k,n} + |\xi|^2)^{-1}(\bar{\xi}\xi^n - q_{k,n}\xi^n - q_{k,n}^n\bar{\xi} + q_{k,n}^{n+1})$$
$$= (q_{k,n}^2 - 2\operatorname{Re}(\xi)q_{k,n} + |\xi|^2)^{-1}(\bar{\xi}\xi^n - q_{k,n}\xi^n - \bar{\xi} + q_{k,n})$$
$$= (q_{k,n}^2 - 2\operatorname{Re}(\xi)q_{k,n} + |\xi|^2)^{-1}(\bar{\xi} - q_{k,n})(\xi^n - 1)$$
$$= (\xi - q_{k,n})^{-*}(\xi^n - 1).$$

Using these relations and recalling that $L_{n-1}(f)$ can be computed via an integral which does not depend on the plane of integration, we obtain

$$L_{n-1}(f)(q_{k,n}) = \frac{1}{2\pi} \int_{\Gamma_{R,J_{n,k}}} ((\xi - q)^{-*} * (\xi^n - q^n)) \,|_{q=q_{k,n}} (\xi^n - 1)^{-1} d\xi_{J_{n,k}} f(\xi)$$

$$= \frac{1}{2\pi} \int_{\Gamma_{R,J_{n,k}}} (\xi - q_{k,n})^{-*}(\xi^n - 1)(\xi^n - 1)^{-1} d\xi_{J_{n,k}} f(\xi)$$

$$= \frac{1}{2\pi} \int_{\Gamma_{R,J_{n,k}}} (\xi - q_{k,n})^{-*} d\xi_{J_{n,k}} f(\xi)$$

$$= f(q_{k,n}),$$

where the last equality follows from the Cauchy formula (1.13). □

Remark 5.2.8. When the function f slice regular on $B(0; \rho)$ is quaternionic intrinsic, i.e., $f(\bar{q}) = \overline{f(q)}$, then by the splitting lemma, its restriction to the complex plane $\mathbb{C}_I$ is of the form $f|_{\mathbb{C}_I}(\zeta) = F(\zeta)$, where F is a $\mathbb{C}_I$-valued holomorphic function on $B(0; \rho) \cap \mathbb{C}_I$. Reasoning as in the proof of Proposition 5.2.5, we see that the Lagrange polynomial

$$L_{n-1}(f|_{\mathbb{C}_I})(\zeta) = L_{n-1}(F)(\zeta)$$

of the holomorphic function F is of the form (5.9), where z_k are the roots of the unity belonging to the complex plane $\mathbb{C}_I$. By the representation formula this polynomial extends to a quaternionic polynomial of the form (5.9).

The partial sum $s_{n-1}(f)(q)$ of the Taylor series at the origin of a function f that is slice regular in a ball centered at the origin can be written in integral form. This integral representation is very useful and will be used in the sequel.

Let us write the power series expansion $f(q) = \sum_{k=0}^{\infty} q^k a_k$, and set

$$s_{n-1}(f)(q) := \sum_{k=0}^{n-1} q^k a_k, \quad \text{where } a_k = \frac{1}{k!} \partial_s^k f(0).$$

Proposition 5.2.9. *Let* $f : B(0; \rho) \to \mathbb{H}$ *be slice regular and let*

$$\Gamma_{R,I} = \{u \in \mathbb{H}, \ u = x + Iy; \ |u| = R\}.$$

We have

$$s_{n-1}(f)(q) = \frac{1}{2\pi} \int_{\Gamma_{R,I}} (\xi - q)^{-*} * (\xi^n - q^n)\xi^{-n} d\xi_I f(\xi), \quad 0 < R < \rho, \quad (5.12)$$

where the integral does not depend on the choice of $I \in \mathbb{S}$.

Proof. When $|q| < |\xi|$, we use the power series expansion of the Cauchy kernel, see (1.11), and we recall that $(\xi - q)^{-*}$ is defined outside its singularities and not only in the ball $|q| < |\xi|$.

Using the Cauchy integral formula, which is independent of the choice of $I \in \mathbb{S}$, see Theorem 1.3.14, and (1.11), we get

$$
\begin{aligned}
f(q) &= \frac{1}{2\pi} \int_{\Gamma_{R,I}} (\xi - q)^{-*} d\xi_I f(\xi) \\
&= \sum_{k=0}^{n-1} q^k \left(\frac{1}{2\pi} \int_{\Gamma_{R,I}} \xi^{-k-1} d\xi_I f(\xi) \right) + \frac{1}{2\pi} \int_{\Gamma_{R,I}} q^n (\xi - q)^{-*} \xi^{-n} d\xi_I f(\xi) \\
&= s_{n-1}(f)(z) + \frac{1}{2\pi} \int_{\Gamma_{R,I}} q^n (\xi - q)^{-*} \xi^{-n} d\xi_I f(\xi),
\end{aligned}
$$

(5.13)

where we used the relation (see Theorem 1.3.14)

$$
\frac{1}{k!} \partial_s^k f(0) = \frac{1}{2\pi} \int_{\Gamma_{R,I}} \xi^{-k-1} d\xi_I f(\xi).
$$

(5.14)

The equality (5.13) gives

$$
s_{n-1}(f)(q) = \frac{1}{2\pi} \int_{\Gamma_{R,I}} (\xi - q)^{-*} d\xi_I f(\xi) - \frac{1}{2\pi} \int_{\Gamma_{R,I}} q^n (\xi - q)^{-*} \xi^{-n} d\xi_I f(\xi).
$$

To complete the proof, we note that the function $(\xi - q)^{-*} - q^n (\xi - q)^{-*} \xi^{-n}$, which is slice regular in the variable q, can be written as

$$
\begin{aligned}
(\xi - q)^{-*} - q^n (\xi - q)^{-*} \xi^{-n} &= (\xi - q)^{-*} - (\xi - q)^{-*} * q^n \xi^{-n} \\
&= (\xi - q)^{-*} * (1 - q^n \xi^{-n}) \\
&= (\xi - q)^{-*} * (\xi^n - q^n) \xi^{-n}. \qquad \square
\end{aligned}
$$

The following result is the quaternionic version of Walsh equiconvergence theorem.

Theorem 5.2.10 (Walsh equiconvergence). *Let $\rho > 1$ and $f : B(0; \rho) \to \mathbb{H}$ be slice regular, i.e., $f(q) = \sum_{k=0}^{\infty} q^k a_k$ with $a_k \in \mathbb{H}$. Then*

(i) *$L_{n-1}(f)$ given by (5.11) converges uniformly with geometric rate to f in any closed ball $\overline{B(0; R)}$ with $1 < R < \rho$;*

(ii) *for all $|q| < \rho^2$,*

$$
\lim_{n \to \infty} [s_{n-1}(f)(q) - L_{n-1}(f)(q)] = 0,
$$

the convergence being again uniform and with geometric rate in any $\overline{B(0; R)}$ with $1 < R < \rho^2$;

(iii) *the result is the best possible, in the sense that for the slice regular function $g : B(0; \rho) \to \mathbb{H}$, given by $g(q) = (q - \rho)^{-1}$, for $q = \rho^2$ we have*

$$
\lim_{n \to \infty} [s_{n-1}(g)(q) - L_{n-1}(g)(q)] \neq 0.
$$

Proof. (i) Using the Cauchy integral formula, see Theorem 1.3.14, we have

$$f(q) - L_{n-1}(f)(q) = \frac{1}{2\pi} \int_{\Gamma_{R,I}} (\xi - q)^{-*} * [1 - (\xi^n - q^n)(\xi^n - 1)^{-1}] d\xi_I f(\xi),$$

for $q \in B(0; R)$. Put

$$A := 1 - (\xi^n - q^n)(\xi^n - 1)^{-1}.$$

Then

$$1 - A = (\xi^n - q^n)(\xi^n - 1)^{-1},$$

whence

$$(1 - A)(\xi^n - 1) = \xi^n - q^n.$$

A simple calculation gives

$$A = (q^n - 1)(\xi^n - 1)^{-1},$$

which when replaced in the above integral yields

$$f(q) - L_{n-1}(f)(q) = \frac{1}{2\pi} \int_{\Gamma_{R,I_q}} (\xi - q)^{-*} * (q^n - 1)(\xi^n - 1)^{-1} d\xi_I f(\xi).$$

Since the maximum modulus principle holds, see [101], we may set

$$M = \max\{|f(\xi)|; \ \xi \in \partial B(0; R)\} < \infty.$$

Take an arbitrary $1 < \mu < R$. Then for any ξ with $|\xi| = R$ and any $|q| \le \mu$ we immediately get

$$|\xi^n - 1| \ge R^n - 1, \quad |q^n - 1| \le \mu^n + 1.$$

Recalling (1.8), we have

$$(\xi - q)^{-*} = [(\bar{\xi} - q) * (\xi - q)]^{-1}(\bar{\xi} - q) = [(\bar{\xi} - q)(\bar{\xi} - q)]^{-1}(\bar{\xi} - q) = (\bar{\xi} - q)^{-1},$$

and so we obtain that

$$|(\xi - q)^{-*}| = |(\xi - \tilde{q})^{-1}| \ge |\,|\xi| - |q|\,| = R - |q| \ge R - \mu,$$

whence

$$|(\xi - q)^{-*} * (q^n - 1)(\xi^n - 1)^{-1}| = |(\xi - \tilde{q})^{-1}| \, |\hat{q}^n - 1| \, |(\xi^n - 1)^{-1}|.$$

Since $\tilde{q}$, $\dot{q}$ belong to the sphere defined by q, as it can be easily seen from (1.8), we have $|\tilde{q}| = |\hat{q}| = |q|$. Thus

$$|(\xi - \tilde{q})^{-1}| \, |\hat{q}^n - 1| \, |(\xi^n - 1)^{-1}| \le \frac{\mu^n + 1}{(R^n - 1)(R - \mu)} \le \frac{4}{R - \mu} \cdot \left(\frac{\mu}{R}\right)^n$$

and

$$|f(q) - L_{n-1}(f)(q)| \le MR \cdot \frac{\mu^n + 1}{(R^n - 1)(R - \mu)} \le C\left(\frac{\mu}{R}\right)^n,$$

for all $|q| \le \mu < R$, and where $C = \frac{4MR}{R-\mu}$ is independent of n and q.

As $1 < R < \rho$ arbitrary, it follows that $L_{n-1}(f)$ converges to f at the geometric rate $\left(\frac{\mu}{R}\right)^n$, uniformly in any closed disk included in $\mathbb{D}_\rho$. This proves (i).

(ii) To prove this item, we consider for any $q \in B(0; R)$ the difference $L_{n-1}(f)(q) - s_{n-1}(f)(q)$, which can be computed as

$$
\begin{aligned}
L_{n-1}&(f)(q) - s_{n-1}(f)(q) \\
&= \frac{1}{2\pi} \int_{\Gamma_{R,I}} (\xi - q)^{-*} * (\xi^n - q^n)[(\xi^n - 1)^{-1} - \xi^{-n}] d\xi_I f(\xi) \\
&= \frac{1}{2\pi} \int_{\Gamma_{R,I}} (\xi - q)^{-*} * (\xi^n - q^n)(\xi^n - 1)^{-1} \xi^{-n} d\xi_I f(\xi).
\end{aligned}
\tag{5.15}
$$

Since $\xi - q$ is a left factor, with respect to the *-product, of $\xi^n - q^n$, the integrand in the right-hand side is slice regular for all (finite) values of q, and in particular for $|q| = \mu < \rho^2$. The estimate

$$|(\xi - q)^{-*} * (\xi^n - q^n)(\xi^n - 1)^{-1} \xi^{-n}| \le \frac{R^n + \mu^n}{(R - \mu)(R^n - 1)R^n}$$

implies

$$
\begin{aligned}
|L_{n-1}(f)(q) - s_{n-1}(f)(q)| &\le \frac{MR(R^n + \mu^n)}{R^n(R^n - 1)(R - \mu)} \\
&\le \frac{2MR}{R - \mu} \cdot \left(\frac{R^n + \mu^n}{R^{2n}}\right) \\
&= C\left(\frac{1}{R^n} + \left(\frac{\mu}{R^2}\right)^n\right),
\end{aligned}
$$

with $C = \frac{2MR}{R-\mu}$. Since the last term in the above inequality tends to zero (as $n \to \infty$) for any $\mu < R^2$, and so part (ii) of the theorem is proved.

(iii) Let $g(q) = (q - \rho)^{-*} = (q - \rho)^{-1}$, $q \in B(0; 1)$. Thus, for $|q| < \rho$, we can write

$$g(q) = -(\rho - q)^{-1} = -\frac{1}{\rho} \cdot \sum_{k=0}^{\infty} \left(\frac{q}{\rho}\right)^k.$$

It is immediate that for the function f the partial sum $s_{n-1}(g)(q)$ is

$$s_{n-1}(g)(q) = -\frac{1}{\rho} \cdot \sum_{k=0}^{n-1} \left(\frac{q}{\rho}\right)^k.$$

Accordingly,

$$f(q) - s_{n-1}(g)(q) = -\frac{1}{\rho} \cdot \sum_{k=n}^{+\infty} \left(\frac{q}{\rho}\right)^k$$

$$= \sum_{j=0}^{+\infty} \left(\frac{q}{\rho}\right)^j \cdot -\frac{1}{\rho} \cdot \frac{q^n}{\rho^n}$$

$$= \left(1 - \frac{q}{\rho}\right)^{-1} \cdot \frac{-q^n}{\rho^{n+1}}$$

$$= (q - \rho)^{-1} \cdot \frac{q^n}{\rho^n}.$$

The above formula also yields

$$g(q) - L_{n-1}(g)(q) = \frac{1}{2\pi} \int_{\Gamma_{R,I}} (\xi - q)^{-*}(q^n - 1)(\xi^n - 1)^{-1}(-I)d\xi \cdot (\xi - \rho)^{-1}.$$

Since the integral does not depend on the choice of the complex plane $\mathbb{C}_I$, we can choose to integrate on the plane $\mathbb{C}_{I_q}$ containing q, and therefore compute exactly as in the complex case. Note that this is true because $\rho \in \mathbb{R}$. As a consequence, again following the computations in the complex case (see, e.g., [193], p. 154), we get

$$g(q) - L_{n-1}(g)(q) = (q - \rho)^{-1}(q^n - 1)(\rho^n - 1)^{-1}.$$

It follows that

$$L_{n-1}(g)(q) - s_{n-1}(g)(q) = (q - \rho)^{-1} \cdot \frac{\rho^n - q^n}{(\rho^n - 1)\rho^n},$$

and taking $q = \rho^2$, we obtain

$$L_{n-1}(g)(q) - s_{n-1}(g)(q) = \frac{1}{\rho - \rho^2},$$

which obviously does not tend to zero. $\qquad\square$

As it happens in the complex case, there are several extensions of the Walsh's theorem. The result below is a generalization of [32, Theorem 1] (for items (i) and (ii)) and of [174, Theorem 1] (for item (iii)).

Theorem 5.2.11. *Let $\rho > 1$ and $f : B(0; \rho) \to \mathbb{H}$ be slice regular, so that $f(q) = \sum_{k=0}^{\infty} q^k a_k$, with $a_k \in \mathbb{H}$. For $l \in \mathbb{N}$ and $j = 0, 1, \ldots, l - 1$, define the polynomials*

$$P_{n-1,j}(f)(q) = \sum_{k=0}^{n-1} q^k a_{k+jn}$$

and

$$S_{n-1,l}(f)(q) = \sum_{j=0}^{l-1} P_{n-1,j}(f)(q).$$

(i) *We have*

$$\lim_{n\to\infty}[L_{n-1}(f)(q) - S_{n-1,l}(f)(q)] = 0, \ \textit{for all } |q| < \rho^{l+1},$$

the convergence being uniform and with geometric rate in any closed ball $B(0; R)$ *with* $1 < R < \rho^{l+1}$;

(ii) *the convergence is the best possible, in the sense that for* $g(q) = (q - \rho)^{-1}$ *and for* $q = \rho^{l+1}$, *we have*

$$\lim_{n\to\infty}[L_{n-1}(g)(q) - S_{n-1,l}(g)(q)] \neq 0;$$

(iii) *if* f *has a singularity on* $|q| = \rho$, *then the sequence*

$$(L_{n-1}(f)(q) - S_{n-1,l}(f)(q))_{n\in\mathbb{N}}$$

can be bounded in at most l *distinct points* $q_k, k = 1, \ldots, l$, *such that no three of them belong to the same sphere and satisfy* $|q_k| > \rho^{l+1}, k = 1, \ldots, l$.

Proof. (i) Using formula (5.14), and with calculations similar to those in the complex case, see the proof of Theorem 1 in [32], we arrive at the formula

$$P_{n-1,j}(f)(q) = \frac{1}{2\pi}\int_{\Gamma_{R,I_q}}(\xi - q)^{-*} * (\xi^n - q^n)\left(\xi^{n(j+1)}\right)^{-1}d\xi_{I_q}f(\xi), \ j = 0, 1, \ldots.$$

Therefore, in view of the formula in Definition 3.1, for each integer $l \geq 1$, we obtain

$$L_{n-1}(f)(q) - S_{n-1,l}(f)(q)$$

$$= \frac{1}{2\pi}\int_{\Gamma_{R,I_q}}(\xi - q)^{-*}(\xi^n - q^n)(\xi^n - 1)^{-1}d\xi_{I_q}f(\xi)$$

$$- \frac{1}{2\pi}\int_{\Gamma_{R,I_q}}(\xi - q)^{-*} * (\xi^n - q^n)\sum_{j=0}^{l-1}(\xi^{n(j+1)})^{-1}d\xi_{I_q}f(\xi)$$

$$= \frac{1}{2\pi}\int_{\Gamma_{R,I_q}}(\xi - q)^{-*} * (\xi^n - q^n)\left[(\xi^n - 1)^{-1} - \sum_{j=0}^{l-1}(\xi^{n(j+1)})^{-1}\right]d\xi_{I_q}f(\xi).$$

Simple properties of the powers of quaternions yield

$$(\xi^n - 1)^{-1} - \sum_{j=0}^{l-1}(\xi^{n(j+1)})^{-1} = (\xi^n - 1)^{-1}(\xi^{nl})^{-1},$$

therefore

$$L_{n-1}(f)(q) - S_{n-1,l}(f)(q)$$

$$= \frac{1}{2\pi}\int_{\Gamma_{R,I_q}}(\xi - q)^{-*} * (\xi^n - q^n)(\xi^n - 1)^{-1}(\xi^{nl})^{-1}d\xi_{I_q}f(\xi).$$

Let $1 < \mu < R$, $|t| = R$, $|q| \le \mu$ and $M = \max\{|f(q)|; \ |q| = R\}$. Reasoning as in the proof of Theorem 5.2.10, we obtain

$$|(\xi - q)^{-*} * (\xi^n - q^n)(\xi^n - 1)^{-1}(\xi^{nl})^{-1}f(\xi)| \le \frac{M(R^n + \mu^n)}{R^{nl}(R^n - 1)(R - \mu)},$$

which implies

$$|L_{n-1}(f)(q) - S_{n-1,l}(f)(q)| \le \frac{MR(R^n + \mu^n)}{R^{nl}(R^n - 1)(R - \mu)}$$

$$\le \frac{2MR}{R - \mu} \cdot \left(\frac{R^n + \mu^n}{R^{n(l+1)}}\right)$$

$$= C\left(\frac{1}{R^{nl}} + \left(\frac{\mu}{R^{l+1}}\right)^n\right),$$

with $C = \frac{2MR}{R-\mu}$. The last term in the above inequality tends to zero (as $n \to \infty$) for any $\mu < R^{l+1}$, and this proves (i).

(ii) Reasoning as in the proof of Theorem 5.2.10, (iii) (but see also [32], or the book [125] p. 9) we see that the choices $g(q) = (q - \rho)^{-1}$ and $q = \rho^{l+1}$ again give

$$\lim_{n \to \infty} [L_{n-1}(g)(q) - S_{n-1,l}(g)(q)] \ne 0.$$

(iii) By contradiction, suppose that there exist $l + 1$ distinct points q_k with $|q_k| > \rho^{l+1}$, $k = 1, \ldots, l+1$, such that no three of them belong to the same sphere and

$$|L_{n-1}(f)(q_k) - S_{n-1,l}(f)(q_k)| \le M, \qquad (5.16)$$

for all $k = 1, \ldots, l+1$ and $n \ge 1$. Since f is slice regular in $B(0; \rho)$ with a singularity on $|q| = \rho$, we have

$$\limsup_{n \to \infty} |a_n|^{1/n} = \frac{1}{\rho}. \qquad (5.17)$$

The rest of the proof follows by adapting the arguments in the proof of Theorem 1 in [174]. We repeat, for the reader's convenience, the part that needs some adaptation of the computations.

For each $\varepsilon > 0$ such that $\rho - \varepsilon > 1$ and $(\rho - \varepsilon)^{l+2} > \rho^{l+1}$ there exists $N \in \mathbb{N}$ depending on ε, such that $|a_n| \le 1/(\rho - \varepsilon)^n$, for all $n \ge N$. By our hypothesis,

$$|q_k| > \rho^{l+1}, \quad k = 1, \ldots, l+1,$$

and so we can set

$$\sigma_1 = \min_{k=1,\ldots,l+1} |q_k| \quad \text{and} \quad \sigma_2 = \min_{k=1,\ldots,l+1} |q_k|,$$

where

$$\rho^{l+1} < \sigma_1 \le \sigma_2 < \rho^{m+1}, \qquad (5.18)$$

and m is the least integer for which the inequality holds. By item (ii) and (5.18), there exists M_1 such that

$$|L_{n-1}(f)(q_k) - \sum_{j=0}^{m-1} P_{n-1,j}(f)(q_k)| \leq M_1, \quad n \geq 1, \ k = 1, \ldots, l+1$$

which, by (5.16), implies

$$|\sum_{j=0}^{m-1} P_{n-1,j}(f)(q_k)| \leq M_2, \quad n \geq 1, \ k = 1, \ldots, l+1. \tag{5.19}$$

Using the estimate $|a_n| \leq 1/(\rho - \varepsilon)^n$ we obtain

$$|P_{n-1,j}(f)(q)| \leq \frac{n|q|^n}{(\rho - \varepsilon)^{(j+1)n}}, \quad \forall n \geq N, \ |q| > \rho, \ j \geq 1.$$

Thus, for $l + 1 \leq m - 1$ we have

$$|\sum_{j=l+1}^{m-1} P_{n-1,j}(f)(q)| \leq \frac{(m-l-1)n|q|^n}{(\rho - \varepsilon)^{(l+2)n}}, \quad \forall n \geq N, \ |q| > \rho.$$

From this inequality and (5.19) we get

$$|P_{n-1,j}(f)(q_k)| \leq M_2 + \frac{(m-l-1)n|q_k|^n}{(\rho - \varepsilon)^{(l+2)n}}, \quad \forall n \geq N, \ k = 1, \ldots, l+1.$$

For any polynomial $p(q)$ we have $q * p(q) = qp(q)$ and $(q * p(q))|_{q=q_k} = q_k p(q_k)$, thus the previous inequality yields

$$|q_k^l P_{n,l}(q_k) - P_{n-1,l}(q_k)| \leq M_3 + \frac{M_4 n|q_k|^n}{(\rho - \varepsilon)^{(l+2)n}}, \quad \forall n \geq N, \ k = 1, \ldots, l+1, \tag{5.20}$$

and

$$q^l P_{n,l}(f)(q) - P_{n-1,l}(f)(q) = \sum_{j=n}^{l+n} q^j a_{ln+j} - \sum_{j=0}^{l-1} q^j a_{ln+j}. \tag{5.21}$$

The rightmost sum in (5.21) is bounded at the points q_k, while the first sum evaluated at q_k satisfies

$$\left| \sum_{j=n}^{l+n} q_k^j a_{ln+j} \right| \leq M_5 + \frac{M_4 n|q_k|^n}{(\rho - \varepsilon)^{(l+2)n}}.$$

Dividing by $|q_k|^n$ and using the definition of σ_1 we get

$$|\sum_{j=0}^{l} q_k^j a_{(l+1)n+j}| \leq \frac{M_5}{\sigma_1^n} + \frac{M_4 n}{(\rho - \varepsilon)^{(l+2)n}}, \quad \forall n \geq N, \ k = 1, \ldots, l+1.$$

Now set
$$\tau := \max\{1/\sigma_1;\ 1/(\rho-\varepsilon)^{l+2})\}.$$

Then by (5.18) we have $\tau < 1/\rho^{l+1}$. Consider the system of $l+1$ linear equations in which the unknowns are the coefficients $a_{(l+1)n+j} \in \mathbb{H}$, i.e.,

$$\sum_{j=0}^{l} q_k^j a_{(l+1)n+j} := w_{k,n}, \quad k = 1, 2, \ldots, l+1,$$

where
$$|w_{k,n}| \le M_6 n\tau^n, \quad \forall n \ge N,\ k = 1, \ldots, l+1. \tag{5.22}$$

Now write the system in the form (5.10) and note that the matrix of coefficients is a Vandermonde matrix of the distinct quaternions $q_k, k = 1, \ldots, l+1$, such that no three of them belong to the same sphere. Therefore, by Theorem 5.2.3, the system has a unique solution for $a_{(l+1)n+j}, j = 0, \ldots, l$. The solution can be computed using the analog of the Cramer's method, see [34, Theorem 2.3]. Thus each $a_{(l+1)n+j}$ is a linear combination (with coefficients depending only on q_k, $k = 1, \ldots, l+1$) of $w_{k,n}$.

Taking into account (5.22), this immediately implies that

$$|a_{(l+1)n+j}| \le M_7 n\tau^n, \quad n \ge N,\ k = 1, \ldots, l+1.$$

This last inequality yields

$$\limsup |a_n|^{1/n} \le \tau^{1/(l+1)} < \frac{1}{\rho},$$

which contradicts (5.17). □

We note that for $l = 1$ we recover Theorem 5.2.10.

To prove the converse of Walsh's result in the quaternionic setting, we need some technical lemmas, exactly as in the proof of the analogous result in the complex case, see [184].

Lemma 5.2.12. *Let $f : \overline{B(0;1)} \to \mathbb{H}$ be slice regular on $B(0;1)$, i.e., $f(q) = \sum_{k=0}^{\infty} q^k a_k$ for all $q \in B(0;1)$, and continuous in $\overline{B(0;1)}$. Then we have*

$$L_{n-1}(f)(q) - S_{n-1,\ell}(f)(q) = L_{n-1}\left(\sum_{k=\ell n}^{\infty} q^k a_k\right)(q), \quad \ell, n \in \mathbb{N},\ q \in B(0;1).$$

Proof. The proof is by induction. For $\ell = 1$ we get

$$S_{n-1,1}(f)(q) = P_{n-1,0}(f)(q) = \sum_{k=0}^{n-1} q^k a_k.$$

According to Proposition 1.4.3 and Theorem 5.2.3, $L_{n-1}(f)(q)$ is the unique polynomial of degree at most $n-1$ coinciding with f on the n roots of unity. This fact gives $L_{n-1}(P)(q) = P(q)$, for any polynomial P of degree $\leq n-1$ and therefore

$$S_{n-1,1}(f)(q) = \sum_{k=0}^{n-1} q^k a_k = L_{n-1}\left(\sum_{k=0}^{n-1} q^k a_k\right)(q).$$

The linearity of L_{n-1} follows from its definition. Thus the case $\ell = 1$ is proved.

We now assume that the assertion is true for $\ell - 1$ and prove it for ℓ. Since $L_{n-1}(g)(q)$ depends only on the n roots of unity and on the values of g taken on the n roots of unity, it is obvious that

$$L_{n-1}(g)(q) = L_{n-1}(e_{nm}g)(q),$$

for all $m, n \in \mathbb{N}$ (here $e_{nm}(q) = q^{nm}$). So, again by the linearity of L_{n-1}, we have

$$L_{n-1}(f)(q) - S_{n-1,\ell}(f)(q) = L_{n-1}(f)(q) - S_{n-1,\ell-1}(f)(q) - P_{n-1,\ell-1}(f)(q)$$

$$= L_{n-1}\left(\sum_{k=(\ell-1)n}^{\infty} q^k a_k\right)(q) - \sum_{k=0}^{n-1} q^k a_{k+(\ell-1)n}$$

$$= L_{n-1}\left(\sum_{k=(\ell-1)n}^{\infty} q^k a_k\right)(q) - L_{n-1}\left(\sum_{k=(\ell-1)n}^{\ell n-1} q^k a_k\right)(q)$$

$$= L_{n-1}\left(\sum_{k=\ell n}^{\infty} q^k a_k\right)(q),$$

and the assertion follows. $\qquad\square$

Lemma 5.2.13. *If $a_k \in \mathbb{H}$, $k = 0, 1, \ldots$, then the following relation holds:*

$$a_{(3\ell^2+1)m+p} - a_{(6\ell^2+1)m+p} = \sum_{j=\ell}^{2\ell-1} a_{(3j\ell+1)m+p}$$

$$+ \sum_{s=\ell+1}^{2\ell-1}\sum_{j=\ell}^{2\ell-1} a_{(3js+2\ell+s+1)m+p} - \sum_{s=\ell}^{2\ell-1}\sum_{j=\ell}^{2\ell-1} a_{(3js+2\ell+j+1)m+p} \quad (5.23)$$

for $\ell = 2, 3, \ldots$, $p = 0, 1, \ldots$, $m = 1, 2, \ldots$.

Proof. The proof is identical to that in the complex case, see [184], in fact it is a manipulation of the sums of the coefficients a_k and the fact that the coefficients are quaternions does not affect the calculations. $\qquad\square$

We are now ready to prove the converse of Walsh's result:

Theorem 5.2.14. *Let $\rho > 1$, l a positive integer, $f : \overline{B(0;1)} \to \mathbb{H}$ be a slice regular function on $B(0;1)$, i.e., $f(q) = \sum_{k=0}^{\infty} q^k a_k$, for all $q \in B(0;1)$, and continuous in $\overline{B(0;1)}$. If the sequence*

$$(L_{n-1}(f)(q) - S_{n-1,l}(f)(q))_{n \in \mathbb{N}},$$

considered in the statement of Theorem 5.2.11 is uniformly bounded in every closed subset of $B(0;\rho^{l+1})$, then f can be slice regularly prolonged to $B(0;\rho)$, i.e., $f : B(0;\rho) \to \mathbb{H}$ is a slice regular function on $B(0;\rho)$.

Proof. The proof mimics the one in the complex case, which can be found, e.g., in [184], Theorem 1, pp. 272–276. First of all, one has to replace everywhere the complex modulus $| \cdot |$ with the quaternionic modulus. We then use the above Lemmas 5.2.12, 5.2.13, the properties of the interpolating polynomials L_{n-1}, and the representations of the coefficients of a slice regular function, in particular, of a polynomial, using Cauchy integrals. This reasoning shows that

$$\limsup_{n \to \infty} |a_n|^{1/n} \leq r^{-1/(3l^2+1)}.$$

This means that there exists $\rho' > 1$ such that f can be prolonged analytically in $B(0;\rho')$ and has a singularity on $\{|q| = \rho'\}$. Then, by Theorem 5.2.11 (iii), the sequence

$$(L_{n-1}(f)(q) - S_{n-1,\ell}(f)(q))_{n \in \mathbb{N}}$$

can be bounded in at most ℓ points $q_k, k = 1, \ldots, \ell$, such that no three of them belong to the same sphere and satisfy $|q_k| > \rho^{l+1}$, $k = 1, \ldots, \ell$. Comparing this with the hypothesis of Theorem 5.2.14, it necessarily implies that $\rho \leq \rho'$, that is f is slice regular on $B(0;\rho)$. $\qquad \square$

5.3 Universality Properties of Power Series and Entire Functions

This section collects results on universality properties of power series originally obtained in the paper of Gal and Sabadini [90].

We start by giving some definitions in the complex case. To this end, we introduce the notation $\mathbb{D}_R = \{z \in \mathbb{C}; |z| < R\}$ and

$$\mathcal{M}_R = \{K; \ K \text{ compact}, K \subset (\mathbb{C} \setminus \overline{\mathbb{D}_R}) \text{ with } \mathbb{C} \setminus K \text{ connected}\}.$$

As it is customary, $\mathbb{D}$ denotes the unit disc.

Definition 5.3.1. A power (Taylor) series $g(z) = \sum_{k=0}^{\infty} a_k z^k$ with radius of convergence R is said to have the *universal approximation property*, if for any $K \in \mathcal{M}_R$

and any $f : K \to \mathbb{C}$, continuous on K and analytic on the interior of K, there exists a subsequence $(S_{n_k}(z))_{k \in \mathbb{N}}$ of the partial sum sequence $(S_n(z))_{n \in \mathbb{N}}$,

$$S_n(z) = \sum_{k=0}^{n} a_k z^k,$$

such that $\lim_{k \to \infty} S_{n_k}(z) = f(z)$ uniformly for $z \in K$.

If there exists a power series with radius of convergence $R > 0$ which has the above universal approximation property in $\mathbb{D}_R$, we will say that the phenomenon of universal power series holds.

The approximation property of an universal power series is evidently related to the so-called *overconvergence property* of a power series:

Definition 5.3.2. The series $\sum_{k=0}^{\infty} a_k z^k$, $a_k \in \mathbb{C}$, of radius of convergence $R > 0$ is said to be overconvergent, if there exist a compact subset $K \subset \{z \in \mathbb{C};\ |z| \geq R\}$ and a subsequence $(S_{n_k}(z))_{k \in \mathbb{N}}$ of the partial sum sequence $(S_n(z))_{n \in \mathbb{N}}$, $S_n(z) = \sum_{k=0}^{n} a_k z^k$, such that, for $k \to \infty$, $S_{n_k}(z)$ converges uniformly on K to a function $f : K \to \mathbb{C}$ (continuous on K and analytic on the interior of K).

The first Taylor series with universal approximation properties was introduced on the real line by Fekete (see, e.g., [157]) and in the complex plane with radius of convergence zero by Seleznev [178].

The existence of a universal power series with positive radius of convergence was established by Chui and Parnes [37] and, independently, by Luh [137]. It is crucial to note that the proof given by Chui and Parnes in [37] is based on the Mergelyan approximation theorem by polynomials.

Since the Mergelyan approximation theorem is not available in the quaternionic setting in its full generality, the proofs in this section will be based on the Runge's approximation theorem by polynomials, see Theorem 4.1.8. What we obtain is a slightly less general phenomenon, that we call the phenomenon of *almost universal power series*.

Generally speaking, we can introduce this notion in the complex setting as well:

Definition 5.3.3. We say that the phenomenon of *almost universal power series* holds, if there exists a power series with radius of convergence 1, with the property that for any open set Ω such that $\Omega \cap \overline{\mathbb{D}} = \emptyset$, and any compact set $K \subset \Omega$ such that $\mathbb{C} \setminus K$ is connected, given any function $f : \Omega \to \mathbb{C}$ analytic in Ω, there exists a subsequence of the partial sums sequence which converges uniformly to f on K.

This weaker notion of almost universal power series is redundant in the complex case. In fact, if we consider f analytic only on the interior of K and continuous on K, then we get the universal power series phenomenon, which obviously imply the "almost universal" power series phenomenon. However, as we shall see, this is not the case in the quaternionic setting, and Theorem 5.3.7 shows that in the quaternionic setting the almost universal power series phenomenon holds. Also, by Theorems 5.3.9 and 5.3.11 one proves the existence of an "almost universal" entire function.

5.3.1 Almost universal power series

Consider a power series $\sum_{k=0}^{\infty} q^k a_k$, $a_k \in \mathbb{H}$, and denote its partial sums by $S_n(q) = \sum_{k=0}^{n} q^k a_k$, $n \in \mathbb{N}$. Observe that $S_n(q)$ is a polynomial with coefficients on the right, hence it is slice regular.

Definition 5.3.4. Let $\Omega \subseteq \mathbb{H}$ be an open set and let $\mathcal{F}(\Omega)$ be the set of all axially symmetric compact sets $K \subset \Omega$, such that $\mathbb{C}_I \setminus (K \cap \mathbb{C}_I)$ is connected for some (and hence for all) $I \in \mathbb{S}$.

To prove our main result we need the following lemmas.

Lemma 5.3.5. *Let $K \in \mathcal{F}(\mathbb{H} \setminus \overline{B(0;1)})$ and let Ω be an axially symmetric open subset of $\mathbb{H}$ containing K. Then, every $f \in \mathcal{R}(\Omega)$ can be uniformly approximated on K by polynomials*

$$p_n(q) = a_{n,0} + q a_{n,1} + \cdots + q^n a_{n,n}, \quad n \in \mathbb{N},$$

such that

$$\lim_{n \to \infty} \sum_{k=0}^{n} |a_{n,k}|^2 = 0.$$

In particular, every $f \in \mathcal{R}(\Omega)$ can be uniformly approximated on K by polynomials

$$p_n(q) = a_{n,0} + q a_{n,1} + \cdots + q^n a_{n,n}$$

whose coefficients $a_{n,j}$ satisfy $|a_{n,j}| \leq 1$, $j = 0, 1, \ldots, n$.

Proof. By hypothesis, $K \cap \overline{B(0;1)} = \emptyset$, where both K and $\overline{B(0;1)}$ are closed bounded sets. It easily follows that the distance between them, in the metric of $\mathbb{R}^4$, is positive. Therefore, the open set Ω can be chosen so small around K that there exists a number $\varepsilon > 0$ with $\Omega \cap B(0; 1 + \varepsilon) = \emptyset$.

Let $f \in \mathcal{R}(\Omega)$ and $\Omega' = \Omega \cup B(0; 1 + \varepsilon)$. If we define $F : \Omega' \to \mathbb{H}$ by $F(q) = f(q)$, for all $q \in \Omega$, $F(q) = 0$, for all $q \in B(0; 1+\varepsilon)$, then clearly $F \in \mathcal{R}(\Omega')$, $K' = K \cup \overline{B(0;1)} \subset \Omega'$ is an axially symmetric compact set and $\mathbb{C}_I \setminus (K \cap \mathbb{C}_I)$ is connected. By the Runge type Theorem 4.1.8, F can be uniformly approximated on K' by polynomials $p_n(q) = a_{n,0} + q a_{n,1} + \cdots + q^n a_{n,n}$, $n \in \mathbb{N}$. Setting

$$|f|_K = \max\{|f(q)|; \ q \in K\},$$

where K is a compact set, we get

$$|p_n - f|_K = |p_n - F|_K \to 0, \text{ as } n \to \infty$$

and for all $I \in \mathbb{S}$ we have, for $n \to \infty$ (see [4], Section 3):

$$\sum_{k=0}^{n} |a_{n,k}|^2 = \frac{1}{2\pi} \int_0^{2\pi} |p_n(e^{I\theta})|^2 d\theta$$

$$\le |p_n|^2_{\overline{B(0;1)}} = |p_n - F|^2_{\overline{B(0;1)}} \to 0,$$

(5.24)

which proves the lemma. $\qquad \square$

Lemma 5.3.6. *Let $K \in \mathcal{F}(\mathbb{H} \setminus \overline{B(0;1)})$, Ω be an axially symmetric open subset of $\mathbb{H}$ containing K, and $(k_n)_n$ be a sequence of natural numbers such that $\lim_{n \to \infty} k_n = \infty$. Then, every $f \in \mathcal{R}(\Omega)$ can be uniformly approximated on K by polynomials of the form*

$$p_n(q) = q^{k_n} + q^{k_n+1} b_{k_n+1} + \cdots + q^{m_n} b_{m_n},$$

where $k_n \le m_n$ and $|b_j| \le 1$, for all $j = k_n + 1, \ldots, m_n$.

Proof. For $K \in \mathcal{F}(\mathbb{H} \setminus \overline{B(0;1)})$, let $d = \max\{|q|; \ q \in K\} > 1$. Since $0 \notin \Omega$, the function

$$g(q) = q^{-(k_n+1)}(f(q) - q^{k_n}), \quad q \in \Omega$$

is in $\mathcal{R}(\Omega)$. Therefore, by the previous Lemma 5.3.5, there exists a polynomial $P_n(q)$ whose coefficients a_j of q^j satisfy $|a_j| \le 1$ and such that

$$|P_n(q) - q^{-(k_n+1)}(f(q) - q^{k_n})| \le \left(\frac{1}{2d} \right)^{k_n+1}, \quad \text{for all } n \in \mathbb{N}, \ q \in K.$$

Multiplying on the left by $d^{k_n+1} = \max\{|q^{k_n+1}|; \ q \in K\}$ and using that $|a \cdot b| = |a| \cdot |b|$, we immediately get

$$|q^{k_n+1} P_n(q) + q^{k_n} - f(q)| \le \left(\frac{1}{2} \right)^{k_n+1}, \quad q \in K.$$

In conclusion, the sought-for $p_n(q)$ is $p_n(q) = q^{k_n} + q^{k_n+1} P_n(q)$, and from Lemma 5.3.5 it is immediate that it has the requisite properties. $\qquad \square$

We can now prove our main theorem:

Theorem 5.3.7. *There exists a quaternionic power series $S(q) = \sum_{k=0}^{\infty} q^k a_k$, with radius of convergence 1 such that, denoting by $S_n(q)$ the n-th partial sum $\sum_{k=0}^{n} q^k a_k$ of S, the following property holds: for every $K \in \mathcal{F}(\mathbb{H} \setminus \overline{B(0;1)})$, for every axially symmetric open subset Ω of $\mathbb{H}$ containing K and for every $f \in \mathcal{R}(\Omega)$, there exists a subsequence $(S_{n_k}(q))_{k \in \mathbb{N}}$ such that $S_{n_k}(q) \to f(q)$ uniformly on K, as $k \to \infty$.*

Proof. We provide a general scheme to construct the series S. To this end, we consider a sequence $K_m \in \mathcal{F}(\mathbb{H} \setminus \overline{B(0;1)})$, $m \in \mathbb{N}$, such that for any $K \in \mathcal{F}(\mathbb{H} \setminus \overline{B(0;1)})$, there exists $m \in \mathbb{N}$ with $K \subset K_m$. Such a sequence $(K_m)_{m \in \mathbb{N}}$ exists:

for instance we can choose those sets in $\mathcal{F}(\mathbb{H} \setminus \overline{B(0;1)})$ which are finite unions of closed balls with rational quaternion centers and rational radii (a quaternion $q = x_0 + ix_1 + jx_2 + kx_3$ is called rational if x_ℓ, $\ell = 0, \ldots, 3$ are rational numbers). We can associate with $(K_m)_{m \in \mathbb{N}}$ a sequence of open sets $(\Omega_m)_{m \in \mathbb{N}}$ such that $K_m \subset \Omega_m$ and $\Omega_m \cap \overline{B(0;1)} = \emptyset$, for all $m \in \mathbb{N}$ (this is based on the well-known fact that in a Hausdorff space two disjoint compact sets can be separated by two open sets).

Now, for each $m = 1, 2, \ldots$, let $f_{m,j}$, $j = 1, 2, \ldots$, be a countable dense set in the space $\mathcal{R}(\Omega_m)$ with respect to the seminorm $|f|_{K_m} = \max\{|f(q)|; \ q \in K_m\}$. By the Runge type result in Theorem 4.1.8, we can take for it the set of polynomials with rational quaternionic coefficients.

Consider the countable list $\mathcal{P}$ of the triplets

$$(f_{m,j}; K_m; \Omega_m)_{m,j \in \mathbb{N}},$$

where for technical reasons we require that each triplet occurs infinitely often in $\mathcal{P}$. Observe that since the functions $f_{m,j}$ are polynomials, they are defined on the whole $\mathbb{H}$ and since in all the next lines in the proof, we use only the approximation properties on the sets K_m, $m \in \mathbb{N}$ (without using the sets Ω_m), in all these triplets we can take $\Omega_m = \mathbb{H}$.

For the sake of simplicity, we write $\mathcal{P}$ in the form

$$\mathcal{P} = \{(g_1; L_1), (g_2; L_2), \ldots\}.$$

We now repeatedly apply Lemma 5.3.6.

Take g_1. By Lemma 5.3.6 there exists a polynomial

$$s_1(q) = 1 + qb_1 + \cdots + q^{k_1}b_{k_1},$$

with coefficients bounded by 1 such that

$$|s_1 - g_1|_{L_1} \le 1.$$

Next g_2. Again by Lemma 5.3.6, there exists a polynomial

$$s_2(q) = q^{k_1+1} + q^{k_1+2}b_{k_1+2} + \cdots + q^{k_2}b_{k_2},$$

with coefficients bounded by 1 such that

$$|s_2 - (g_2 - s_1)|_{L_2} \le \frac{1}{2}.$$

Iterating the procedure, we obtain that when considering g_n there exists a polynomial

$$s_n(q) = q^{k_{n-1}+1} + q^{k_{n-1}+2}b_{k_{n-1}+2} + \cdots + q^{l_{\ell n}}b_{k_n},$$

with coefficients bounded by 1 such that

$$|s_n - (g_n - s_1 - \cdots - s_{n-1})|_{L_n} \le \frac{1}{n}.$$

We set

$$S(q) = \sum_{j=1}^{\infty} s_j(q),$$

and we claim that $S(q)$ is the sought-for power series, and since its coefficients are bounded by 1, it converges in the open unit ball.

To complete the proof, take $f \in \mathcal{R}(\Omega)$. By Theorem 4.1.8, there exists a sequence of polynomials $\{p_j\}$ such that

$$|f - p_j|_K < \frac{1}{j},$$

for any $K \in \mathcal{F}(\mathbb{H} \setminus \overline{B(0;1)})$. Assume that $K \subset K_m$ where $K_m \in \mathcal{F}(\mathbb{H} \setminus \overline{B(0;1)})$. Then for any $j \in \mathbb{N}$ select f_{m,k_j} such that

$$|f_{m,k_j} - p_j|_{K_m} < \frac{1}{j}.$$

The list of pairs $(f_{m,k_j}; K_m)$ occurs infinitely often in $\mathcal{P}$ and so, using the notation introduced above, we have

$$g_{n_1} = g_{n_2} = \cdots = f_{m,j_n},$$

and

$$L_{n_1} = L_{n_2} = \cdots = K_m.$$

Let S_{N_j} be the partial sum up to N_j of S, namely $S_{N_j}(q) = s_1(q) + \cdots + s_{n_j}(q)$. We have, following [37], that

$$|S_{N_j} - f_{m,k_j}|_{K_m} = |S_{N_j} - g_{n_j}|_{L_{n_j}} = |s_{n_j} - (g_{n_j} - s_1 - \cdots - s_{n_j-1})| \le \frac{1}{n_j}.$$

Moreover, by the triangle inequality,

$$\begin{aligned}
|S_{N_j} - f|_K &\le |S_{N_j} - f_{m,k_j}|_K + |f_{m,k_j} - p_j|_K + |p_j - f|_K \\
&\le |S_{N_j} - f_{m,k_j}|_{K_m} + |f_{m,k_j} - p_j|_{K_m} + |p_j - f|_K \\
&< \frac{1}{n_j} + \frac{1}{j} + \frac{1}{j}
\end{aligned}$$

and since all the terms on the right-hand side tend to zero when $j \to \infty$, this completes the proof. $\square$

5.3.2   Almost universal entire functions

In this section we extend to the quaternionic case a result originally proved by Birkhoff [16] in the complex plane and then generalized by Seidel and Walsh to simply connected sets. The result is stated next and can be found, e.g., in [177]:

Theorem 5.3.8. *There exists an entire function $F(z)$ such that given an arbitrary function $f(z)$ analytic in a simply connected region $R \subseteq \mathbb{C}$, for suitably chosen $a_1, a_2, \ldots$ the relation*

$$\lim_{n \to \infty} F(z + a_n) = f(z)$$

holds for $z \in R$ uniformly on any compact set in R.

As explained in Chapter 1, in the quaternionic setting by entire function we mean a slice regular function whose power series expansion in Theorem 1.1.5 converges in $\mathbb{H}$.

Further, if g is a polynomial with quaternionic coefficients, the composition $f \circ g$ is not slice regular in general. However, if g is quaternionic intrinsic, i.e., if g is a polynomial with real coefficients, then $f \circ g$ is slice regular.

In the next result we consider functions of the form $F(q + a_n)$, where F is quaternionic entire and a_n is a real number for every $n \in \mathbb{N}$. Recalling the definition of $\mathfrak{R}(\mathbb{H})$, see Definition 4.2.8, we have the following:

Theorem 5.3.9. *There exists a quaternionic entire function $F(q)$ such that given an arbitrary function $f(q)$ that is slice regular in a region $\Omega \in \mathfrak{R}(\mathbb{H})$, for suitably chosen $a_1, a_2, \ldots \in \mathbb{R}$ the relation*

$$\lim_{n \to \infty} F(q + a_n) = f(q) \tag{5.25}$$

holds for $q \in \Omega$ uniformly on any compact set in $\mathcal{F}(\mathbb{H})$ contained in Ω.

Proof. Following the proof of the result in the complex case given in [177], we define the spheres

$$S(4^n, 2^n) = \{q \in \mathbb{H}; \ |q - 4^n| = 2^n\} \quad \text{for } n = 1, 2, \ldots$$

and the spheres

$$\Sigma(0, 4^n + 2^n + 1) = \{q \in \mathbb{H}; \ |q| = 4^n + 2^n + 1\} \quad \text{for } n = 1, 2, \ldots.$$

It is immediate that $S(4^n, 2^n)$ and $S(4^m, 2^m)$ are mutually disjoint if $n \neq m$ and that $\Sigma(0, 4^n + 2^n + 1)$ contains in its interior $S(4^j, 2^j)$ for $j = 1, 2, \ldots, n$, but it does not contain $S(4^j, 2^j)$, nor its interior points if $j > n$. We will construct the function F as a limit of polynomials with rational quaternionic coefficients, i.e., with coefficients of the form $a_0 + ia_1 + ja_2 + ka_3$ with $a_i \in \mathbb{Q}$. Our assumption on the coefficients is not restrictive since it does not change properties like (uniform) convergence to a given function. Thus, let us consider all the polynomials with rational quaternionic coefficients and arrange these polynomials in a sequence (P_n).

Let $\pi_1(q)$ be a polynomial such that

$$|P_1(q - 4) - \pi_1(q)| < 1/2$$

for q on or inside $S(4,2)$. Such a polynomial exists: for example, we can take $\pi_1(q) = P_1(q-4)$. Let us define the function

$$\tilde{P}_2(q) = \begin{cases} P_2(q-4^2), & \text{for } q \text{ on or inside } S(4^2, 2^2), \\ \pi_1(q), & \text{for } q \text{ on or inside } \Sigma(0, 4+2+1). \end{cases}$$

We apply Theorem 4.1.8 to $\tilde{P}_2$ to deduce that there exists a polynomial $\pi_2(q)$ such that

$$|P_2(q-4^2) - \pi_2(q)| < 1/2^2, \quad \text{on or inside } S(4^2, 2^2),$$
$$|\pi_1(q) - \pi_2(q)| < 1/2^2, \quad \text{on or inside } \Sigma(0, 4+2+1).$$

Continuing, we inductively define

$$\tilde{P}_n(q) = \begin{cases} P_n(q-4^n), & \text{for } q \text{ on or inside } S(4^n, 2^n), \\ \pi_{n-1}(q), & \text{for } q \text{ on or inside } \Sigma(0, 4^{n-1} + 2^{n-1} + 1). \end{cases}$$

Using Theorem 4.1.8, we can find a polynomial $\pi_n(q)$ such that

$$|P_n(q-4^n) - \pi_n(q)| < 1/2^n, \quad \text{on or inside } S(4^n, 2^n),$$
$$|\pi_{n-1}(q) - \pi_n(q)| < 1/2^n, \quad \text{on or inside } \Sigma(0, 4^{n-1} + 2^{n-1} + 1).$$

The sequence $(\pi_n(q))$ converges uniformly inside the spheres Σ, and hence uniformly on every bounded set. Thus the function $F(q) = \lim_{n\to\infty} \pi_n(q)$ converges at any point in $\mathbb{H}$ and it is entire. We now show that it satisfies (5.25), see Proposition 4.1 in [50]. Let f be a slice regular function in a region $\Omega \in \mathfrak{R}(\mathbb{H})$. By Theorem 4.1.8, there exists a subsequence $(P_{n_k}(q))$ of the polynomials considered above, such that

$$\lim_{k\to\infty} P_{n_k}(q) = f(q) \tag{5.26}$$

in Ω, uniformly on every $K \in \mathcal{F}(\mathbb{H})$ contained in Ω. We apply the above construction and consider $q \in S(4^n, 2^n)$. We have

$$F(q) = \pi_n(q) + (\pi_{n+1}(q) - \pi_n(q)) + (\pi_{n+2}(q) - \pi_{n+1}(q)) + \cdots$$

and

$$|F(q) - P_n(q-4^n)| \leq |P_n(q-4^n) - \pi_n(q)| + |\pi_{n+1}(q) - \pi_n(q)|$$
$$+ |\pi_{n+2}(q) - \pi_{n+1}(q)| + \cdots$$
$$< \frac{1}{2^n} + \frac{1}{2^{n+1}} + \cdots = \frac{1}{2^{n-1}},$$

and so

$$\lim_{n\to\infty} [F(q+4^n) - P_n(q)] = 0$$

for q in any bounded set. Thus, by (5.26), we finally have

$$\lim_{k \to \infty} |F(q + 4^{n_k}) - f(q)|$$
$$\leq \lim_{k \to \infty} |F(q + 4^{n_k}) - P_{n_k}(q)| + \lim_{k \to \infty} |P_{n_k}(q) - f(q)| = 0,$$

which concludes the proof. □

Remark 5.3.10. The construction in the proof shows that the sequence $(a_n)_{n \in \mathbb{N}}$ depends on the approximated function f.

We now prove another result which, in the complex case, was originally obtained by MacLane [138], and which shows that there exists a slice regular function whose set of derivatives is dense in $\mathcal{R}(\mathbb{H})$. To state it, we recall that $F^{(n)}(x + Iy) = \frac{\partial^n}{\partial x^n} F(x + Iy)$ (see Section 1.1).

Theorem 5.3.11. *There exists an entire slice regular function F such that the set $\{F^{(n)}\}_{n \in \mathbb{N}}$ is dense in $\mathcal{R}(\mathbb{H})$.*

Proof. We follow the lines of the proof given in [12]. Let us define a linear operator I on the monomials q^n by

$$I(q^n) = \frac{q^{n+1}}{n+1}$$

and extend it by linearity to the set of all polynomials. Note that

$$I^k(q^n) = \frac{q^{n+k}}{(n+k) \cdots (n+1)},$$

and for $|q| \leq r$,

$$|I^k(q^n)| \leq \frac{r^{n+k}}{(n+k) \cdots (n+1)} \leq \frac{r^{n+k}}{k!}.$$

As a consequence, $\max_{|q| \leq r} |I^k(q^n)| \to 0$ for $k \to \infty$, and so for any $\delta > 0$ and any $r > 0$ there exists $k_0 \in \mathbb{N}$ such that

$$\max_{|q| \leq r} |I^k(q^n)| < \delta, \quad \text{for } k \geq k_0.$$

Now observe that, using the Cauchy estimates, given $f \in \mathcal{R}(\mathbb{H})$, $\varepsilon > 0$, and $m \in \mathbb{N}$, if $|f(q)| \leq \delta$ for $|q| \leq r$, then for $|q| \leq r/2$ we have

$$|f^{(j)}(q)| \leq \frac{j! \max_{|q| \leq r} |f(q)|}{(r/2)^j} \leq j! 2^j / r^j \delta < \varepsilon,$$

for any $j = 0, \dots, m$, provided δ is sufficiently small. In particular, if we consider a polynomial $P(q)$ and any $r > 0$, $\varepsilon > 0$, and $m \in \mathbb{N}$, then there exists $k_0 \in \mathbb{N}$ such that for $k \geq k_0$

$$\max_{|q| \leq r} |(I^k(P))^{(j)}| < \varepsilon, \tag{5.27}$$

for $j = 0, \ldots, m$.

Let $\{P_n\}_{n\in\mathbb{N}}$ be a sequence of polynomials dense in $\mathcal{R}(\mathbb{H})$, and consider

$$\sum_{j=1}^{\infty} I^{k_j}(P_j)(q),$$

where the integers k_j are provided below. Here we denote $I^{k_j}(P_j)(q)$ by $Q_j(q)$.

We set $k_1 = 0$, so $Q_1 = P_1$; then we choose $k_2 > k_1 + \deg(P_1)$ and, since we set $Q_2 = I^{k_2}(P_2)$, by (5.27) we can select k_2 to be such that

$$|Q_2(q)| < 1/2^2 \quad \text{for } |q| \le 2.$$

Then, inductively, for $n \ge 3$, we take $k_n > k_{n-1} + \deg(P_{n-1})$, we set $Q_n = I^{k_n}(P_n)$ and we select k_n to be such that

$$|Q_n(q)| \le \frac{1}{2^n}, \quad |Q'_n(q)| \le \frac{1}{2^n}, \ldots, \quad |Q_n^{(k_{n-1})}(q)| \le \frac{1}{2^n} \tag{5.28}$$

for $|q| \le n$. The series

$$\sum_{j=1}^{\infty} I^{k_j}(P_j)(q) = \sum_{j=1}^{\infty} Q_j(q)$$

converges uniformly on the bounded subsets in $\mathbb{H}$ to a slice regular function $F(q)$, because (5.28) implies that:

$$\max_{|q|\le j} |Q_j(q)| \le \frac{1}{2^j}.$$

We claim that the function F is such that $\{F^{(n)}\}_{n\in\mathbb{N}}$ is dense in $\mathcal{R}(\mathbb{H})$. Indeed, let $f \in \mathcal{R}(\mathbb{H})$ and let $r > 0$, $\varepsilon > 0$ be arbitrary numbers. Let $n_0 \in \mathbb{N}$ be such that $n_0 > r$ and $1/2^{n_0-1} < \varepsilon$. Since the sequence $\{P_n\}$ is dense in $\mathcal{R}(\mathbb{H})$, there exists $n \in \mathbb{N}$, $n > n_0$, such that

$$\max_{|q|\le n_0} |f(q) - P_n(q)| < \varepsilon.$$

The conditions on the numbers k_n imply that $Q_j^{(k_n)}(q) = 0$ for $j = 1, \ldots, n-1$ and $Q_n^{(k_n)}(q) = P_n(q)$, thus

$$|f(q) - F^{(k_n)}(q)| = |f(q) - P_n(q) - \sum_{j=n+1}^{\infty} Q_j^{(k_n)}(q)|$$

$$\le |f(q) - P_n(q)| + \sum_{j=n+1}^{\infty} |Q_j^{(k_n)}(q)| < 2\varepsilon.$$

Hence, $\max_{|q|\le n_0} |f(q) - F^{(k_n)}(q)| < 2\varepsilon$ and the claim is established. $\qquad\square$

Chapter 6

Inequalities for Quaternionic Polynomials

Many results in approximation theory depend on the fact that the derivative of a polynomial cannot be, in general, too large. In this spirit, this chapter studies the Bernstein type inequality, its Erdös–Lax improvement and its opposite inequality, namely the Turán inequality together with some of its generalizations, in the case of quaternionic polynomials.

The results presented in this chapter, except those where the authors are explicitly mentioned, were obtained in the papers of Gal and Sabadini [94], [88].

6.1 The complex case

One of the most known polynomial inequality with important applications in approximation theory is the following Bernstein inequality for complex polynomials.

Theorem 6.1.1 (Bernstein [15], Riesz [169]). *If $P(z)$ is an algebraic polynomial of degree n with complex coefficients, then $\|P'\| \le n \cdot \|P\|$, where the norm of P is defined by*

$$\|P\| = \max_{|z| \le 1} |P(z)| = \max_{|z|=1} |P(z)|.$$

A refinement of the Bernstein inequality conjectured by Erdös and proved by Lax [133] can be stated as follows:

Theorem 6.1.2 (Erdős–Lax [133]). *If $P(z)$ is an algebraic polynomial of degree n with complex coefficients which does not have a zero in the open unit disk of $\mathbb{C}$, then*

$$\|P'\| \le \frac{n}{2} \cdot \|P\|.$$

A kind of opposite to the Erdös–Lax inequality was proved by Turán in [188] and can be stated as follows.

© Springer Nature Switzerland AG 2019
S. G. Gal, I. Sabadini, *Quaternionic Approximation*, Frontiers in Mathematics,
https://doi.org/10.1007/978-3-030-10666-9_6

Theorem 6.1.3 (Turán [188]). *If $P(z)$ is an algebraic polynomial of degree n with complex coefficients having all its zeros in $|z| \leq 1$, then*

$$\|P'\| \geq \frac{n}{2} \cdot \|P\|.$$

The inequality is sharp and is an equality if P has all of its zeroes on the unit circle $|z| = 1$.

Theorem 6.1.3 was generalized in several ways. The following is a version in which instead of the unit disk one considers a disk of smaller radius:

Theorem 6.1.4 (Malik [139]). *If $P(z)$ is an algebraic polynomial of degree n with complex coefficients having all its zeros in the disk $|z| \leq K$, with $0 < K \leq 1$, then*

$$\|P'\| \geq \frac{n}{1+K} \cdot \|P\|.$$

The inequality is sharp and equality holds for $P(z) = \left(\dfrac{z+K}{1+K}\right)^n$.

For disks larger than the unit disk we have:

Theorem 6.1.5 (Govil [112]). *If $P(z)$ is an algebraic polynomial of degree n with complex coefficients having all its zeros in the disk $|z| \leq K$, with $K \geq 1$, then*

$$\|P'\| \geq \frac{n}{1+K^n} \cdot \|P\|.$$

The inequality is sharp and equality holds for $P(z) = \dfrac{z^n + K^n}{1 + K^n}$.

Among other generalizations, we mention the following:

Theorem 6.1.6 (Giroux, Rahman and Schmeisser [108]). *If*

$$P_n(z) = \prod_{j=1}^{n}(z - z_j), \quad z, z_j \in \mathbb{C},$$

$|z_j| \leq 1$, $j = 1, \ldots, n$ and $\rho = \frac{1}{n} \cdot \sum_{j=1}^{n} |z_j|$, *then*

$$\|P_n'\| \geq \frac{n}{1+\rho} \cdot \|P_n\|.$$

Theorem 6.1.7 (Giroux, Rahman and Schmeisser [108]). *Let*

$$P_n(z) = \prod_{j=1}^{n}(z - z_j), \quad z, z_j \in \mathbb{C}$$

be such that $\mathrm{Re}(z_j) \geq 1$, $j = 1, \ldots, n$. *Then*

$$\|P'_n\| \geq n \cdot \prod_{j=1}^{n} (1 + \mathrm{Re}(z_j))^{1/n} \cdot \|P_n\|^{1-2/n}.$$

Equality holds if, in addition, $P_n(z)$ takes real values for real z and if there is a $\lambda \geq 1$ such that all the zeroes z_j lie in the set

$$\{z \in \mathbb{C}; \; |z - \lambda + 1| \leq \lambda\} \cup \{z \in \mathbb{C}; \; \mathrm{Re}(z) \geq 1\}.$$

The main aim of this chapter is to extend Theorems 6.1.1–6.1.5 to quaternionic polynomials.

If P is a quaternionic polynomial, then the norm of P is defined by

$$\|P\| = \max_{|q| \leq 1} |P(q)| = \max_{|q| = 1} |P(q)|.$$

Note that we have:

Proposition 6.1.8. *Let P be a quaternionic polynomial of the form $P(q) = \sum_{k=0}^{n} q^k a_k$ and let $P^c(q) = \sum_{k=0}^{n} q^k \overline{a_k}$. Then $\|P\| = \|P^c\|$.*

Proof. First of all, we observe that some lengthy but easy computations show that if $p \in \mathbb{H}$ and $I \in \mathbb{S}$, there exists $J \in \mathbb{S}$ such that $pI = Jp$. Then, since a polynomial P is a special case of a slice regular function, we write it as $P(q) = P(x + Iy) = \alpha(x, y) + I\beta(x, y)$ and $P^c(q) = P^c(x + Iy) = \overline{\alpha(x, y)} + I\overline{\beta(x, y)}$. We then have

$$\|P\| = \max_{|q|=1} |P(q)| = \max_{I \in \mathbb{S}} \max_{x^2+y^2=1} |\alpha(x, y) + I\beta(x, y)|$$

$$- \max_{I \in \mathbb{S}} \max_{x^2+y^2=1} |\overline{\alpha(x, y) + I\beta(x, y)}| = \max_{I \in \mathbb{S}} \max_{x^2+y^2=1} |\overline{\alpha(x, y)} - \overline{\beta(x, y)}I|$$

$$= \max_{J \in \mathbb{S}} \max_{x^2+y^2=1} |\overline{\alpha(x, y)} + J\overline{\beta(x, y)}| = \|P^c\|. \qquad \square$$

6.2 Bernstein's Inequality

As we shall see below, the Bernstein inequality extends to the quaternionic case. We will provide two proofs. The first is based on the following quaternionic version of the Gauss–Lucas type theorem (see [190] and [106]) which was stated for arbitrary $n \in \mathbb{N}$ for the first time in [190], but later proved to actually be valid only for $n = 0, 1, 2$ in [106]. Note that according to [106], the Gauss–Lucas type result is not valid, in general, for $n \geq 3$.

Theorem 6.2.1. *Let $P(q) = \sum_{k=0}^{n} q^k a_k$, with $0 \leq n \leq 2$ and quaternionic coefficients a_k. Then the zeros of $P'(q)$ are in the axially symmetric completion of the convex hull $\mathcal{K}(Z_{P^s})$ of the zero set Z_{P^s} of P^s, where $P^s = P^c * P = P * P^c$ and $P^c(q) = \sum_{k=0}^{n} q^k \overline{a_k}$.*

The main result of this section is the Bernstein inequality in the quaternionic setting. The result was obtained for the first time by the authors of this book in [88] for $0 \leq n \leq 2$ and later, for arbitrary $n \in \mathbb{N}$ in Xu [195].

Theorem 6.2.2. *If P is a quaternionic polynomial of degree n, then*

$$\|P'\| \leq n \cdot \|P\|.$$

Proof. We will give two different proofs, one for the case when $0 \leq n \leq 2$ based on the Theorem 6.2.1 and on the authors' paper [88], and the other for arbitrary n, but based on the paper [195].

Case $0 \leq n \leq 2$. This proof will be divided into four steps.

Step 1. Define $Q(q) = Mq^n$ and $f(q) = Q^{-1}(q) * P(q) = \frac{1}{M}q^{-n}P(q)$, where $M = \max_{|q|=1}|P(q)|$. We show that

$$|f(q)| \leq 1 \quad \text{for all } |q| \geq 1. \tag{6.1}$$

Indeed, f is slice regular as function of q for $|q| > 1$ and

$$|f(q)| = \frac{1}{M}|P(q)| \leq 1,$$

for all $|q| = 1$. Then, denoting $P(q) = \sum_{k=0}^{n} q^k a_k$ and $\tilde{q} = q^{-1}$, after a simple calculation we can write $f(q) = h(\tilde{q})$, where

$$h(\tilde{q}) = \frac{1}{M}\sum_{k=0}^{n}(\tilde{q})^{n-k}a_k$$

is slice regular for $|\tilde{q}| < 1$, and since $|q| = 1$ if and only if $|\tilde{q}| = 1$, we also have $|h(\tilde{q})| = |f(q)| \leq 1$, for all $|\tilde{q}| = 1$.

Applying the maximum modulus theorem (see Theorem 3.4 in [99]), it follows that $|h(\tilde{q})| \leq 1$ for all $|\tilde{q}| \leq 1$, which in view of the equality $h(\tilde{q}) = f(q)$ yields (6.1).

Step 2. We prove that all the zeroes of the (left) regular polynomial of degree n,

$$g(q) = P(q) - Q(q)\lambda = P(q) - q^n\lambda M,$$

where $\lambda \in \mathbb{H}$ satisfies $|\lambda| > 1$, belong to the open unit ball $B(0; 1)$.

Indeed, let $q_0 \in \mathbb{H}$ be such that $g(q_0) = P(q_0) - Q(q_0)\lambda = 0$. We have two subcases: $Q(q_0) \neq 0$ (i.e., $q_0 \neq 0$) and $Q(q_0) = 0$ (i.e., $q_0 = 0$).

In the first subcase,

$$|P(q_0)| = |\lambda| \cdot |Q(q_0)| > |Q(q_0)| = M \cdot |q_0^n|,$$

which immediately implies $|f(q_0)| = |\frac{1}{M}q_0^{-n}P(q_0)| > 1$, and by (6.1) we necessarily get $|q_0| < 1$.

In the second subcase, we trivially have $|q_0| < 1$ since $q_0 = 0$.

Step 3. We show that all the zeroes of g^s also are included in $B(0;1)$, where g^s is defined as in the statement of Theorem 6.2.1. Indeed, since $g^s = g * g^c$ with

$$g^c(q) = q^n(\overline{a_n} - M\overline{\lambda}) + \sum_{k=0}^{n-1} q^k \overline{a_k}$$

and $(g * g^c)(q) = 0$ if and only if $g(q) = 0$ or, if $g(q) \neq 0$ then $g^c(g(q)^{-1}qg(q)) = 0$, the conclusion of the Step 2 immediately yields the conclusion of Step 3.

Step 4. Denoting by Z_{g^s} the zero set of g^s and by $\mathcal{K}(Z_{g^s})$ the convex hull of Z_{g^s}, from Step 3 we have $Z_{g^s} \subset B(0;1)$, which implies that the axially symmetric completion of $\mathcal{K}(Z_{g^s})$ is included in the closed ball $\overline{B(0;1)}$. Applying now Theorem 6.2.1, it follows that $g'(q) = P'(q) - Q'(q)\lambda$ has all its zeroes in $|q| \leq 1$; in other words, for all $|\lambda| > 1$ and $|q| > 1$ we have $g'(q) \neq 0$, which is equivalent to $\left|[Q'(q)]^{-1}P'(q)\right| \neq |\lambda| > 1$.

This clearly implies $\left|[Q'(q)]^{-1}P'(q)\right| \leq 1$, for all $|q| > 1$, i.e., $|P'(q)| \leq |Q'(q)|$, for all $|q| > 1$. From the continuity of P' and Q', for any $|q_0| = 1$ and taking a sequence q_m with $|q_m| > 1$, $\lim_{m \to \infty} q_m = q_0$, we easily get that $|P'(q_0)| \leq |Q'(q_0)|$. This implies

$$|P'(q)| \leq |Q'(q)| = nM|q|^{n-1}, \quad \text{for all } |q| \geq 1,$$

and taking the maximum over all q with $|q| = 1$ we get $\max_{|q|=1} |P'(q)| \leq nM$, which proves the theorem for $0 \leq n \leq 2$.

Case of arbitrary $n \in \mathbb{N} \cup \{0\}$. This second proof is different, in fact it is completely based on the ideas of a proof in the complex case in Rahman and Schmeisser's book [165]. In the quaternionic setting, the proof was originally given by Xu in the paper [195]. Thus, by following exactly the lines of the proof of Theorem 14.1.1, on pp. 508–509 in [165], we consider the Fejér kernel

$$K_n(\theta) = \frac{1}{n+1}\left(\frac{\sin((n+1)\theta/2)}{\sin(\theta/2)}\right)^2$$

$$= 1 + 2\sum_{\mu=1}^{n}\left(1 - \frac{\mu}{n+1}\right) \cdot \cos(\mu\theta)$$

$$= \sum_{\mu=-n}^{n}\left(1 - \frac{|\mu|}{n+1}\right) \cdot e^{i\mu\theta}$$

and for $g : \mathbb{R} \to \mathbb{C}$, 2π-periodic, i.e., of the form $g(\theta) = \sum_{\nu=-\infty}^{\infty} e^{i\nu\theta} c_\nu$, we consider the arithmetic mean of the partial sums of the Fourier series of q, $s_0(g), s_1(g), \ldots, s_n(g)$, i.e.,

$$\sigma_n(g)(\theta) = \frac{1}{2\pi}\int_{-\pi}^{\pi} g(\theta - \varphi)K_n(\varphi)d\varphi = \sum_{\mu=-n}^{n}\left(1 - \frac{|\mu|}{n+1}\right)e^{i\mu\theta}c_\mu$$

(here $i^2 = -1$). Also note that if $|g(\theta)| \leq M$ for all $\theta \in \mathbb{R}$, then $|\sigma_n(g)(\theta)| \leq M$, for all $\theta \in \mathbb{R}$ and $n \in \mathbb{N}$.

Now, let $P(q) = \sum_{\nu=0}^n q^\nu a_n$ be a quaternionic polynomial of degree at most n. For an arbitrary $I \in \mathbb{S}$, define the 2π-periodic function $g : \mathbb{R} \to \mathbb{H}$, by the formula

$$g(\theta) = e^{In\theta} \cdot P(e^{-I\theta}) = \sum_{\mu=0}^n e^{I\mu\theta} a_{n-\mu}.$$

Reasoning exactly as in the complex case, we get

$$\sigma_{n-1}(g)(-\theta) = \frac{1}{n} e^{-I(n-1)\theta} P'(e^{I\theta}) = \frac{1}{n} q^{-(n-1)} P'(q)|_{q=e^{I\theta}}. \qquad (6.2)$$

Using the maximum modulus theorem we have

$$\max_{\theta \in \mathbb{R}} |g(\theta)| = \max_{q \in \mathbb{C}_I, |q| \leq 1} |P_I(q)|$$

$$\leq \max_{|q| \leq 1} |P_I(q)| \leq \max_{|q| \leq 1} |P(q)|$$

$$= \max_{|q|=1} |P(q)| = \|P\|,$$

and combining with the inequality $|\sigma_{n-1}(g)(\theta)| \leq \|P\|$, for all $n \in \mathbb{N}$, $\theta \in \mathbb{R}$ and $I \in \mathbb{S}$, passing to the norm $\|\cdot\|$ in (6.2), we immediately obtain

$$|P'(q)| \leq n\|P\|, \quad \text{for all } |q| = 1,$$

which yields the required inequality. $\qquad \square$

Remark 6.2.3. It is evident that a similar inequality holds for polynomials of degree n of the form $P(q) = \sum_{k=0}^n a_k q^k$, with $a_k \in \mathbb{H}$, i.e., with coefficients on the left.

6.3   Erdős–Lax's Inequality

The Erdős–Lax inequality, see Theorem 6.1.2, cannot be, in general, extended to quaternionic setting. In fact, we have :

Theorem 6.3.1. *The Erdős–Lax inequality is not valid, in general, for quaternionic polynomials.*

Proof. We provide a counterexample. Let

$$P(q) = (q - i) * (q - j) = q^2 - q(i + j) + k.$$

The only root of this polynomial of degree two is $q = i$, namely $q = i$ has multiplicity 2, thus $P(q)$ has no other roots in the open unit ball $\mathbb{B}$. We claim that for this polynomial the Erdős–Lax inequality does not hold.

Let $q = e^{I\theta}$, where $I = ai + bj + ck$, $a^2 + b^2 + c^2 = 1$. The proof is divided into two steps.

Step 1. Clearly, $P'(q) = 2q - (i + j)$, so that

$$P'(e^{I\theta}) = 2\cos\theta + i(2a\sin\theta - 1) + j(2b\sin\theta - 1) + k(2c\sin\theta).$$

It is then readily verified that

$$|P'(e^{I\theta})|^2 = 6 - 4(a + b)\sin\theta$$

and

$$\max_{I \in \mathbb{S}, \; \theta \in [0, 2\pi)} |P'(e^{I\theta})|^2 = \max(6 - 4(a + b)\sin\theta) \geq 6 + 4\sqrt{2}, \tag{6.3}$$

since $\max_{a^2+b^2+c^2=1}(a + b) = \max_{a^2+b^2=1}(a + b) = \max(a + \sqrt{1 - a^2}) = \sqrt{2}$.

Step 2. We have:

$$P(e^{I\theta}) = e^{2I\theta} - e^{I\theta}(i + j) + k$$
$$= \cos(2\theta) + a\sin\theta + b\sin\theta + i(a\sin(2\theta) - \cos\theta + c\sin\theta)$$
$$+ j(b\sin(2\theta) - \cos\theta - c\sin\theta) + k(c\sin(2\theta) - a\sin\theta + b\sin\theta + 1).$$

Some lengthy but simple computations show that

$$|P(e^{I\theta})|^2 = 4 - 4a\sin\theta + 4c\cos\theta\sin\theta,$$

and so the maximum of $|P(e^{I\theta})|$ is attained when the maximum of $-4a\sin\theta + 4c\cos\theta\sin\theta$ is. We have

$$\sin\theta(-4a + 4c\cos\theta) \leq |\sin\theta(-4a + 4c\cos\theta)|$$
$$\leq 4|(-a + c\cos\theta)| \leq 4(|a| + |c|)$$
$$\leq 4\sqrt{2},$$

where the last inequality follows from

$$\max_{a^2+b^2+c^2=1}(|a| + |c|) = \max_{a^2+c^2=1}(|a| + |c|) = \max(|a| + \sqrt{1 - a^2}) = \sqrt{2}.$$

Thus

$$\max_{I \in \mathbb{S}, \; \theta \in [0, 2\pi)} |P(e^{I\theta})|^2 \leq 4 + 4\sqrt{2}. \tag{6.4}$$

From inequalities (6.3) and (6.4) it follows that

$$\|P'\| \geq (6 + 4\sqrt{2})^{1/2}$$
$$> (4 + 4\sqrt{2})^{1/2}$$
$$\geq \|P\| = \frac{2}{2} \cdot \|P\|$$

and the statement follows. $\qquad\qquad\square$

The Erdős–Lax inequality holds true, however, at least for a class of polynomials specified in the following result.

Proposition 6.3.2. *Suppose the polynomial* $P(q) = \sum_{k=0}^{n} q^k a_k$, *with* $a_k \in \mathbb{C}_I$ *for some fixed* $I \in \mathbb{S}$, $a_n \neq 0$, *has no zeros in the ball* $|q| < 1$. *Then*

$$\|P'\| \leq \frac{n}{2} \cdot \|P\|.$$

Proof. If P has all its coefficients in $\mathbb{C}_I$ for some fixed $I \in \mathbb{S}$ then it factorises as

$$(q - \alpha_1) * \cdots * (q - \alpha_r)(q^2 - 2\mathrm{Re}(a_1)\overset{\bullet}{q} + |a_1|^2) \cdots (q^2 - 2\mathrm{Re}(a_s)q + |a_s|^2).$$

In fact, if we restrict $P(q)$ to the complex plane $\mathbb{C}_I$, whose variable is denoted by z, the polynomial $P(z)$ has exactly n zeros on $\mathbb{C}_I$ and so

$$P(z) = \sum_{k=0}^{n} z^n a_n = (z - \alpha_1) \cdots (z - \alpha_r)(z - \beta_1)(z - \bar{\beta}_1) \cdots (z - \beta_s)(z - \bar{\beta}_s),$$

where $\alpha_i \in \mathbb{R}$, $i = 1, \ldots, r$, $\beta_i \in \mathbb{C}_I \setminus \mathbb{R}$, $i = 1, \ldots, s$. Now consider the polynomial P_1 in the quaternionic variable q defined by

$$P_1(q) = (q - \alpha_1) * \cdots * (q - \alpha_r) * (q - \beta_1) * (q - \bar{\beta}_1) * \cdots * (q - \beta_s) * (q - \bar{\beta}_s).$$

We have $P_1(z) = P(z)$ and thus by the identity principle for slice regular functions (and, in particular, for polynomials) we have $P_1(q) = P(q)$. By hypothesis, all the roots of P are outside the unit ball and so are the roots of the restriction P_I of P to $\mathbb{C}_I$ are outside the unit disk. The Erdős–Lax inequality in the complex case is valid for P_I, and so $\|P_I'\| \leq \frac{n}{2}\|P_I\|$ where $\|P_I\| = \max_{|z|=1} |P_I(z)|$. We now note that P is such that $P(\mathbb{C}_I) \subset \mathbb{C}_I$ so that Corollary 3.4 in [168] implies that

$$\|P\| = \max_{x^2+y^2=1, \ J \in \mathbb{S}} |P(x + Jy)| = \max_{x^2+y^2=1} |P(x + Iy)| = \|P_I\|,$$

and, similarly, $\|P'\| = \|P_I'\|$. Thus we have

$$\|P'\| = \|P_I'\| \leq \frac{n}{2}\|P_I\| = \frac{n}{2}\|P\|$$

and the assertion follows.   □

Remark 6.3.3. The bound in Proposition 6.3.2 is optimal, as it can be seen by taking the polynomial $P(q) = (1 + q^n)/2$.

Remark 6.3.4. We note that Proposition 6.3.2 holds, in particular, when $P(q) = (q - a)^{*n}$, with $a \in \mathbb{H}$ satisfying $|a| \geq 1$. In fact, $a \in \mathbb{C}_I$ some $I \in \mathbb{S}$ and thus all the coefficients of P are in $\mathbb{C}_I$.

Proposition 6.3.2 can be generalized as follows.

Proposition 6.3.5. *Let $P(q)$ be a polynomial of degree n with coefficients in a complex plane $\mathbb{C}_I$ for some $I \in \mathbb{S}$, that has no zeros in the ball $|q| < K$ with $K \geq 1$. Then*

$$\|P'\| \leq \frac{n}{1+K} \cdot \|P\|.$$

Proof. We will use considerations presented in the proof of Proposition 6.3.2 and the validity of the result in the complex case, see the paper of Malik [139].

Since, by hypothesis, P has no zeros in the ball $|q| < K$, according to the proof of Proposition 6.3.2 the restriction $P_I(z)$ of $P(q)$ to the complex plane $\mathbb{C}_I$ has no zeros in the disk $|z| < K$, thus it satisfies the inequality

$$\|P_I'\| \leq \frac{n}{1+K} \cdot \|P_I\|.$$

The fact that P has coefficients in the complex plane $\mathbb{C}_I$ implies that $\|P\| = \|P_I\|$ and $\|P'\| = \|P_I'\|$, and the conclusion follows. $\qquad\square$

6.4 Turán's Inequality

The main aim of this section is to extend Theorem 6.1.3 and then its generalizations in Theorems 6.1.4–6.1.7, to quaternionic polynomials.

All the known methods of proof for Turán's inequality in the complex case are based on the calculation and lower estimates of the expression $|\text{Re}[P'(z)/P(z)]|$. But due to noncommutativity, in the quaternionic setting, the expression $P'(q)/P(q)$, for $n \geq 2$, is not the right tool to consider and in fact it does not serve well even when P has multiple zeros.

Nevertheless, by using other methods of proofs we can extend the Turán inequality for a subset of quaternionic polynomials. Specifically the method of Lagrange multipliers allows us to show that Turán's inequality remains valid for all quaternionic polynomials of degree $n \leq 2$ and for some classes of quaternionic polynomials of degree $n \geq 3$.

Theorem 6.4.1. (i) *Let $P(q)$ be a polynomial of degree $n \leq 2$ with quaternionic coefficients that has all its zeros in the closed ball $|q| \leq 1$. Then*

$$\|P'\| \geq \frac{n}{2} \cdot \|P\|.$$

(ii) *Let $P(q) = (q - \alpha) * (q - \beta) * (q - \gamma)$ be a polynomial of degree 3 with $|\alpha| \leq 1$, $|\beta| \leq 1$, $|\gamma| \leq 1$. If $\alpha + \beta + \gamma = 0$ and*

$$\left\{ |\alpha| \leq \frac{-3 + \sqrt{17}}{2}, \quad or \ |\beta| \leq \frac{-3 + \sqrt{17}}{2}, \quad or \ |\gamma| \leq \frac{-3 + \sqrt{17}}{2} \right\},$$

then $\|P'\| \geq \frac{3}{2}\|P\|$.

(iii) *Let $P(q) = (q - \alpha) * (q - \beta) * (q - \gamma)$ be a polynomial of degree 3 with $|\alpha| \leq 1$,*
$|\beta| \leq 1$, $|\gamma| \leq 1$. If $\alpha\beta + \alpha\gamma + \beta\gamma = 0$, then $\|P'\| \geq \frac{3}{2}\|P\|$.

Proof. (i) Since the case $n = 1$ is immediate, let us consider a quaternionic polynomial $P(q)$ of degree 2 with roots in $|q| \leq 1$, that is,

$$P(q) = (q - \alpha) * (q - \beta) = q^2 - q(\alpha + \beta) + \alpha\beta, \tag{6.5}$$

with $\alpha, \beta \in \mathbb{H}$, $|\alpha|, |\beta| \leq 1$. Obviously,

$$P'(q) = 2q - (\alpha + \beta). \tag{6.6}$$

Let us write $\alpha = \alpha_0 + i\alpha_1 + j\alpha_2 + k\alpha_3$, $\beta = \beta_0 + i\beta_1 + j\beta_2 + k\beta_3$, $\gamma = \alpha + \beta = \gamma_0 + i\gamma_1 + j\gamma_2 + k\gamma_3$, where $\gamma_i = \alpha_i + \beta_i$, $i = 0, 1, 2, 3$. Take $q = e^{I\theta}$, with $I \in \mathbb{S}$ and $\theta \in [0, 2\pi)$, arbitrary. Write $I = ix + jy + zk$, where $x^2 + y^2 + z^2 = 1$. We first prove the equality

$$|P'(e^{I\theta})|^2 = -4\gamma_0 \cos(\theta) - 4(x\gamma_1 + y\gamma_2 + z\gamma_3)\sin(\theta) + 4 + |\alpha + \beta|^2. \tag{6.7}$$

Indeed, we have

$$
\begin{aligned}
P'(e^{I\theta}) &= 2\cos(\theta) + 2(ix + jy + kz)\sin(\theta) - (\gamma_0 + i\gamma_1 + j\gamma_2 + k\gamma_3) \\
&= (2\cos(\theta) - \gamma_0) + i(2x\sin(\theta) - \gamma_1) + j(2y\sin(\theta) - \gamma_2) + k(2z\sin(\theta) - \gamma_3),
\end{aligned}
$$

whence

$$
\begin{aligned}
|P'(e^{I\theta})|^2 &= (2\cos(\theta) - \gamma_0)^2 + (2x\sin(\theta) - \gamma_1)^2 + (2y\sin(\theta) - \gamma_2)^2 \\
&\quad + (2z\sin(\theta) - \gamma_3)^2 \\
&= (4\cos^2(\theta) - 4\gamma_0\cos(\theta) + \gamma_0^2) + (4x^2\sin^2(\theta) - 4x\gamma_1\sin(\theta) + \gamma_1^2) \\
&\quad + (4y^2\sin^2(\theta) - 4y\gamma_2\sin(\theta) + \gamma_2^2) + (4z^2\sin^2(\theta) - 4z\gamma_3\sin(\theta) + \gamma_3^2) \\
&= 4\cos^2(\theta) + 4\sin^2(\theta)(x^2 + y^2 + z^2) - 4\gamma_0\cos(\theta) \\
&\quad - 4\sin(\theta)(x\gamma_1 + y\gamma_2 + z\gamma_3) + (\gamma_0^2 + \gamma_1^2 + \gamma_2^2 + \gamma_3^2).
\end{aligned}
$$

Since $\gamma_0^2 + \gamma_1^2 + \gamma_2^2 + \gamma_3^2 = |\alpha + \beta|^2$, this yields (6.7),

We now fix $\theta \in (0, 2\pi)$, with $\sin(\theta) \neq 0$. By using the method of Lagrange multipliers, let us introduce the Lagrange function

$$
\begin{aligned}
f(x, y, z) = {}& 4 + |\alpha + \beta|^2 - 4(\gamma_0\cos(\theta) + x\gamma_1\sin(\theta) \\
&+ y\gamma_2\sin(\theta) + z\gamma_3\sin(\theta)) + \lambda(x^2 + y^2 + z^2).
\end{aligned}
$$

The system of equations

$$\frac{\partial f}{\partial x} = 0, \quad \frac{\partial f}{\partial y} = 0, \quad \frac{\partial f}{\partial z} = 0, \quad x^2 + y^2 + z^2 = 1,$$

becomes

$$-4\gamma_1 \sin(\theta) + 2\lambda x = 0, \quad -4\gamma_2 \sin(\theta) + 2\lambda y = 0,$$
$$-4\gamma_3 \sin(\theta) + 2\lambda z = 0, \quad x^2 + y^2 + z^2 = 1,$$

which immediately leads to the solutions

$$x = \frac{2\gamma_1 \sin(\theta)}{\lambda}, \quad y = \frac{2\gamma_2 \sin(\theta)}{\lambda}, \quad z = \frac{2\gamma_3 \sin(\theta)}{\lambda},$$

$$\lambda^2 = 4\sin^2(\theta)(\gamma_1^2 + \gamma_2^2 + \gamma_3^2).$$

We get for λ two possibilities:

Case 1. $\lambda = -2\sin(\theta)\sqrt{\gamma_1^2 + \gamma_2^2 + \gamma_3^2}$.

Then replacing x, y, z in the formula (6.7), we easily get

$$|P'(e^{I\theta})|^2 = -4\gamma_0 \cos(\theta) + 4\sqrt{\gamma_1^2 + \gamma_2^2 + \gamma_3^2}\,\sin(\theta) + 4 + |\alpha + \beta|^2.$$

Since in general for $E(x) = A\cos(x) + B\sin(x) + C$, we have $E(x) = \sqrt{A^2 + B^2}\cos(x - \eta) + C$, where $\eta = \arctan\frac{B}{A}$, we obtain

$$|P'(e^{I\theta})|^2 = \sqrt{16\gamma_0^2 + 16\gamma_1^2 + 16\gamma_2^2 + 16\gamma_3^2}\,\cos(\theta - \eta) + 4 + |\alpha + \beta|^2$$
$$= 4|\alpha + \beta| \cdot \cos(\theta - \eta) + 4 + |\alpha + \beta|^2.$$

We conclude that for $\theta = \eta$ and x, y, z given as above,

$$|P'(e^{I\theta})|^2 = (2 + |\alpha + \beta|)^2,$$

which immediately implies that $\|P'\| \geq 2 + |\alpha + \beta|$.

Note that since by (6.6) we easily get $\|P'\| \leq 2 + |\alpha + \beta|$, we actually have $\|P'\| = 2 + |\alpha + \beta|$. Here recall that $\|Q\| = \sup_{|q|\leq 1}\{|Q(q)|\}$.

Case 2. $\lambda = 2\sin(\theta)\sqrt{\gamma_1^2 + \gamma_2^2 + \gamma_3^2}$.

Reasoning exactly as in the Case 1, we obtain

$$|P'(e^{I\theta})|^2 = -4\gamma_0 \cos(\theta) - 4\sqrt{\gamma_1^2 + \gamma_2^2 + \gamma_3^2}\,\sin(\theta) + 4 + |\alpha + \beta|^2,$$

which leads to the same conclusion as in Case 1.

Therefore,

$$\|P\| \leq 1 + |\alpha + \beta| + |\alpha\beta| \leq 2 + |\alpha + \beta| = \|P'\|,$$

which proves the assertion (i).

(ii) We can write

$$P(q) = q^3 - q^2(\alpha + \beta + \gamma) + q(\alpha\beta + \alpha\gamma + \beta\gamma) - \alpha\beta\gamma,$$

which by hypothesis immediately implies that

$$P(q) = q^3 + q(\alpha\beta + \alpha\gamma + \beta\gamma) - \alpha\beta\gamma.$$

We get

$$\|P\| \leq 1 + |\alpha\beta + \alpha\gamma + \beta\gamma| + |\alpha\beta\gamma|. \tag{6.8}$$

On the other hand,

$$P'(q) = 3q^2 + (\alpha\beta + \alpha\gamma + \beta\gamma). \tag{6.9}$$

In what follows we could apply the method used for $n = 2$ to the polynomial in (6.9).

Thus, denoting $\alpha\beta + \alpha\gamma + \beta\gamma = A_0 + iA_1 + jA_2 + kA_3$ and taking $q = e^{I\theta}$, a simple calculation shows that

$$\begin{aligned} P'(e^{I\theta}) &= (3\cos(2\theta) + A_0) + i(3x\sin(2\theta) + A_1) \\ &\quad + j(3y\sin(2\theta) + A_2) + k(3z\sin(2\theta) + A_3) \end{aligned}$$

and

$$\begin{aligned} |P'(e^{I\theta})|^2 &= (3\cos(2\theta) + A_0)^2 + (3x\sin(2\theta) + A_1)^2 \\ &\quad + (3y\sin(2\theta) + A_2)^2 + (3z\sin(2\theta) + A_3)^2 \\ &= 9 + 6[A_0\cos(2\theta) + xA_1\sin(2\theta) + yA_2\sin(2\theta) \\ &\quad + zA_3\sin(2\theta)] + A_0^2 + A_1^2 + A_2^2 + A_3^2, \end{aligned}$$

where $A_0^2 + A_1^2 + A_2^2 + A_3^2 = |\alpha\beta + \alpha\gamma + \beta\gamma|^2$.

Introducing the Lagrange function

$$f(x, y, z) = |P'(e^{I\theta})|^2 + \lambda(x^2 + y^2 + z^2),$$

the system of equations

$$\frac{\partial f}{\partial x} = 0, \quad \frac{\partial f}{\partial y} = 0, \quad \frac{\partial f}{\partial z} = 0, \quad x^2 + y^2 + z^2 = 1$$

reduces to

$$6A_1\sin(2\theta) + 2\lambda x = 0, \quad 6A_2\sin(2\theta) + 2\lambda y = 0, \quad 6A_3\sin(2\theta) + 2\lambda z = 0,$$

$$x^2 + y^2 + z^2 = 1,$$

which immediately leads to the solutions

$$x = -\frac{3A_1\sin(2\theta)}{\lambda}, \quad y = -\frac{3A_2\sin(2\theta)}{\lambda}, \quad z = -\frac{3A_3\sin(2\theta)}{\lambda},$$

with

$$\lambda^2 = 9\sin^2(2\theta)(A_1^2 + A_2^2 + A3^2).$$

As in the case of $n = 2$, we have two possibilities:

1. $\lambda = -3\sin(2\theta)\sqrt{A_1^2 + A_2^2 + A_3^2}$;

2. $\lambda = 3\sin(2\theta)\sqrt{A_1^2 + A_2^2 + A_3^2}$.

Case 1. Replacing the above values obtained for x, y and z in the expression of $|P'(e^{I\theta})|^2$, and reasoning exactly as in the case $n = 2$, we immediately obtain

$$|P'(e^{I\theta})|^2 = 9 + |\alpha\beta + \alpha\gamma + \beta\gamma|^2 + 6A_0\cos(2\theta) + 6\sin(2\theta)\sqrt{A_1^2 + A_2^2 + A_3^2}$$

$$= 9 + |\alpha\beta + \alpha\gamma + \beta\gamma|^2 + 6\cos(2\theta - \eta)\cdot|\alpha\beta + \alpha\gamma + \beta\gamma|.$$

Taking here $2\theta = \eta$, we get

$$|P'(e^{I\theta})|^2 = 9 + |\alpha\beta + \alpha\gamma + \beta\gamma|^2 + 6\cdot|\alpha\beta + \alpha\gamma + \beta\gamma|$$

$$= (3 + |\alpha\beta + \alpha\gamma + \beta\gamma|)^2,$$

i.e.,

$$3 + |\alpha\beta + \alpha\gamma + \beta\gamma| = |P'(e^{I\theta})| \leq \|P'\|.$$

Since by (6.9) we also obviously have $\|P'\| \leq 3 + |\alpha\beta + \alpha\gamma + \beta\gamma|$, we conclude that $3 + |\alpha\beta + \alpha\gamma + \beta\gamma| = \|P'\|$.

Case 2. is analogous.

Now, taking into account (6.8), it follows that if

$$\frac{3}{2}(1 + |\alpha\beta + \alpha\gamma + \beta\gamma| + |\alpha\beta\gamma|) \leq 3 + |\alpha\beta + \alpha\gamma + \beta\gamma|,$$

then Turán's inequality holds.

The above inequality is obviously equivalent to

$$|\alpha\beta + \alpha\gamma + \beta\gamma| + 3\cdot|\alpha\beta\gamma| \leq 3.$$

From $\alpha + \beta + \gamma = 0$ it follows that $|\beta + \gamma| = |\alpha|$ and

$$|\alpha\beta + \alpha\gamma + \beta\gamma| + 3\cdot|\alpha\beta\gamma| = |\alpha(\beta + \gamma) + \beta\gamma| + 3\cdot|\alpha|\cdot|\beta\gamma|$$

$$\leq |\alpha|\cdot|\beta + \gamma| + |\beta\gamma| + 3\cdot|\alpha|\cdot|\beta\gamma| \leq |\alpha|^2 + 3|\alpha| + 1.$$

The condition $|\alpha|^2 + 3|\alpha| + 1 \leq 3$ is equivalent to $|\alpha| \leq \frac{-3+\sqrt{17}}{2}$, which immediately implies that for $|\alpha| \leq \frac{-3+\sqrt{17}}{2}$ we have $\|P'\| \geq \frac{3}{2}\|P\|$.

For reasons of symmetry, we easily arrive at the same conclusion if $|\gamma| \leq (-3 + \sqrt{17})/2$.

Now, in the case of β, we arrive at the same conclusion, but with a slightly different proof. Indeed, as in the above cases, we start from the condition $|\alpha\beta + \alpha\gamma + \beta\gamma| + 3|\alpha\beta\gamma| \leq 3$, and putting here $\beta = -(\alpha + \gamma)$, we get

$$|-\alpha(\alpha + \gamma) + \alpha\gamma - (\alpha + \gamma)\gamma| + 3|\alpha|\cdot|\gamma|\cdot|\beta| = |\alpha^2 + \alpha\gamma + \gamma^2| + 3|\alpha|\cdot|\gamma|\cdot|\beta|$$

$$= |(\alpha + \gamma)\cdot(\alpha + \gamma) - \gamma\alpha| + 3|\alpha|\cdot|\gamma|\cdot|\beta| \leq |\beta|^2 + |\gamma\alpha| + 3|\beta| \leq |\beta|^2 + 3|\beta| + 1,$$

hence, we arrive at the same type of inequality for β, that is, $|\beta|^2 + 3|\beta| + 1 \leq 3$.

(iii) We can write

$$P(q) = q^3 - q^2(\alpha + \beta + \gamma) + q(\alpha\beta + \alpha\gamma + \beta\gamma) - \alpha\beta\gamma.$$

In order to get a flavor of the difficulties in using the method from item (ii), we will perform below the calculations in the most general case, which will lead even to the results for (ii).

To this end, write $\alpha + \beta + \gamma = B_0 + iB_1 + jB_2 + kB_3$, $\alpha\beta + \alpha\gamma + \beta\gamma = A_0 + iA_1 + jA_2 + kA_3$ and $I = ix + jy + kz$ with $x^2 + y^2 + z^2 = 1$. Since $P'(q) = 3q^2 - 2q(\alpha + \beta + \gamma) + (\alpha\beta + \alpha\gamma + \beta\gamma)$, we get

$P'(e^{I\theta})$

$= (3\cos(2\theta) - 2B_0\cos(\theta) + 2xB_1\sin(\theta) + 2yB_2\sin(\theta) + 2zB_3\sin(\theta) + A_0)$
$\quad + i(3x\sin(2\theta) - 2B_1\cos(\theta) - 2B_3y\sin(\theta) + 2B_2z\sin(\theta) - 2B_0x\sin(\theta) + A_1)$
$\quad + j(3y\sin(\theta) - 2B_2\cos(\theta) - 2B_1z\sin(\theta) + 2B_3x\sin(\theta) - 2B_0y\sin(\theta) + A_2)$
$\quad + k(3z\sin(\theta) - 2B_3\cos(\theta) - 2B_2x\sin(\theta) + 2B_1y\sin(\theta) - 2B_0z\sin(\theta) + A_3).$

This implies

$|P'(e^{I\theta})|^2$

$= (3\cos(2\theta) - 2B_0\cos(\theta) + 2xB_1\sin(\theta) + 2yB_2\sin(\theta) + 2zB_3\sin(\theta) + A_0)^2$
$\quad + (3x\sin(2\theta) - 2B_1\cos(\theta) - 2B_3y\sin(\theta) + 2B_2z\sin(\theta) - 2B_0x\sin(\theta) + A_1)^2$
$\quad + (3y\sin(\theta) - 2B_2\cos(\theta) - 2B_1z\sin(\theta) + 2B_3x\sin(\theta) - 2B_0y\sin(\theta) + A_2)^2$
$\quad + (3z\sin(\theta) - 2B_3\cos(\theta) - 2B_2x\sin(\theta) + 2B_1y\sin(\theta) - 2B_0z\sin(\theta) + A_3)^2$
$= \{9[\cos^2(2\theta) + \sin^2(2\theta)] + 4\cos^2(\theta)|\alpha + \beta + \gamma|^2 + 4x^2\sin^2(\theta)|\alpha + \beta + \gamma|^2$
$\quad + 4y^2\sin^2(\theta)|\alpha + \beta + \gamma|^2 + 4z^2\sin^2(\theta)|\alpha + \beta + \gamma|^2 + |\alpha\beta + \alpha\gamma + \beta\gamma|^2\}$
$\quad + \{[-12B_0\cos(2\theta)\sin(\theta) + 12B_1x\cos(2\theta)\sin(\theta) + 12B_2y\cos(2\theta)\sin(\theta)$
$\quad + 12B_3z\cos(2\theta)\sin(\theta) + 6A_0\cos(2\theta)]$
$\quad + [-8B_0B_1x\cos(\theta)\sin(\theta) - 8B_0B_2y\cos(\theta)\sin(\theta) - 8B_0B_3z\cos(\theta)\sin(\theta)$
$\quad - 4A_0B_0\cos(\theta)]$
$\quad + [8B_1B_2xy\sin^2(\theta) + 8B_1B_3xz\sin^2(\theta) + 4A_0B_1x\sin(\theta)]$
$\quad + [8B_2B_3yz\sin^2(\theta) + 4A_0B_2y\sin(\theta) + 4A_0B_3z\sin(\theta)]\}$
$\quad + \{[-12B_1x\sin(2\theta)\cos(\theta) - 12B_3xy\sin(2\theta)\cos(\theta) + 12B_2xz\sin(2\theta)\cos(\theta)$
$\quad - 12B_0x^2\sin(2\theta)\cos(\theta) + 6A_1x\sin(2\theta)]$
$\quad + [8B_1B_3y\cos(\theta)\sin(\theta) - 8B_1B_2z\cos(\theta)\sin(\theta) + 8B_0B_1x\cos(\theta)\sin(\theta)$
$\quad - 4A_1B_1\cos(\theta)]$
$\quad + [-8B_2B_3zy\sin^2(\theta) + 8B_0B_3xy\sin^2(\theta) - 4A_1B_3y\sin(\theta)$

$$-8B_0B_2xz\sin^2(\theta)+4A_1B_2z\sin(\theta)-4A_1B_0x\sin(\theta)]\}$$
$$+\{[-12B_2y\sin(2\theta)\cos(\theta)-12B_1yz\sin(2\theta)\cos(\theta)+12B_3xy\sin(2\theta)\cos(\theta)$$
$$-12B_0y^2\sin(2\theta)\cos(\theta)+6A_2y\sin(2\theta)]$$
$$+[8B_1B_2z\cos(\theta)\sin(\theta)-8B_2B_3x\cos(\theta)\sin(\theta)+8B_0B_2y\cos(\theta)\sin(\theta)$$
$$-4A_2B_2\cos(\theta)]$$
$$+[-8B_1B_3xz\sin^2(\theta)+8B_0B_1yz\sin^2(\theta)-4A_2B_1z\sin(\theta)$$
$$-8B_0B_3xy\sin^2(\theta)+4A_2B_3x\sin(\theta)-4A_2B_0y\sin(\theta)]\}$$
$$+\{[-12B_3z\sin(2\theta)\cos(\theta)-12B_2xz\sin(2\theta)\cos(\theta)+12B_1yz\sin(2\theta)\cos(\theta)$$
$$-12B_0z^2\sin(2\theta)\cos(\theta)+6A_3z\sin(2\theta)]$$
$$+[8B_2B_3x\cos(\theta)\sin(\theta)-8B_1B_3y\cos(\theta)\sin(\theta)+8B_0B_3z\cos(\theta)\sin(\theta)$$
$$-4A_3B_3\cos(\theta)]$$
$$+[-8B_1B_2xy\sin^2(\theta)+8B_0B_2xz\sin^2(\theta)-4A_3B_2x\sin(\theta)$$
$$-8B_0B_1yz\sin^2(\theta)+4A_3B_1y\sin(\theta)-4A_3B_0z\sin(\theta)]\}\,.$$

After simplifications and some simple calculations, we easily arrive at

$$|P'(e^{I\theta})|^2$$
$$=9+4|\alpha+\beta+\gamma|^2+|\alpha\beta+\alpha\gamma+\beta\gamma|^2-12B_0\cos(\theta)$$
$$-12\sin(\theta)(B_1x+B_2y+B_3z)+6A_0\cos(2\theta)+6\sin(2\theta)[A_1x+A_2y+A_3z]$$
$$-4\cos(\theta)[A_0B_0+A_1B_1+A_2B_2+A_3B_3]$$
$$+4\sin(\theta)[(A_0B_1-A_1B_0+A_2B_3-A_3B_2)x$$
$$+(A_0B_2-A_2B_0+A_3B_1-A_1B_3)y+(A_0B_3-A_3B_0+A_1B_2-A_2B_1)z].$$

Let us observe that if $\alpha+\beta+\gamma=0$, that is, if $B_0=B_1=B_2=B_3=0$, then we get for $|P'(e^{I\theta})|^2$ exactly the expression from (ii).

Now, let us suppose that $\alpha\beta+\alpha\gamma+\beta\gamma=0$, that is $A_0=A_1=A_2=A_3=0$. It follows that

$$P'(e^{I\theta})=9+4|\alpha+\beta+\gamma|^2-12B_0\cos(\theta)-12\sin(\theta)(B_1x+B_2y+B_3z).$$

In what follows, we reason as in item (ii).

Firstly, we note that

$$\|P\|\le 1+|\alpha+\beta+\gamma|+|\alpha\beta\gamma|.$$

Then, denoting $f(x,y,z)=|P'(e^{I\theta})+\lambda(x^2+y^2+z^2)$, by the Lagrange multiplier method, we consider the system

$$\frac{\partial f}{\partial x} = -12B_1 \sin(\theta) + 2\lambda x = 0,$$

$$\frac{\partial f}{\partial y} = -12B_2 \sin(\theta) + 2\lambda y = 0,$$

$$\frac{\partial f}{\partial z} = -12B_3 \sin(\theta) + 2\lambda z = 0, \quad x^2 + y^2 + z^2 = 1,$$

which immediately leads to the solution

$$x = \frac{6B_1 \sin(\theta)}{\lambda}, \quad y = \frac{6B_2 \sin(\theta)}{\lambda}, \quad z = \frac{6B_3 \sin(\theta)}{\lambda},$$

and

$$\lambda^2 = 36 \sin^2(\theta)[B_1^2 + B_2^2 + B_3^2].$$

Choose, for example, $\lambda = -6\sin(\theta)\sqrt{B_1^2 + B_2^2 + B_3^2}$ (the case $\lambda = 6\sin(\theta)\sqrt{B_1^2 + B_2^2 + B_3^2}$ is similar). Inserting this value in the expression of $|P'(e^{I\theta})|^2$ and reasoning as in items (i) and (ii), we get

$$|P'(e^{I\theta})|^2 = 9 + 4|\alpha + \beta + \gamma|^2 - 12B_0 \cos(\theta) + 12 \sin(\theta)\sqrt{B_1^2 + B_2^2 + B_3^2}$$

$$= 9 + 4|\alpha + \beta + \gamma|^2 + 12 \cos(\theta - \eta) \cdot |\alpha + \beta + \gamma|.$$

This yields

$$\max_{|q|=1} |P'(e^{I\theta})|^2 = 9 + 4|\alpha + \beta + \gamma|^2 + 12|\alpha + \beta + \gamma|$$

$$= (2|\alpha + \beta + \gamma| + 3)^2,$$

that is, $\|P'\| = 2|\alpha + \beta + \gamma| + 3$.

But we note that

$$\|P\| \leq 1 + |\alpha + \beta + \gamma| + |\alpha\beta\gamma|,$$

whence

$$\frac{3}{2}\|P\| \leq \frac{3}{2}(1 + |\alpha + \beta + \gamma| + |\alpha\beta\gamma|) \leq 2|\alpha + \beta + \gamma| + 3 = \|P'\|,$$

valid for all α, β, γ with $|\alpha| \leq 1$, $|\beta| \leq 1$, $|\gamma| \leq 1$. This proves (iii) and the theorem. $\qquad\square$

Remark 6.4.2. The additional hypothesis in the statement of Theorem 6.4.1, (ii) that { $|\alpha| \leq \frac{-3+\sqrt{17}}{2}$ or $|\beta| \leq \frac{-3+\sqrt{17}}{2}$ or $|\gamma| \leq \frac{-3+\sqrt{17}}{2}$ }, seems to be generated by the estimate (6.8), which is far from being the best upper estimate for $\|P\|$. It is natural to expect that finding the best (sharp) upper estimate for $\|P\|$ will eliminate this additional hypothesis.

In what follows, the method in the proof of Theorem 6.4.1 will be used to extend Turan's inequality for subclasses of quaternionic polynomials of a higher order $n \in \mathbb{N}$, $n \geq 4$.

For $n = 4$, we can consider quaternionic polynomials of the form

$$P_4(q) = (q - \alpha) * (q - \beta) * (q - \gamma) * (q - \lambda),$$

with $|\alpha|, |\beta|, |\gamma|, |\lambda| \leq 1$ satisfying, in addition, the conditions $\alpha + \beta + \gamma + \lambda = 0$ and $\alpha\beta + \alpha\gamma + \alpha\lambda + \beta\gamma + \beta\lambda + \gamma\lambda = 0$, namely

$$P(q) = q^4 - q(\alpha\beta\gamma + \alpha\beta\lambda + \alpha\gamma\lambda + \beta\gamma\lambda) + \alpha\beta\gamma\lambda.$$

For arbitrary $n \in \mathbb{N}$, the method of proof of Theorem 6.4.1 can be used for quaternionic polynomials of the form

$$P_n(q) = (q - \alpha_1) * (q - \alpha_2) * \cdots * (q - \alpha_n)$$
$$= q^n + (-1)^{n-1}q \cdot \sum_{1 \leq i_1 < \cdots < i_{n-1} \leq n} \alpha_{i_1} \cdots \alpha_{i_{n-1}} + (-1)^n \cdot \alpha_1\alpha_2 \cdots \alpha_n,$$

with $|\alpha_i| \leq 1$, for all $i = 1, \ldots, n$ satisfying, in addition, the conditions

$$\sum_{i=1}^{n} \alpha_i = 0, \qquad \sum_{1 \leq i_1 < i_2 \leq n} \alpha_{i_1}\alpha_{i_2} = 0, \ldots, \qquad \sum_{1 \leq i_1 < \cdots < i_{n-2} \leq n} \alpha_{i_1} \cdots \alpha_{i_{n-2}} = 0.$$

We will do the calculation for the $n = 4$ case, but the general case $n \geq 5$ is dealt with in a similar way.

Theorem 6.4.3. *Let $P(q) = (q - \alpha) * (q - \beta) * (q - \gamma) * (q - \lambda)$ be a polynomial of degree 4 with $|\alpha| \leq 1$, $|\beta| \leq 1$, $|\gamma| \leq 1$, $|\lambda| \leq 1$. If*

$$\alpha + \beta + \gamma + \lambda = 0 \quad and \quad \alpha\beta + \alpha\gamma + \alpha\lambda + \beta\gamma + \beta\lambda + \gamma\lambda = 0 \qquad (6.10)$$

and $|\alpha| \leq x^*$ or ($|\beta| \leq x^*$ and $\beta\alpha = \alpha\beta$) or ($|\gamma| \leq x^*$ and $\gamma\lambda = \lambda\gamma$) or $|\lambda| \leq x^*$, where

$$x^* = \left(\frac{1}{2} + \sqrt{\frac{59}{108}}\right)^{1/3} + \left(\frac{1}{2} - \sqrt{\frac{59}{108}}\right)^{1/3},$$

then Turán's inequality $\|P'\| \geq 2\|P\|$ holds.

Proof. First it is immediate that

$$\|P\| \leq 1 + |\alpha\beta\gamma + \alpha\beta\lambda + \alpha\gamma\lambda + \beta\gamma\lambda| + |\alpha\beta\gamma\lambda| \qquad (6.11)$$

and

$$P'(q) = 4q^3 - (\alpha\beta\gamma + \alpha\beta\lambda + \alpha\gamma\lambda + \beta\gamma\lambda).$$

Writing $q = e^{I\theta}$, $I = ix + jy + kz \in \mathbb{S}$, setting

$$\alpha\beta\gamma + \alpha\beta\lambda + \alpha\gamma\lambda + \beta\gamma\lambda = A_0 + iA_1 + jA_2 + kA_3,$$

and inserting in $P'(q)$ we obtain

$$P'(e^{I\theta}) = (4\cos(3\theta) - A_0) + i(4x\sin(3\theta) - A_1) + j(4y\sin(3\theta) - A_2)$$
$$+ k(4z\sin(3\theta) - A_3).$$

Therefore,

$$|P'(e^{I\theta})|^2 = (4\cos(3\theta) - A_0)^2 + (4x\sin(3\theta) - A_1)^2 + (4y\sin(3\theta) - A_2)^2$$
$$+ (4z\sin(3\theta) - A_3)^2$$
$$= 16 - 8[A_0\cos(3\theta) + A_1x\sin(3\theta) + A_2y\sin(3\theta)$$
$$+ A_3z\sin(3\theta)] + A_0^2 + A_1^2 + A_2^2 + A_3^2,$$

where

$$A_0^2 + A_1^2 + A_2^2 + A_3^2 = |\alpha\beta\gamma + \alpha\beta\lambda + \alpha\gamma\lambda + \beta\gamma\lambda|^2.$$

Introducing the Lagrange function

$$f(x, y, z) = |P'(e^{I\theta})|^2 + \mu(x^2 + y^2 + z^2),$$

and applying the method used for $n = 2$ and $n = 3$, we obtain the equations

$$-8A_1\sin(3\theta) + 2\mu x = 0, \quad -8A_2\sin(3\theta) + 2\mu y = 0, \quad -8A_3\sin(3\theta) + 2\mu z = 0,$$
$$x^2 + y^2 + z^2 = 1,$$

with the solutions

$$x = \frac{4A_1\sin(3\theta)}{\mu}, \quad y = \frac{4A_2\sin(3\theta)}{\mu}, \quad z = \frac{4A_3\sin(3\theta)}{\mu},$$
$$\mu^2 = 16\sin^2(3\theta)(A_1^2 + A_2^2 + A_3^2).$$

We have two possibilities:

1. $\mu = -4\sin(3\theta)\sqrt{A_1^2 + A_2^2 + A_3}$;

2. $\mu = 4\sin(3\theta)\sqrt{A_1^2 + A_2^2 + A_3}$.

In what follows we treat only Case 1 since Case 2 is similar.

Case 1. Replacing x, y, z and μ, after simple calculation and arguing as in the case $n = 2$, we obtain

$$|P'(e^{I\theta})|^2 = 16 + |\alpha\beta\gamma + \alpha\beta\lambda + \alpha\gamma\lambda + \beta\gamma\lambda|^2 - 8A_0\cos(3\theta)$$
$$+ 8\sin(3\theta)\sqrt{A_1^2 + A_2^2 + A_3^2}$$
$$= 16 + |\alpha\beta\gamma + \alpha\beta\lambda + \alpha\gamma\lambda + \beta\gamma\lambda|^2$$
$$+ \sqrt{64(A_0^2 + A_1^2 + A_2^2 + A_3^2)}\cos(3\theta - \eta).$$

Choosing $\eta = 3\theta$, we get

$$\|P'\|^2 \geq |P'(e^{I\theta})|^2 = (4 + |\alpha\beta\gamma + \alpha\beta\lambda + \alpha\gamma\lambda + \beta\gamma\lambda|)^2,$$

that is,

$$\|P'\|^2 \geq 4 + |\alpha\beta\gamma + \alpha\beta\lambda + \alpha\gamma\lambda + \beta\gamma\lambda|.$$

Combining this inequality with (6.11), it follows that if

$$4 + |\alpha\beta\gamma + \alpha\beta\lambda + \alpha\gamma\lambda + \beta\gamma\lambda| \geq 2(1 + |\alpha\beta\gamma + \alpha\beta\lambda + \alpha\gamma\lambda + \beta\gamma\lambda| + |\alpha\beta\gamma\lambda|),$$

then Turan's inequality holds.

But the last inequality is obviously equivalent with

$$|\alpha\beta\gamma + \alpha\beta\lambda + \alpha\gamma\lambda + \beta\gamma\lambda| + 2|\alpha\beta\gamma\lambda| \leq 2. \tag{6.12}$$

From (6.10) we get $|\alpha| = |\beta + \gamma + \lambda|$ and $|\alpha\beta + \alpha\gamma + \alpha\lambda| = |\alpha|^2 = |\beta\gamma + \beta\lambda + \gamma\lambda|$, whence

$$|\alpha\beta\gamma + \alpha\beta\lambda + \alpha\gamma\lambda + \beta\gamma\lambda| + 2|\alpha\beta\gamma\lambda| \leq |\alpha|^3 + |\beta\gamma\lambda| + 2|\alpha| \cdot |\beta\gamma\lambda|$$
$$\leq |\alpha|^3 + 2|\alpha| + 1.$$

To ensure (6.12) holds, we look for α satisfying $|\alpha|^3 + 2|\alpha| + 1 \leq 2$, that is,

$$|\alpha|^3 + 2|\alpha| - 1 \leq 0.$$

Denoting $S(x) = x^3 + 2x - 1$, since $S'(x) = 3x^2 + 2 > 0$ for all $x \geq 0$, it follows that S is (strictly) increasing on $[0, 1]$, and therefore the only real root x^* (which is between 0 and 1) of the cubic equation $x^3 + 2x - 1 = 0$ will represent the upper bound for $|\alpha|$ satisfying (6.12).

By the Cardano formula,

$$x^* = \left(-\frac{1}{2} + \sqrt{\left(\frac{-1}{2}\right)^2 + \left(\frac{2}{3}\right)^3}\right)^{1/3} + \left(-\frac{1}{2} - \sqrt{\left(\frac{-1}{2}\right)^2 + \left(\frac{2}{3}\right)^3}\right)^{1/3}$$

$$= \left(\frac{1}{2} + \sqrt{\frac{59}{108}}\right)^{1/3} + \left(\frac{1}{2} - \sqrt{\frac{59}{108}}\right)^{1/3}.$$

Then the hypothesis $|\alpha| \leq x^*$ immediately implies the required inequality (6.12).

By symmetry, a similar conclusion holds if $|\lambda| \leq x^*$. Then, since β commutes with α, it easily follows that we can argue in much the same way as for α and λ. Finally, the same conclusion holds if γ commutes with λ, and the theorem is proved. $\square$

Remark 6.4.4. Since in the proof of the above theorem we have $S(1/2) > 0$ and $S(1/3) < 0$, it follows that $x^* \in \left(\frac{1}{3}, \frac{1}{2}\right)$.

Another method of proof for Turán's inequality in complex analysis relies on the Erdős–Lax inequality. This technique can be used for the class of polynomials for which the latter inequality holds, as follows.

Theorem 6.4.5. *Let $P(q)$ be a polynomial of degree $n \geq 3$ with coefficients in $\mathbb{C}_I$ for some $I \in \mathbb{S}$ that has all its zeros in the closed ball $|q| \leq 1$. Then*

$$\|P'\| \geq \frac{n}{2} \cdot \|P\|.$$

Proof. We borrow here the simple idea used in the complex case in [139]: if P is a quaternionic polynomial such that $P(q)$ has all its zeros in the closed ball $|q| \leq 1$, then the polynomial of degree $\leq n$, defined as $Q(q) = q^n * P(1/q)$ satisfies $\|Q\| = \|P\|$ and all its zeros lie in the domain $|q| > 1$.

By the formula for Q, it is immediate that its coefficients belong to the same complex plane $\mathbb{C}_I$ to which the coefficients of P belong.

Therefore, according to the Proposition 6.3.2, we get

$$\|Q'\| \leq \frac{n}{2}\|Q\| \leq \frac{n}{2}\|P\|.$$

But, since

$$Q'(q) = nq^{n-1} * P(1/q) - q^{n-2} * P'(1/q),$$

denoting $e_k(q) = q^k$ we immediately get

$$\|e_{n-2} * P'(e_{-1})\| \geq n\|e_{n-1} * P(e_{-1})\| - \|Q'\|,$$

whence

$$\|P'\| \geq n\|P\| - \frac{n}{2}\|Q\| \geq \frac{n}{2}\|P\|. \qquad \square$$

Remark 6.4.6. The only obstacle which prevents us from extending Turan's inequality to quaternionic polynomials of degree $n \geq 3$, is that given a quaternionic polynomial of degree $n \geq 3$, P_n, we don't know how to perform a suitable rotation to always ensure that $\|P\| = \max_{|q|=1} |P(q)|$ is attained at the real point $q_0 = 1$ (which holds in the complex case).

Remark 6.4.7. If we consider the polynomial of degree 2 given by $P(q) = (q - i) * (q - j) = q^2 - q(i + j) + k$, as in the proof of Theorem 6.3.1, we have $\|P'\| > \frac{2}{2} \cdot \|P\|$. Since the only root i has modulus 1, we see that the equality claimed in the second part of Theorem 6.1.3 does not hold, in general, for quaternionic polynomials.

In what follows, we prove some generalizations of Theorem 6.4.1, the first main result being an analogue of Theorem 6.1.5.

Theorem 6.4.8. *Let $P(q)$ be a polynomial of degree n with coefficients in a complex plane $\mathbb{C}_I$ for some $I \in \mathbb{S}$, that has all its zeros in the closed ball $|q| \leq K$ with $K \geq 1$. Then*

$$\|P'\| \geq \frac{n}{1 + K^n} \cdot \|P\|.$$

The inequality is sharp and equality holds for $P(q) = \dfrac{q^n + K^n}{1 + K^n}.$

Proof. First we prove that if $T(q)$ is a slice regular function, in particular a polynomial, for $|q| > \rho > 0$, then for any $r > \rho$ we have

$$\max_{|q| \geq r} |T(q)| = \max_{|q| = r} |T(q)|. \tag{6.13}$$

Indeed, denoting $F(q) = T(q^{-1})$, since $F(q)$ is slice regular in the ball $|q^{-1}| \geq r$, i.e., $|q| \leq \frac{1}{r}$, the maximum modulus theorem shows that

$$\max_{|q| \leq 1/r} |F(q)| = \max_{|q| = 1/r} |F(q)|,$$

which is obviously equivalent with

$$\max_{|q^{-1}| \geq r} |T(q^{-1})| = \max_{|q^{-1}| = r} |T(q^{-1})|.$$

Substituting here $q^{-1} = q'$ one obtains the required relationship (6.13).

From (6.13) and for $K \geq 1$, we immediately obtain that

$$\max_{|q| = K} |T(q)| \leq \max_{|q| = 1} |T(q)|. \tag{6.14}$$

Now, choosing in (6.14) $T(q) = q^{-n} * S(q)$, where $S(q)$ is an arbitrary polynomial of degree n, we immediately obtain

$$\max_{|q| = K} |S(q)| \leq K^n \cdot \max_{|q| = 1} |S(q)|. \tag{6.15}$$

Moreover, the polynomial $R(q) = P(Kq)$ is of degree n and has all its zeros in $|q| \leq 1$. By the hypothesis on P, it is immediate that all its coefficients belong to the complex plane $\mathbb{C}_I$. Therefore, by Theorem 6.4.5,

$$\|R'\| \geq \frac{n}{2} \cdot \|R\|. \tag{6.16}$$

Denoting $Q(q) = q^n * R\left(q^{-1}\right)$, it is immediate that Q is a polynomial of degree n which has all its zeros in $|q| > 1$, and that the coefficients of Q are in $\mathbb{C}_I$. Moreover, $\|Q\| = \|R\|$.

In what follows, we will prove the following inequality, valid for all $K \geq 1$:

$$\max_{|q| = K} |Q(q)| \leq \frac{K^n + 1}{2} \cdot \max_{|q| = 1} |Q(q)|. \tag{6.17}$$

Indeed, by Proposition 6.3.2, $\|Q'\| \leq \frac{n}{2}\|Q\|$, which in conjunction with the inequality (6.15) applied to the polynomial $Q'(q)/(n/2)$ of degree $n - 1$, yields

$$\max_{|q| = K} |Q'(q)| \leq \frac{n}{2} K^{n-1} \cdot \|Q\|.$$

Since any $q \in \mathbb{H} \setminus \mathbb{R}$ with $|q| = K$ can be (uniquely) written as $q = Ke^{I_q\varphi}$, $I_q \in \mathbb{S}$, and any $q \in \mathbb{R}$ with $|q| = K$ can be written as $q = Ke^{I0}$, if $q = K$, and as $q = Ke^{I\pi}$ with an arbitrary $I \in \mathbb{S}$, the above inequality yields

$$|Q'(Ke^{I\varphi})| \leq \frac{n}{2}K^{n-1} \cdot \|Q\|, \text{ for all } \varphi \in [0, 2\pi) \text{ and } I \in \mathbb{S}. \qquad (6.18)$$

But for each $I \in \mathbb{S}$ and $\varphi \in [0, 2\pi)$ fixed, we have

$$Q(Ke^{I\varphi}) - Q(e^{I\varphi}) = \int_1^K e^{I\varphi} * Q'(re^{I\varphi})dr,$$

which in conjunction with (6.18) written for K replaced by $r \geq 1$, implies

$$|Q(Ke^{I\varphi})| \leq |Q(Ke^{I\varphi}) - Q(e^{I\varphi})| + |Q(e^{I\varphi})|$$

$$\leq \int_1^K |Q'(re^{I\varphi})|dr + \|Q\|$$

$$\leq \frac{\|Q\|n}{2} \cdot \int_1^K r^{n-1}dr + \|Q\|$$

$$= \|Q\| \left(\frac{K^n + 1}{2}\right).$$

This immediately yields (6.17).

Now, since $\|Q\| = \|R\|$, inequality (6.17) shows that

$$\|R\| \geq \frac{2}{1 + K^n} \max_{|q|=K} |Q(q)|. \qquad (6.19)$$

Using $Q(q) = q^n R(q^{-1}) = q^n \cdot P(K \cdot q^{-1})$, we deduce that

$$\max_{|q|=K} |Q(q)| = K^n \cdot \max_{|q|=1} |P(q)| = K^n \cdot \|P\|. \qquad (6.20)$$

Then (6.19) combined with (6.20) implies

$$\|R\| \geq \frac{2K^n}{1 + K^n} \cdot \|P\|,$$

which by (6.16) gives

$$\frac{2}{n}\|R'\| \geq \frac{2K^n}{1 + K^n} \cdot \|P\|. \qquad (6.21)$$

Since $R'(q) = KP'(Kq)$, using (6.15) for polynomials of degree $n - 1$ we get

$$\|R'\| = K \cdot \max_{|\xi|=K} |P'(\xi)| \leq K \cdot K^{n-1}\|P'\| = K^n\|P'\|,$$

which combined with (6.21) gives

$$\frac{2}{n} \cdot K^n\|P'\| \geq \frac{2K^n}{1 + K^n}\|P\|.$$

This proves the first assertion of the theorem. The fact that the inequality is sharp and equality holds for $P(q) = \frac{q^n + K^n}{1 + K^n}$ is immediate. $\qquad \square$

The following analogue of Theorem 6.1.4 is also valid.

Theorem 6.4.9. *Let $P(q)$ be a polynomial of degree n with coefficients in the complex plane $\mathbb{C}_I$ for some $I \in \mathbb{S}$, that has all its zeros in the closed ball $|q| \leq K$ with $K \leq 1$. Then*

$$\|P'\| \geq \frac{n}{1+K} \cdot \|P\|.$$

The inequality is sharp and equality holds for $P(q) = \left(\frac{q+K}{1+K}\right)^n$.

Proof. Denoting $Q(q) = q^n * P(q^{-1})$, it is immediate that Q is a polynomial of degree n, having all zeros in the ball $|q| \leq \frac{1}{K} \geq 1$, and that by its definition its coefficients belong to $\mathbb{C}_I$.

Then, since $Q'(q) = nq^{n-1} * P(q^{-1}) - q^{n-2} * P'(q^{-1})$, applying Proposition 6.3.5 to $Q(q)$ with $L = \frac{1}{K}$, we immediately obtain

$$\|P'\| \geq |n\|P\| - \|Q'\| \geq n\|P\| - \|Q'\|$$
$$\geq n\|P\| - \frac{n}{1+1/K}\|P\| = \|P\| \cdot \frac{n}{1+K},$$

which proves the inequality in the statement.

The fact that the inequality is sharp and equality holds for $P(q) = \left(\frac{q+K}{1+K}\right)^n$ is immediate. $\qquad\square$

In the particular case $n = 2$, the analogues of Theorem 6.1.5 and Theorem 6.1.4 hold.

Theorem 6.4.10. *Let $P(q) = (q - \alpha) * (q - \beta)$ be a quaternionic polynomial.*

(i) *If $|\alpha|, |\beta| \leq K$ with $K \geq 1$, then $\|P'\| \geq \frac{2}{1+K^2} \cdot \|P\|$.*

(ii) *If $|\alpha|, |\beta| \leq K$ with $K \leq 1$, then $\|P'\| \geq \frac{2}{1+K} \cdot \|P\|$.*

Proof. The proof of Theorem 6.4.1, (i) shows that

$$\|P\| \leq 1 + |\alpha + \beta| + |\alpha \cdot \beta| \quad \text{and} \quad \|P'\| = 2 + |\alpha + \beta|.$$

(i) We have

$$\frac{2}{1+K^2}\|P\| \leq \frac{2}{1+K^2} + \frac{2}{1+K^2}|\alpha + \beta| + \frac{2|\alpha\beta|}{1+K^2} \leq \frac{2+2K^2}{1+K^2} + \frac{2}{1+K^2}|\alpha + \beta|$$
$$= 2 + \frac{2}{1+K^2}|\alpha + \beta| \leq 2 + |\alpha + \beta| = \|P'\|,$$

since $\frac{2}{1+K^2} \leq 1$.

(ii) We have

$$\frac{2}{1+K} \cdot \|P\| \leq \frac{2}{1+K} + \frac{2}{1+K}|\alpha + \beta| + \frac{2|\alpha\beta|}{1+K} \leq \frac{2+2K^2}{1+K} + \frac{2}{1+K}|\alpha + \beta|.$$

We claim that

$$\frac{2 + 2K^2}{1 + K} + \frac{2}{1 + K}|\alpha + \beta| \leq 2 + |\alpha + \beta| = \|P'\|.$$

Indeed, this inequality is obviously equivalent to

$$\frac{1 - K}{1 + K}|\alpha + \beta| \leq \frac{2K(1 - K)}{1 + K},$$

which reduces to the valid inequality $|\alpha + \beta| \leq 2K$. This concludes the proof. $\square$

Remark 6.4.11. In the case $n = 3$, the estimates in the proof of Theorem 6.4.1, (ii) lead to only partial generalizations of the above types, as follows.

Theorem 6.4.12. Let $P(q) = (q - \alpha) * (q - \beta) * (q - \gamma)$ be a quaternionic polynomial with $\alpha + \beta + \gamma = 0$.

  (i) If $|\alpha|, |\beta|, |\gamma| \leq K$ with $K \geq 2^{1/3}$, then $\|P'\| \geq \frac{3}{1 + K^3} \cdot \|P\|$.

  (ii) If $|\alpha|, |\beta|, |\gamma| \leq K$ with $K \leq \frac{1}{2}$, then $\|P'\| \geq \frac{3}{1 + K} \cdot \|P\|$.

Proof. The proof of Theorem 6.4.1, (ii) shows that

$$\|P\| \leq 1 + |\alpha\beta + \alpha\gamma + \beta\gamma| + |\alpha\beta\gamma| \text{ and } \|P'\| \geq 3 + |\alpha\beta + \alpha\gamma + \beta\gamma|.$$

  (i) We have

$$\frac{3}{1 + K^3} \cdot \|P\| \leq \frac{3}{1 + K^3} + \frac{3}{1 + K^3}|\alpha\beta + \alpha\gamma + \beta\gamma| + \frac{3}{1 + K^3}|\alpha\beta\gamma|$$
$$\leq \frac{3 + 3K^3}{1 + K^3} + \frac{3}{1 + K^3}|\alpha\beta + \alpha\gamma + \beta\gamma| = 3 + \frac{3}{1 + K^3}|\alpha\beta + \alpha\gamma + \beta\gamma|.$$

Then, the inequality

$$3 + \frac{3}{1 + K^3}|\alpha\beta + \alpha\gamma + \beta\gamma| \leq 3 + |\alpha\beta + \alpha\gamma + \beta\gamma|$$

(which guarantees the required inequality), is equivalent to $\frac{3}{1 + K^3} \leq 1$, i.e., to $K \geq 2^{1/3}$.

  (ii) We have

$$\frac{3}{1 + K} \cdot \|P\| \leq \frac{3}{1 + K} + \frac{3}{1 + K}|\alpha\beta + \alpha\gamma + \beta\gamma| + \frac{3}{1 + K}|\alpha\beta\gamma|$$
$$\leq \frac{3 + 3K^3}{1 + K} + \frac{3}{1 + K}|\alpha\beta + \alpha\gamma + \beta\gamma| = 3(K^2 - K + 1) + \frac{3}{1 + K}|\alpha\beta + \alpha\gamma + \beta\gamma|.$$

Then, the inequality

$$3(K^2 - K + 1) + \frac{3}{1 + K}|\alpha\beta + \alpha\gamma + \beta\gamma| \leq 3 + |\alpha\beta + \alpha\gamma + \beta\gamma|$$

(which guarantees the required inequality), is equivalent to

$$3K(K-1) \le \left(1 - \frac{3}{1+K}\right)|\alpha\beta + \alpha\gamma + \beta\gamma|,$$

which in turn is obviously equivalent to

$$|\alpha\beta + \alpha\gamma + \beta\gamma| \le \frac{3K(1-K^2)}{2-K}.$$

Since $|\alpha\beta + \alpha\gamma + \beta\gamma| \le 3K^2$ and $3K^2 \le \frac{3K(1-K^2)}{2-K}$ is equivalent to $K \le \frac{1}{2}$, we arrive at the desired conclusion. $\square$

6.5 Notes on Turan Type Inequalities

Another obstacle in extending Turan's inequality to quaternionic polynomials of degree $n \ge 3$, in addition to the one indicated in Remark 6.4.6, is that the method of Lagrange multipliers used in the proof of Theorem 6.4.1 for the particular cases $n = 2$ and $n = 3$, for general $n \ge 4$ leads to extremely complicated calculations.

Also, the quaternionic extension of Theorems 6.1.6 and 6.1.7, under hypotheses similar to those in Theorems 6.4.8 and 6.4.9 or to those for the particular cases $n = 2$ and $n = 3$ in the Theorems 6.4.10 and 6.4.12, remains an open question.

Chapter 7

Approximation of nullsolutions of generalized Cauchy–Riemann operators

In this chapter we present some results concerning approximation theory in the setting of nullsolutions of generalized Cauchy–Riemann operators in the quaternionic and in the Clifford algebra setting. This last case was considered only marginally in this work, but in this chapter we deal also with this more general situation when it is the framework of the original sources. It is obvious that one can always specialize the results to quaternions.

As we shall see, the approximation results are different in spirit from the result we proved in the case of slice regular functions. The only exceptions are some Runge-type results proved by Brackx and Delanghe in [61] and which have been generalized to the complex setting by Ryan in [173] and to spherical monogenics of complex degree by Van Lancker in [191].

Then we introduce an important topic, namely Appell systems, which are of crucial importance for approximating monogenic functions. In another section, we consider results in L^2 and Sobolev spaces obtained in [27] by Cação, Gürlebeck and Malonek using monogenic polynomials. Here we also present an overview of the studies on complete orthonormal systems of monogenic polynomials, see [25], [26], [148], [149], [150]. We then discuss the case of axially monogenic functions treated by Common [55], who constructed their Padé approximants. In the case of axially and biaxially monogenic functions, one can also obtain some toroidal series expansion, as shown in the fifth section.

Finally, in the last section, following the work [143] by Falcão and Malonek we present the polynomial approximation of quasi-conformal monogenic maps.

It is worth mentioning that there are various other papers containing elements of approximation in Clifford analysis. For example, the problem of approximation

© Springer Nature Switzerland AG 2019
S. G. Gal, I. Sabadini, *Quaternionic Approximation*, Frontiers in Mathematics,
https://doi.org/10.1007/978-3-030-10666-9_7

and interpolation through the use of generalized analytic functions was treated by Gürlebeck, see [115]. Also relationships between the interpolation of monogenic functions in the unit ball of the $\mathbb{R}^3$ and the best approximation of their boundary values were considered by Gürlebeck in [116].

7.1   Runge Type Results

This section describes Runge type theorems for the solutions of the equation $Df = 0$, where D is the generalized Cauchy–Riemann operator (1.23) acting on functions f defined on subsets of $\mathbb{R}^{n+1}$, and with values in $\mathbb{R}_n$ (or $\mathbb{C}_n$). Accordingly, in the sequel $\mathcal{A}_n$ denotes either the Clifford algebra $\mathbb{R}_n$ or $\mathbb{C}_n$. The results in this section hold more generally for functions in the kernel of D^k, $k \geq 1$ and defined on subsets of $\mathbb{R}^m$, $m \leq n$, but since these functions are outside the scopes of this book, we refer the interested reader to the original paper [61].

The first result asserts that if K is a compact subset of $\mathbb{R}^{n+1}$ and f satisfies the equation $Df = 0$ in an open neighborhood of K, then f can be uniformly approximated on K by a sequence of polynomial solutions of the equation (if $\mathbb{R}^{n+1}\backslash K$ is connected) or by a sequence of "rational" solutions having their singularities outside K (with K arbitrary). Then, based on these results, generalizations of the well-known Runge and Hartogs–Rosenthal theorems and a Runge type theorem for regular solutions at infinity of the operator D are considered.

We denote by $\mathcal{M}(U, \mathcal{A}_n)$ the set of monogenic functions on the open set $U \subseteq \mathbb{R}^{n+1}$ with values in $\mathcal{A}_n$. It is known that, as it happens for holomorphic functions, i.e., nullsolutions of the Cauchy–Riemann operator, the set $\mathcal{M}(U, \mathcal{A}_n)$ equipped with the topology of uniform convergence on compact sets is a right $\mathcal{A}_n$-Fréchet module.

If $K \subset \mathbb{R}^{n+1}$ is a compact set, then we denote by $\mathcal{M}(K, \mathcal{A}_n)$ the set of all functions f for which there exists an open neighborhood U of K, such that $f \in \mathcal{M}(U, \mathcal{A}_n)$. In a more refined way, $\mathcal{M}(K, \mathcal{A}_n)$ is the inductive limit of $\mathcal{M}(U, \mathcal{A}_n)$ where U runs over all open sets containing K.

In Chapter 1 we have introduced the symmetric polynomials (1.24) which belong to $\mathcal{M}(\mathbb{R}^{n+1}, \mathcal{A}_n)$ and we have stated the Taylor expansion of a monogenic function in a given point $a \in \mathbb{R}^{n+1}$. Let us consider the analogs of the outer spherical monogenics (1.27) at the point a:

$$W^{(a)}_{\ell_1,\dots,\ell_p}(x) = (-1)^p \frac{\partial^p}{\partial_{y_{\ell_1}} \cdots \partial_{y_{\ell_p}}} E(x - a) = \left[\frac{\partial^p}{\partial_{u_{\ell_1}} \cdots \partial_{u_{\ell_p}}} E(x - u) \right]_{|u=a} , \quad (7.1)$$

and set

$$\mathcal{M}(a) = \{W^{(a)}_{\ell_1,\dots,\ell_p}(x) \mid (\ell_1,\dots,\ell_p) \in \{1,\dots,n\}^p, \ p \in \mathbb{N}\}.$$

If $A = (a_i)_{i \in I}$ is a finite or countable subset of $\mathbb{R}^{n+1}$, we put

$$\mathcal{M}(A) := \bigcup_{i \in I} \mathcal{M}(a_i)$$

and denote by $\mathscr{M}_r(A)$ the right-linear span of $\mathscr{M}(A)$. Then $\mathscr{M}_r(A)$ is a right $\mathcal{A}_n$-module containing functions whose singularities (k-poles) lie outside K, namely in A.

Let K be an arbitrary compact subset of $\mathbb{R}^{n+1}$ and let $U_0, U_1, \ldots,$ be the components of $\mathbb{R}^{n+1} \setminus K$, with U_0 unbounded. Moreover, let $A = (a_i)$ be a subset of $\mathbb{R}^{n+1} \setminus K$, such that $a_i \in U_i$, $i = 1, 2, \ldots$.

The following Runge type results hold, see the original source [61] and also [20].

Theorem 7.1.1 (Delanghe and Brackx [61]). *Let K be a compact subset of $\mathbb{R}^{n+1}$ and let A be a subset of $\mathbb{R}^{n+1} \setminus K$ having one point in each bounded component of $\mathbb{R}^{n+1} \setminus K$. Then each function left monogenic in a neighborhood of K can be approximated uniformly on K by functions in $\mathscr{M}(\mathbb{R}^{n+1}, \mathcal{A}_n) \oplus \mathscr{M}_r(A)$, i.e., $\mathscr{M}(\mathbb{R}^{n+1}, \mathcal{A}_n) \oplus \mathscr{M}_r(A)$ is uniformly dense in $\mathscr{M}(K, \mathcal{A}_n)$.*

The following refinement is valid when K satisfies some additional conditions:

Theorem 7.1.2 (Delanghe and Brackx [61]). *Let K be a compact subset of U such that $U \setminus K$ has no component relatively compact in U. Then every function left monogenic in a neighborhood of K can be uniformly approximated on K by functions in $\mathscr{M}(U, \mathcal{A}_n)$, i.e., $\mathscr{M}(U, \mathcal{A}_n)$ is uniformly dense in $\mathscr{M}(K, \mathcal{A}_n)$.*

In the case of an open subset U of $\mathbb{R}^{n+1}$ we have:

Theorem 7.1.3 (Delanghe and Brackx [61]). *Let U be an open subset of $\mathbb{R}^{n+1}$ and α be a subset of $\mathbb{R}^{n+1} \setminus U$ having one point in each component of $\mathbb{R}^{n+1} \setminus U$. Then, every function left monogenic in U can be approximated uniformly on each compact subset K of U by functions in $\mathscr{M}(\mathbb{R}^{n+1}, \mathcal{A}_n) \oplus \mathscr{M}_r(\alpha)$, i.e., $\mathscr{M}(\mathbb{R}^{n+1}, \mathcal{A}_n) \oplus \mathscr{M}_r(\alpha)$ is dense in $\mathscr{M}(U, \mathcal{A}_n)$ for the topology of uniform compact convergence. In particular, if $\mathbb{R}^{n+1} \setminus U$ is connected and unbounded, then $\mathscr{M}(\mathbb{R}^{n+1}, \mathcal{A}_n)$ is dense in $\mathscr{M}(U, \mathcal{A}_n)$.*

Moreover, the following Hartogs–Rosenthal type result holds.

Theorem 7.1.4 (Delanghe and Brackx [61]). *Let K be a compact subset of $\mathbb{R}^{n+1}$ with Lebesgue measure zero and A be a subset of $\mathbb{R}^{n+1} \setminus K$ having one point in each bounded component of $\mathbb{R}^{n+1} \setminus K$. Then, $\mathscr{M}(\mathbb{R}^{n+1}, \mathcal{A}_n) \oplus \mathscr{M}_r(A)$ is dense in $C^0(K, \mathcal{A}_n)$.*

Remark 7.1.5. The Runge type approximation results in the above theorems were extended by Ryan in [173] for nullsolutions of the complex Dirac operator $\sum_{j=1}^n e_j \dfrac{\partial}{\partial z_j}$, namely to functions f which are holomorphic in a domain $U \subseteq \mathbb{C}^n$ and such that

$$\sum_{j=1}^n e_j \frac{\partial}{\partial z_j} f(z) = 0, \quad z = e_1 z_1 + \cdots + e_m z_m \in U.$$

Approximation results have been also proved for nullsolutions to the complex Laplacian

$$\Delta_z = \sum_{j=1}^{n} e_j \frac{\partial^2}{\partial z_j^2}$$

on special domains in $\mathbb{C}^n$. Thus, for some special classes of holomorphic functions which satisfy the complex, homogeneous Dirac equation, or the complex Laplace equation, Runge approximation theorems on open subsets, in $\mathbb{C}^n$, of the complement of a compact set were obtained. Details of the pretty complicated techniques used can be found in [173].

Remark 7.1.6. Another Runge type approximation was considered in [191] for spherical monogenics of complex degree κ, i.e., functions $f : U \to \mathbb{C}_n$ of class $\mathcal{C}^1$ satisfying $(\Gamma + \kappa)f = 0$ on U, where $\Gamma = -x \wedge D = \sum_{i<j} e_i e_j (x_i \frac{\partial}{\partial x_j} - x_j \frac{\partial}{\partial x_i})$ is the Gamma operator. For special values of κ these functions may admit a geometric interpretation; for example, for $\kappa = \frac{1}{2}(1-n)$ they belong to the kernel of the Atiyah–Singer–Dirac operator used in differential geometry.

7.2   Appell systems

Quaternionic (or Clifford algebra-valued) functions in the kernel of a generalized Cauchy–Riemann operator can be locally expressed in terms of the symmetric polynomials (1.21) (resp. (1.24)). This local expansion characterizes the hyperholomorhic functions, in the sense that it is also equivalent to the hypercomplex derivability if this is suitably understood, see [118, 140]. Thus the three approaches to hyperholomorphic functions, namely, the one of Riemann, based on a suitable differential operator for which a function is a nullsolution, the one of Weierstrass, based on series expansions, and the one of Cauchy, based on a hypercomplex derivative, are all equivalent in this framework. However, a disadvantage of the symmetric polynomials is that they are not orthogonal in the L^2-inner product. The Gram–Schmidt algorithm is not easy to apply and the calculations are not numerically stable. Taylor and Fourier series, i.e., local and global approximations, can be obtained, for practical purposes (for example, for constructing explicit representations for solutions to boundary value problems in three or four dimensions), if orthogonal bases are considered. The first construction in the case of dimension 3 was done in Caçao's dissertation [23], and was further developed by Bock, Caçao, Gürlebeck and Malonek in [24, 25, 26, 27]. In particular, in [26] the authors overcome some of these problems by constructing complete orthonormal systems of quaternionic valued polynomials in L^2. The authors established an orthogonal decomposition of the space of square integrable monogenic functions (from $\mathbb{R}^3$ to itself or to $\mathbb{H}$) with respect to the derivatives of arbitrary order. This is a generalization of earlier results in [25], where some of the derivatives were already calculated.

To explain in more detail the problem, let us recall that, in general, when dealing with special systems of functions, an important property is that the derivative of a function in the system is a multiple of another function in the same system. Moreover, it is expected that the factor is given by the total degree of the original polynomial. This property, which trivially holds for powers z^k of a complex variable, was extended by Appell [9] to rather general polynomial systems, nowadays called Appell systems.

Following this idea, a sequence $(P_k(x))_{k\geq 0}$ of monogenic polynomials $P_k(x)$ is called an *Appell sequence* if it satisfies the following two conditions: the total degree of P_k is k, and $(d/dx)P_k(x) = kP_{k-1}(x)$, for any $k = 1, 2, \ldots$. This sequence of polynomials is normalized by setting $P_0(x) = 1$.

The literature is rather rich in results about Appell systems of monogenic polynomials in $\mathbb{R}^3$ and more generally in $\mathbb{R}^n$. The first systems of paravector valued monogenic polynomials, orthogonal but not complete, were constructed by Malonek and co-authors in [69, 142]. In particular, in [69] the monogenic polynomials are of the form

$$P_k(x) = \sum_{s=0}^{k} T_s^k x^{k-s} \bar{x}^s, \tag{7.2}$$

where the T_s^k are suitably defined real numbers. In [28] Cação and Malonek constructed an orthogonal Appell basis in L^2 with a real-valued inner product for the solutions to the Riesz system in dimension 3.

Later, in [17], the authors constructed an orthogonal Appell basis of square integrable monogenic polynomials in $\mathbb{R}^3$ that are solutions to the Moisil–Teodorescu system and with respect to a quaternionic-valued inner product. It is worthwhile mentioning that this Appell basis is useful when dealing with approximation of the Lamé–Navier equations in elasticity theory.

Special monogenic polynomials in Clifford analysis of the form (7.2) in terms of x and its conjugate have been considered before in [1] and in other papers of these authors. But the special monogenic polynomials considered by the authors do not take into account the concept of hypercomplex differentiability [118] and the corresponding use of the hypercomplex derivative, which first appeared in Malonek's paper [140] and was not available at that time for the investigation of Appell sequences of monogenic polynomials. In the quaternionic case and using a completely different approach, these polynomials were obtained in [23].

The definition of hypercomplex differentiability in [140] can be explained after a few notations are set. Given $x = x_0 + \sum_{k=1}^{n} x_k e_k \in \mathbb{R}^{n+1}$, we consider $z_k = x_k - x_0 e_k$, $k = 1, \ldots, n$ and $\vec{z} = (z_1, \ldots, z_n)$. The set of these vectors $\vec{z}$ will be denoted by $\mathcal{H}^n$. It is immediate that $\mathcal{H}^n$ can be identified with $\mathbb{R}^{n+1}$ (seen as the set of paravectors of the form $r_0 + e_1 r_1 + \cdots + e_n r_n$); moreover, we can equip $\mathcal{H}^n$ with the inner product $(\vec{z}, \vec{w}) = \sum_{k=1}^{n} \bar{z}_k w_k$ and the induced norm $\|\vec{z}\| = (\sum_{k=1}^{n} \bar{z}_k z_k)^{1/2}$.

Definition 7.2.1 (Malonek [140]). Let f be a continuous map from $\mathbb{R}^{n+1}$ to $\mathcal{A}_n$.

The function $f = f(\vec{z})$ is said to be hyperdifferentiable on the left (right) in $\vec{a}$ if there exists a left (right) $\mathcal{A}_n$-linear mapping $L : \mathcal{H}^n \to \mathcal{A}_n$ such that

$$\lim_{\Delta\vec{z} \to 0} \frac{|f(\vec{a} + \Delta\vec{z}) - f(\vec{a}) - L(\Delta\vec{z})|}{\|\Delta\vec{z}\|} = 0.$$

The linear map L is called the left (right) *hypercomplex derivative* of f in $\vec{a}$.

If a function is left (right) hyperdifferentiable, then the linear map L is uniquely determined. This concept is crucial since it represents the Cauchy approach in the theory of monogenic functions, as the following results show:

Theorem 7.2.2 (Malonek [140]). *Let $f = f(\vec{z})$ be a continuously real differentiable function in $\vec{z} \in \mathcal{H}^n$. Then the function is left (right) hyperdifferentiable if and only if it is left (right) monogenic.*

To give the flavor of the results in the language of Appell sequences, we mention some of the results obtained in [17]. To this purpose we need some notations. First of all, we consider functions $f : U \subseteq \mathbb{R}^3 \to \mathbb{H}$ in the kernel of the operator $D = \partial_{x_0} + e_1 \partial_{x_1} + e_2 \partial_{x_2}$ given in (1.23).

We use spherical coordinates in $\mathbb{R}^3$, so that (x_0, x_1, x_2) can be expressed as

$$x_0 = r\cos\theta, \quad x_1 = r\sin\theta\cos\phi, \quad x_2 = r\sin\theta\sin\phi,$$

where $0 < r < \infty$, $0 < \theta \leq \pi$, $0 < \phi \leq 2\pi$. Moreover, we denote by $\mathbb{B}$ the unit ball in $\mathbb{R}^3$, namely

$$\mathbb{B} = \{x = x_0 + e_1 x_1 + e_2 x_2; \ x_0^2 + x_1^2 + x_2^2 < 1\}.$$

Definition 7.2.3. The *spherical monogenics*

$$X_n^0, \quad X_n^m, \quad Y_n^m, \quad m = 1, \ldots, n+1$$

on the boundary of the unit ball are defined by

$$\begin{aligned}
X_n^0 &= A^{0,n} + B^{0,n}(\cos\phi i + \sin\phi j), \\
X_n^m &= A^{m,n}\cos m\phi + B^{m,n}(\cos\phi i + \sin\phi j)\cos m\phi \\
&\quad + C^{m,n}(-\sin\phi i + \cos\phi j)\sin m\phi, \\
Y_n^m &= A^{m,n}\sin m\phi + B^{m,n}(\cos\phi i + \sin\phi j)\sin m\phi \\
&\quad + C^{m,n}(-\sin\phi i - \cos\phi j)\cos m\phi,
\end{aligned} \tag{7.3}$$

where the coefficients $A^{m,n}$, $B^{m,n}$, $C^{m,n}$ can be computed explicitly in terms of θ via the Legendre polynomials (see [17]).

Furthermore, we set

$$X_{n,s}^{0,\dagger} := r^n X_n^0 e_s, \quad X_{n,s}^{m,\dagger} := r^n X_n^m e_s, \quad Y_{n,s}^{m,\dagger} := r^n Y_n^m e_s, \quad s = 0, 1, 2, 3 \tag{7.4}$$

where $e_0 = 1$, $e_1 = i$, $e_2 = j$, $e_3 = k$. Finally, let $\tilde{X}_{n,s}^{m,\dagger}$, $m = 0, 1, 2, \ldots$, and $\tilde{Y}_{n,s}^{m,\dagger}$ $m = 1, 2, \ldots$, be the normalized elements obtained by dividing the elements (7.4) by their norm in $L^2(\mathbb{B})$. We have:

Theorem 7.2.4 (Bock and Gürlebeck [17]). *For any $n = 0, 1, 2, \ldots$ the solid spherical monogenics*

$$\varphi_n^0 := \tilde{X}_{n,0}^{0,\dagger}, \quad \varphi_n^\ell := c_{n,-\ell}(\tilde{X}_{n,0}^{\ell,\dagger} - \tilde{Y}_{n,3}^{\ell,\dagger}), \tag{7.5}$$

where $c_{n,-\ell} = \sqrt{(n+1)/(2n - 2\ell + 2)}$, $\ell = 1, \ldots, n$, are orthonormal in $L^2(\mathbb{B})$.

The polynomials (φ_n^ℓ) form an Appell system and allow to prove the following expansion for monogenic functions:

Theorem 7.2.5 (Bock and Gürlebeck [17]). *Let $f \in L^2(\mathbb{B})$ be a quaternionic valued function monogenic in the unit ball $\mathbb{B}$ of $\mathbb{R}^3$. Then f can be uniquely represented in terms of the orthonormal system (7.5) as follows*

$$f(x) = \sum_{n=0}^{\infty} \sum_{\ell=0}^{n} \varphi_n^\ell(x) \alpha_{n,\ell}, \quad \alpha_{n,\ell} = \int_{\mathbb{B}} \overline{\varphi_n^\ell(x)} f(x) \, dV,$$

where dV denotes the Lebesgue measure in $\mathbb{R}^3$.

Let us now consider a variation of the system $(\varphi_n^\ell(x))$, specifically, let us introduce

$$\hat{A}_n^\ell := \sqrt{\frac{(2\ell + 3)(2\ell + 1)!}{(2n + 3)(n + \ell +)!(n - \ell)!}} \, \varphi_n^\ell, \quad \ell = 0, \ldots, n. \tag{7.6}$$

We have:

Theorem 7.2.6 (Bock and Gürlebeck [17]). *The system of homogeneous monogenic polynomials $(\hat{A}_n^\ell)$ is a complete orthogonal Appell system for monogenic functions in the unit ball $\mathbb{B}$ that belong to $L^2(\mathbb{B})$, namely*

$$\frac{1}{2}(\partial_{x_0} - e_1 \partial_{x_1} - e_2 \partial_{x_2})\hat{A}_n^\ell(x) = n\hat{A}_{n-1}^\ell(x), \quad \ell = 0, \ldots, n - 1,$$

$$\frac{1}{2}(\partial_{x_0} - e_1 \partial_{x_1} - e_2 \partial_{x_2})\hat{A}_n^\ell(x) = 0, \quad \ell = n.$$

In general, the construction of orthogonal bases for sets of solutions to systems of differential equations is a difficult problem. The construction approach formulated by Gelfand and Tsetlin, see [97], is somewhat easier, but requires the setting of irreducible modules over a simple Lie algebra. This is the approach followed in [18] in dimension 3 and in [131] in arbitrary dimension n. Since it has wider generality, we mention the main result from [131] concerning the approximation of spherical monogenics in dimension n.

By $F_{n,j}^{(k-j)}$ we denote the polynomials

$$F_{n,j}^{(k-j)}(x) := \frac{(j+1)_{k-j}}{(n+2j-2)_{k-j}} |x|^{k-j} C_{k-j}^{n/2+j-1}\left(\frac{x_n}{x}\right), \quad x \in \mathbb{R}^n \tag{7.7}$$

where C_k^ν is the Gegenbauer polynomial

$$C_k^\nu(z) = \sum_{i=0}^{[k/2]} \frac{(-1)^i (\nu)_{k-i}}{i!(k-2i)!}(2z)^{k-2i}$$

and $(\nu)_k = \nu(\nu+1)\cdots(\nu+k-1)$. Set

$$X_{n,j}^{(k-j)}(x) := F_{n,j}^{(k-j)}(x) + \frac{j+1}{n+2j-1}F_{n,j+1}^{(k-j-1)}(x)\underline{x}e_n, \tag{7.8}$$

where $F_{n,j}^{(k-j)}$ are as in (7.7), $F_{n,k+1}^{(-1)} = 0$ and $\underline{x}$ is the 1-vector part of the paravector x. One can construct an orthogonal basis for the space of monogenic polynomials of degree k in $\mathbb{R}^n$ by induction on the dimension n, as explained in [20], pp. 262–264.

This basis, which possesses the Appell property, is formed by the polynomials

$$f_{k,\mu} = X_{n,k_{n-1}}^{(k-k_{n-1})} X_{n-1,k_{n-2}}^{(k_{n-1}-k_{n-2})} \cdots X_{3,k_2}^{(k_3-k_2)}(x_1 - e_1 e_2 x_2)^{k_2} \tag{7.9}$$

where μ is a sequence of integers $(k_{n-1}, \ldots, k_2)$ such that $k \geq k_{n-1} \geq \cdots \geq k_2 \geq 0$. We note that this basis is also a complete orthogonal Appell system for the set of square integrable monogenic functions on the unit ball $\mathbb{B}_n$ of $\mathbb{R}^n$ as proved in the following:

Theorem 7.2.7 (Lavicka, [131]). *Let $n \geq 3$ and for any $k \in \mathbb{N}_0$ denote by J_k^n the set of sequences $(k_{n-1}, \ldots, k_2)$ such that $k \geq k_{n-1} \geq \cdots \geq k_2 \geq 0$.*

1. *An orthogonal basis for monogenic functions in $\mathbb{B}_n$ in $L^2(\mathbb{B}_n)$ is provided by the polynomials $f_{k,\mu}$ for $k \in \mathbb{N}_0$ and $\mu \in J_k^n$.*

2. *Every function f monogenic in $\mathbb{B}_n$ and belonging to $L^2(\mathbb{B}_n)$ has a unique orthogonal series expansion*

$$f = \sum_{k=0}^{\infty} \sum_{\mu \in J_k^n} f_{k,\mu} a_{k,\mu} \tag{7.10}$$

with coefficients $a_{k,\mu} \in \mathbb{R}_n$. Moreover, for $\mu = (k_{n-1}, \ldots, k_2) \in J_k^n$,

$$a_{k,\mu} = \frac{1}{k!}\frac{1}{2}(\partial_{x_1} + e_1 e_2 \partial_{x_2})\partial_{x_3}^{k_3-k_2}\cdots\partial_{x_n}^{k-k_{n-1}}f(x)_{|x=0}.$$

The series expansion (7.10) is called the generalized Taylor series of f. As we pointed out, the particular case $n = 3$ is studied in [18] and the results coincide with those in [17].

7.3 Approximation in L^2

Here we present some special classes of polynomials in several real variables that are useful for approximating monogenic functions in the framework of Clifford analysis, in the L^2 norm.

In this section, we consider the case of the Clifford algebra $\mathbb{R}_2$, namely the skew field of quaternions $\mathbb{H}$ whose elements q will be written in the form $q = x_0 + x_1 e_1 + x_2 e_2 + x_3 e_1 e_2$. Then, for each element $x = (x_0, x_1, x_2) \in \mathbb{R}^3$, we consider the quaternion $z = x_0 + x_1 e_1 + x_2 e_2$ and denote by $\mathrm{Re}(z) = x_0$ its real part and by $\mathrm{Im}(z) = \frac{1}{2}(z - \bar{z})$ its imaginary part. Next, we consider the generalized Cauchy–Riemann operator

$$D = \frac{\partial}{\partial x_0} + e_1 \frac{\partial}{\partial x_1} + e_2 \frac{\partial}{\partial x_2}$$

and its conjugate

$$\overline{D} = \frac{\partial}{\partial x_0} - e_1 \frac{\partial}{\partial x_1} - e_2 \frac{\partial}{\partial x_2}.$$

In order to choose a basis of polynomials, we set

$$H_\mu^k(z) = z_1^{\mu_1} \times z_2^{\mu_2} = \frac{\mu!}{|\mu|!} \sum_{\pi(i_1, \dots, i_{|\mu|})} z_{i_1} \cdots z_{i_{|\mu|}},$$

where $\vec{z} = (z_1, z_2)$, $z_i = x_i - e_i x_0$, $\mu = (\mu_1, \mu_2)$ is a multi-index, $|\mu| = \mu_1 + \mu_2$, $\mu! = \mu_1! \mu_2!$.

As we already know from Chapter 1, $H_\mu^k(\vec{z})$ is a monogenic and homogeneous polynomial of degree k for each $k \in \mathbb{N}$, $|\mu| = k$. As one may expect, the set $S = (H_\mu^k)_{k=1}^\infty \bigcup \{1\}$ is a system of $\mathbb{H}$-linearly independent functions.

Let us define an inner product in $L^2(\mathbb{B})$ by setting

$$\langle f, g \rangle_{L^2(\mathbb{B})} = \int_{\mathbb{B}} \overline{f} g \, dV,$$

where dV is the Lebesgue volume measure in $\mathbb{R}^3$. Using this inner product, rewriting the polynomials in spherical coordinates and using the orthogonality properties of spherical harmonics of different degrees, one obtains

$$\langle H_\mu^k, H_\nu^h \rangle_{L^2(\mathbb{B})} = 0, \quad k \neq h.$$

Thus, for different degrees k the sets $\{H_\mu^k(z) \mid |\mu| = k\}$ are orthogonal in $L^2(\mathbb{B})$. Polynomials having the same degree k can be made orthogonal using the Gram–Schmidt algorithm. Thus we can assume that

$$\{H_\mu^k(z); \ |\mu| = k\}_{k=1}^\infty \cup \{1\}$$

is a complete orthonormal system in in $L^2(\mathbb{B})$.

Therefore, any $f \in L^2(\mathbb{B})$ can be expanded in a Fourier series with respect to this system.

Consider the Sobolev space $W_2^m(\mathbb{B})$, that is, the set of functions with all (multiindexed) partial derivatives up to order m belonging to $L^2(\mathbb{B})$. Let us denote by ∂^α the partial derivative $\dfrac{\partial^{\alpha_1+\alpha_2+\alpha_3}}{\partial_{x_0}^{\alpha_0}\partial_{x_1}^{\alpha_1}\partial_{x_2}^{\alpha_2}}$, where $\alpha = (\alpha_0, \alpha_1, \alpha_2)$. Reasoning as above one obtains

$$\langle H_\mu^k, H_\nu^h \rangle_{W_2^m(\mathbb{B})} = \sum_{|\alpha| \leq m} \langle \partial^\alpha H_\mu^k, \partial^\alpha H_\nu^h \rangle_{L^2(\mathbb{B})} = 0, \quad k \neq h,$$

since $\partial^\alpha H_\mu^k$, $\partial^\alpha H_\nu^h$ are homogeneous, monogenic polynomials of degree $k - |\alpha|$ and $h - |\alpha|$, respectively, where $|\alpha| \leq m$.

A main result in this framework is the following.

Theorem 7.3.1 (Cação, Gürlebeck and Malonek [27]). *Let $S_r(f)$ be the best approximation for $f \in W_2^m(\mathbb{B}) \bigcap \ker D$ in $L^2(\mathbb{B})$. Then*

$$\|f - S_r(f)\|_{L^2(\mathbb{B})} \leq Cn^{-m}\|f\|_{W_2^m},$$

for all $r \in \mathbb{N}$, where $C > 0$ is a positive constant.

For more details, the reader may consult [27].

One may also consider L^2-functions on the unit sphere in $\mathbb{R}^{m+1}$. In this case, the most widely used approximation is the one by spherical harmonics. We recall that a spherical harmonic of degree k is a k-homogeneous polynomial which is harmonic. In [59] the authors consider instead spherical monogenics of degree k, namely k-homogeneous polynomials $P_k(x)$ which are monogenic and, as before, the specific case of functions from $\mathbb{R}^3$ to $\mathbb{R}_2 = \mathbb{H}$. The idea is then to construct an embedding of two complex variables into the unit sphere of $\mathbb{R}^3$ in order to obtain an embedding of holomorphic functions. This is done by associating with each function $f(z_1, z_2) = \sum_{p,q=0}^\infty z_1^p z_2^q a_{p,q}$ of two complex variables z_1, z_2, the function

$$\sum_{p,q=0}^\infty (x_0 + \underline{x})^p(x_1 - e_1e_2x_2)^q a_{p,q}, \tag{7.11}$$

where $\underline{x} = x_1e_1 + x_2e_2$. Using this correspondence, one may use techniques and advantageous features of holomorphic functions to obtain results in the space L^2 of the unit sphere. In particular, in [59] the authors study the orthonormalization of quaternionic polynomials of the form $(x_0 + \underline{x})^p(x_1 - e_1e_2x_2)^q$ and the convergence of the series (7.11).

Remark 7.3.2. The paper [33] also deals with approximation results in the case of the space L^2 of the unit sphere. In this context, spherical harmonics and their derivatives present a clear advantage. But for numerical algorithms, there are several issues arising for spherical monogenics, among others the fast growth of the number of spherical monogenics of the same degree, and the stability of the numerical algorithms for the orthonormalization process. For this reason, the authors make use of monogenic wavelets instead of spherical monogenics as approximating functions on the sphere (and in the ball). Their main result is a Jackson-type theorem which holds under suitable assumptions on the frame chosen, see Theorem 5.2 in [33], for n-points approximation in L^2 of the unit sphere $\mathbb{S}^2$. The techniques hold in any dimension.

In the context of quaternionic and Clifford analysis there are many other results concerning complete orthogonal systems of monogenic polynomials which are of relevance in approximation theory also in the L^2 space. Thus, we summarize some of the results available in the recent literature in a qualitative way and we refer to the original sources for more information.

In [148], the author investigates approximation properties for monogenic functions by suitable polynomial systems defined in $\mathbb{R}^3$ and with values in the reduced quaternions, i.e., the space generated by $1, e_1, e_2$. In this work, monogenic functions are meant to be solutions of the Riesz system $\operatorname{div} f = 0$, $\operatorname{rot} f = 0$. The results are obtained over 3-dimensional prolate spheroids, namely quadric surfaces generated by rotating an ellipse about its major axis. Using elliptic-cylindrical coordinates, the author constructs a system of monogenic polynomials and proves orthonormality properties. These polynomials are explicitly expressed in terms of products of Ferrer's associated Legendre functions of the first kind, multiplied by a pair of trigonometric functions.

The paper [149] continues the study in [148], but for nullsolutions of the Moisil–Teodorescu operator $e_1 \partial_{x_1} + e_2 \partial_{x_2} + e_1 e_2 \partial_{x_3}$ for which a complete orthogonal system of monogenic polynomials is constructed in the framework of square integrable functions. The author provides explicit representations of the L^2-norms of these polynomials. This makes it possible to define the Fourier expansion of a square integrable, quaternionic valued monogenic function.

The study in [148] of nullsolutions of the Riesz system is further developed in [150]. The authors construct a complete system of homogeneous monogenic functions for the domain exterior of a ball in $\mathbb{R}^3$ by means of spherical harmonic functions. Then, they construct a new complete orthogonal system for the monogenic square integrable functions, for the domain exterior of prolate spheroids. The underlying functions are expressed in terms of Ferrer's associated Legendre functions of the second kind, and thus they are not homogeneous functions. Nevertheless, the authors describe their asymptotic behavior, which is exactly the same as in the spherical case. Another main result in the paper, Theorem 3.2, asserts that for a monogenic function the Fourier expansion constructed by the authors is convergent.

7.4 Approximation of axially monogenic functions

In classical complex analysis it is known that the polynomials obtained by truncating a Taylor series of a function f (holomorphic in an open disk which does not contain any singularities of f) provide a good approximation of f. However, when approaching the boundary of the disk these polynomials are slowly converging to f and outside the disk they diverge. A good approximation on the whole domain of holomorphy of f can be obtained via the so-called Padé approximants:

Definition 7.4.1. The $[M/N]$ *Padé approximant* of the holomorphic function $f(z)$ of the complex variable z is the rational function $f_{M/N}$ defined by the condition

$$f_{M/N}(z)\left(\sum_{k=0}^{N} b_k z^k\right) - \sum_{m=0}^{M} a_m z^m = 0,$$

where the coefficients a_m, b_k are given by

$$b_0 = 1, \quad f(z)\left(\sum_{k=0}^{N} b_k z^k\right) - \sum_{m=0}^{M} a_m z^m = O(z^{N+M+1}).$$

This definition was given in the hypercomplex setting by Common, see [55], who defined the Padé approximants in the axially monogenic case.

Monogenic functions on an open set $U \subseteq \mathbb{R}^{n+1}$ which are invariant under the action of $SO(n)$ have an expansion of the form $f(x) = \sum_{k=0}^{+\infty} f_k(x)$, where f_k are called axially monogenic functions of degree k. The functions f_k have the form

$$f_k(x) = A_k(x_0, r, \underline{\omega}) + \underline{\omega} B_k(x_0, r, \underline{\omega}), \quad x = x_0 + \underline{\omega} r, \ \underline{\omega} \in \mathbb{S}^{n-1},$$

and $A_k(x_0, r, \underline{\omega})$, $B_k(x_0, r, \underline{\omega})$ satisfy the following Vekua type system:

$$\frac{\partial}{\partial x_0} A_k - \frac{\partial}{\partial r} B_k = \frac{k+m-1}{r} B_k,$$

$$\frac{\partial}{\partial x_0} B_k + \frac{\partial}{\partial r} A_k = \frac{k}{r} A_k.$$

In what follows, we assume that $A_k(x_0, r, \underline{\omega})$, $B_k(x_0, r, \underline{\omega})$ are of the form

$$A_k(x_0, r, \underline{\omega}) = A_k(x_0, r) P_k(\underline{\omega}), \quad B_k(x_0, r, \underline{\omega}) = B_k(x_0, r) P_k(\underline{\omega}),$$

and $P_k(\underline{\omega})$ is a spherical monogenic of degree k. Note that for $k = 0$ we obtain the axially monogenic functions considered in (1.32).

Definition 7.4.2 (Common, [55]). Given a function $H(z)$, let $H_{N+j/N}(z)$, $N = 0, 1, 2, \ldots$, $j = -1, 0, 1, 2, \ldots$ be its $[N + j/N]$ Padé approximant. We define the axially monogenic Padé approximant of the axially monogenic function of degree k

$$f_k(x) = \frac{1}{2\pi i} \int_C \frac{1 + s\bar{x}}{((1 + s x_0)^2 + s^2 r^2)^{(n+1+2k)/2}} H\left(-\frac{1}{s}\right) \frac{ds}{s} P_k(\underline{x}),$$

where C is a suitable integration contour, as

$$f^j_{k,N}(x) = \frac{1}{2\pi i} \int_C \frac{1+s\bar{x}}{((1+sx_0)^2 + s^2 r^2)^{(n+1+2k)/2}} H_{N+j/N}\left(-\frac{1}{s}\right) \frac{ds}{s} p_k(x)$$

for all x in the set

$$D_n(\delta, R) = \{x_0^2 + r^2 \leq R^2;\ x_0 \leq 0,\ \rho \geq 0\} \cup \{x_0^2 + r^2 \geq \lambda^2;\ x_0 > 0\}.$$

The axially monogenic function f_k of degree k in the variable $x = x_0 + r\underline{\omega}$ can also be written as

$$f_k(x) = \sum_{\ell=0}^{+\infty} \frac{(1)^\ell}{\ell!} \left(k + \frac{n+1}{2}\right)_\ell (p_{2\ell+k}(x_0, r)c_{2\ell} - q_{2\ell+k+1}(x_0, r)) P_k(\underline{\omega}),$$

where $p_{s,k,n}$, $q_{s,k,n}$ are inner and outer power functions, namely special axially monogenic functions in which the functions A_k, B_k are expressed via suitable special functions, see [55] for the precise definition.

With all these notations at hand, we can state the main result on the uniform convergence of the Padé approximants to an axially monogenic function of degree k:

Theorem 7.4.3 (Common, [55]). *For a fixed $j = -1, 0, 1, \ldots$ the approximants $f^j_{k,N}(x)$ converge uniformly to $f_k(x)$ as $N \to +\infty$ for all $x \in D_1(\delta, R)$ and hence $x \in D_m(\delta, R)$, when the c_j's exist and satisfy $0 < c_j \lesssim (2j)!$.*

Common concludes [55] by showing that the Padé approximants $f^j_{k,N}$ do in fact approximate the function f_k in its domain of monogenicity, thus completely mimicking the classical complex case.

7.5 Series expansions of monogenic functions

When a function f is monogenic in a neighborhood of the origin in $\mathbb{R}^{n+1}$, for example in the unit ball $\mathbb{B}_{n+1}$, we can write its Taylor expansion at the origin (1.25), namely

$$f(x) = \sum_{p=0}^{\infty} \left(\sum_{(\ell_1, \ldots, \ell_p)} V^{(a)}_{\ell_1 \ldots \ell_p}(x) \partial_{x_{\ell_1}} \cdots \partial_{x_{\ell_p}} f(a) \right),$$

which converges uniformly and absolutely in $\mathbb{B}_{n+1}$. We have observed in Chapter 1 that (1.25) can be seen as a series ordered by homogeneous polynomials of fixed degree. If we consider the series (1.26) in which the terms of the same degree are not grouped, its domain of convergence is, in general, different from the one of the series in (1.25). In fact, following [141] and using the notations in Section 7.2, for $\overrightarrow{r} = (r_1, \ldots, r_n)$ we can define

$$\mathcal{U}(\overrightarrow{r}) = \{(z_1, \ldots, z_n) \text{ such that } x_0^2 + x_k^2 < r_k,\ k = 1, \ldots, n\}.$$

The natural domain of convergence of the function $f(x)$ written using (1.26), i.e.,

$$f(x) = \sum_{p=0}^{\infty} \sum_{(\ell_1,\ldots,\ell_p)} V^{(a)}_{\ell_1\ldots\ell_p}(x)\partial_{x_{\ell_1}} \cdots \partial_{x_{\ell_p}} f(a),$$

is the polycylinder $\mathcal{U}(\vec{r})$.

In some cases, one can also consider other types of expansions, for example, the toroidal expansions in [158]. To this end we recall that a paravector can be written in the form $x = x_0 + \underline{x} = x_0 + r\underline{\omega}$, $\underline{\omega} \in \mathbb{S}^{n-1}$, $r = \underline{x}$.

Definition 7.5.1. We call *toroidal expansion of axial type* a convergent series of the form

$$f(x_0,\underline{x}) = \sum_{k=0}^{\infty}\sum_{\ell=0}^{\infty} Z^k \overline{Z}^\ell A_{k,\ell}(x_0,\underline{x}), \quad Z = x_0 + (r-1)\underline{\omega}. \tag{7.12}$$

Theorem 7.5.2 (Penã Penã and Sommen [158]). *A sufficient condition for f as in (7.12) to be monogenic is*

$$A_{k,\ell+1} = -\frac{1}{2(\ell+1)}\left(\partial_{x_0} A_{k,\ell} + \underline{\omega}\partial_r A_{k,\ell} + \frac{n-1}{2r}\underline{\omega}(A_{k,\ell} - A_{\ell,k}) + \frac{1}{r}\underline{\omega}\Gamma_{\underline{x}} A_{\ell,k}\right),$$

where all the functions depend on $(x_0,\underline{x})$ and

$$\Gamma_{\underline{x}} = -\sum_{i<k} e_i e_k(x_i \partial_{x_k} - x_k \partial_{x_i}).$$

We note that the preceding result allows one to generate monogenic functions starting from a sequence $(A_{k,0})$ (initial condition) and calculating all the other coefficients via the formula.

One can also generalize the expansion to the biaxial case: if we write $\mathbb{R}^n = \mathbb{R}^{p_1} \oplus \mathbb{R}^{p_2}$, $p_1 + p_2 = n$, then we have $\underline{x} = \underline{x}_1 + \underline{x}_2$ with

$$\underline{x}_1 = \sum_{j=1}^{p_1} x_j e_j, \quad \underline{x}_2 = \sum_{j=1}^{p_2} x_{p_1+j} e_{p_2+j}.$$

Associated with these variables we have the operators $\partial_{\underline{x}} = \partial_{\underline{x}_1} + \partial_{\underline{x}_2}$, $\partial_{\underline{x}_1} = \sum_{j=1}^{p_1} e_j \partial_{x_j}$, $\partial_{\underline{x}_2} = \sum_{j=1}^{p_2} e_{p_2+j}\partial_{x_{p_1+j}}$. Then, by introducing spherical coordinates in $\mathbb{R}^{p_1}$ and $\mathbb{R}^{p_2}$ by writing $\underline{x}_i = r_i\underline{\omega}_i$, $\underline{\omega}_i \in \mathbb{S}^{p_i-1}$, $i = 1,2$ we can give the following definition:

Definition 7.5.3. We call *toroidal expansion of biaxial type* a convergent series of biaxial type around $\mathbb{S}^{p_1-1} \times \mathbb{S}^{p_2-1}$ of the form

$$f(\underline{x}_1,\underline{x}_2) = \sum_{k=0}^{\infty}\sum_{\ell=0}^{\infty} Z^k \overline{Z}^\ell A_{k,\ell}(\underline{x}_1,\underline{x}_2), \quad Z = (r_1-1) + (r_2-1)\underline{\omega}_1\underline{\omega}_2. \tag{7.13}$$

Theorem 7.5.4 (Penã Penã and Sommen [158]). *A sufficient condition for a function f of the form (7.13) to be monogenic is*

$$A_{k+1,\ell} = \frac{\omega}{2(k+1)} \left(\underline{\omega}_1 \left(\partial_{r_1} A_{k,\ell} + \frac{1}{r_1} \Gamma_{\underline{x}_1} A_{\ell,k} \right) + \underline{\omega}_2 \left(\partial_{r_2} A_{k,\ell} + \frac{1}{r_2} \Gamma_{\underline{x}_2} A_{\ell,k} \right) \right.$$
$$\left. + \left(\frac{p_1 - 1}{2r_1} \underline{\omega}_1 + \frac{p_2 - 1}{2r_2} \underline{\omega}_2 \right) \times (A_{k,\ell} - A_{\ell,k}) \right), \quad k, \ell \geq 0.$$

Also in this case, one may construct the toroidal expansion starting from the sequence $(A_{0,\ell})$ as initial condition.

Another interesting expansion that can be found in [158] is

$$f(x_0, \underline{x}_1, \underline{x}_2) = \sum_{k=0}^{\infty} \sum_{\ell=0}^{\infty} (x_0 + \underline{x}_1)^k (x_0 - \underline{x}_1)^\ell A_{k,\ell}(x_0, \underline{x}_1, \underline{x}_2).$$

For this series, the authors provide some recurrence relations on the coefficient functions $A_{k,\ell}$ which are sufficient conditions for f to be monogenic.

7.6 Polynomial approximation of monogenic quasi-conformal maps

In this section we discuss a more geometric problem, namely how to approximate quasi-conformal maps. It is an interesting case, since it shows how the nature of monogenic functions enters at the very simple level of the linear approximation of a function.

We consider monogenic functions $f : U \subseteq \mathbb{R}^3 \to \mathbb{H}$. The generalized Cauchy–Riemann operator D is an areolar derivative in the sense of Pompeiu [161]. If f is monogenic, its areolar derivative vanishes, but this means that the other areolar derivative $\frac{1}{2}\overline{D}$ can be considered as the hypercomplex derivative of f. This derivative of f is itself a monogenic function.

The Taylor expansion of a monogenic function can be expressed in terms of the two complex variables, $z_1 = x_1 - e_1 x_0$ and $z_2 = x_2 - e_2 x_0$. For simplicity, let us denote by $z_1^k \times z_2^h$ the symmetric product

$$\frac{1}{(k+h)!} \sum \pi(i_1, \ldots, i_{k+h}) z_{i_1} \cdots z_{k+h}$$

where k terms are equal to z_1, h terms are equal to z_2, and the sum is over all the possible permutations.

In the sequel we will consider paravector-valued functions, namely functions of the form $f = f_0 + f_1 e_1 + f_2 e_2$, with f_0, f_1, f_2 real-valued. We have:

Theorem 7.6.1. *The Taylor expansion of a paravector-valued function f in a neighborhood of the origin is*

$$f(x) = \sum_{n=0}^{\infty} \sum_{k=0}^{n} \binom{n}{k} z_2^{n-k} \times z_2^k a_{n-k,k},$$

where $x = x_0 + e_1 x_1 + e_2 x_2$,

$$a_{n-k,k} = \frac{1}{n!} \frac{\partial^n f}{\partial x_1^{n-k} \partial x_2^k}(0), \quad k = 0, \ldots, n,$$

and

$$a_{n-k,k} = a_{n-k,k}^0 + a_{n-k,k}^1 e_1 + a_{n-k,k}^2 e_2, \quad k = 0, \ldots, n, \; n \in \mathbb{N} \qquad (7.14)$$

are paravectors.

We have the following result, see [143]:

Theorem 7.6.2. *A homogeneous monogenic polynomial of degree n*

$$P(z) = \sum_{k=0}^{n} \binom{n}{k} z_2^{n-k} \times z_2^k a_{n-k,k}$$

with arbitrary paravector-valued coefficients as in (7.14) is paravector-valued, if and only if

$$a_{n-k,k}^1 = a_{n-k-1,k+1}^2, \quad k = 0, \ldots, n.$$

The constraints on the coefficients $a_{n-k,k}$ of a paravector-valued polynomial P imply, in particular, that P is both left and right monogenic.

The class of functions from $U \subset \mathbb{R}^3$ to $\mathbb{R}^3$ contain, as a particular case, maps which are quasi-conformal from a domain $U \subset \mathbb{R}^3$ containing the origin to the interior of a ball centered at the origin. It follows that the Taylor expansion of such functions is such that $a_{0,0} = 0$ and the degree 1 approximating polynomial is $z_1 a_{1,0} + z_2 a_{0,1}$. In the classical complex case, to impose the value at the derivative at the origin, for example $f'(0) = 1$, would mean to impose a first-order approximation with the polynomial $P_1(z) = z$. However, due to the nature of the monogenic functions, a linear monogenic function with hypercomplex derivative $\frac{1}{2}\overline{D}f(0)$ equal to 1 is such that $-e_1 a_{1,0} - e_2 a_{0,1} = 1$ and using the constraints in Theorem 7.6.1 we deduce that

$$-e_1(a_{1,0}^0 + e_1 a_{1,0}^1) - e_2(a_{0,1}^0 + e_2 a_{0,1}^1) = 1.$$

Thus, a linear monogenic function with hypercomplex derivative equal to 1 at the origin is such that $a_{1,0}^0 = a_{0,1}^0 = 0$ and $a_{1,0}^1 + a_{0,1}^2 = 1$. If we set $a_{1,0}^1 = a_{0,1}^2 = c$, we conclude that the linear approximation is given by

$$f(z) = z_1(a_{1,0}^1 e_1 + c e_2) + z_2(c e_1 + a_{0,1}^2 e_2),$$

which still contains three real parameters. To determine them, one may impose an extra condition namely $Df(0) = 1$, where D is the generalized Cauchy–Riemann operator,

$$D = \frac{\partial}{\partial x_0} + e_1 \frac{\partial}{\partial x_1} + e_2 \frac{\partial}{\partial x_2}.$$

Using this condition we finally conclude that the linear approximation is given by

$$w = f(z) = \frac{1}{2}(z_1 e_1 + z_2 e_2).$$

Remark 7.6.3. From the geometric point of view this result means that at the first step of the approximation, the unit ball in the w-coordinate is obtained from the oblate ellipsoid

$$\{z = x_0 + e_1 x_1 + e_2 x_2; \ x_0^2 + \frac{1}{4} x_1^2 + \frac{1}{4} x_2^2 = 1\}.$$

From the analytic point of view, the extra condition $Df(0) = 1$ imposes a symmetric behaviour with respect to the variables x_1 and x_2.

Bibliography

[1] Abul-ez, M.A. and Constales, D., Basic sets of polynomials in Clifford analysis, *Complex Var. Theory Appl.* **14** (1990), 177–185.

[2] Adler, S.L., *Quaternionic Quantum Mechanics and Quantum Fields*, Oxford University Press, New York, 1995.

[3] Ahlfors, L.V., *Complex Analysis. An Introduction to the Theory of Analytic Functions of One Complex Variable*, Third edition, McGraw-Hill, Inc., New York, 1979.

[4] Alpay, D., Bolotnikov, V.. Colombo, F. and Sabadini, I., Self-mappings of the quaternionic unit ball: multiplier properties, Schwarz–Pick inequality, and Nevanlinna–Pick interpolation problem, *Indiana Univ. Math. J.* **64** (2015), 151–180.

[5] Alpay D., Colombo F., Kimsey D.P., The spectral theorem for quaternionic unbounded normal operators based on the S-spectrum, *J. Math. Phys.* **57** (2016), 023503.

[6] Alpay, D., Colombo, F. and Sabadini, I., Pontryagin–de Branges–Rovnyak spaces of slice hyper-holomorphic functions, *J. Anal. Math.* **121** (2013), 87–125.

[7] Alpay D., Colombo F., Sabadini I., *Slice Hyperholomorphic Schur Analysis*, Operator Theory Advances and Applications 256, Birkhäuser, Basel, 2016.

[8] Andrievskii, V.V., Belyi, V.I. and Dzjadyk, V.K., *Conformal Invariants in Constructive Theory of Functions of Complex Variable*, World Federation Publishers, Atlanta, Georgia, 1995.

[9] Appell, P., Sur une classe de polynômes. *Annals scientifiques de l'Ecole Normale Supérieure* **21** (1880), 9.

[10] Arakelian, N.I., Uniform approximation on closed sets by entire functions (Russian), *Izv. Akad. Nauk. SSSR* **28** (1964), 1187–1206.

[11] Aral, A., Gupta, V. and Agarwal, R.P., *Applications of q-Calculus in Operator Theory*, Springer, New York, 2013.

© Springer Nature Switzerland AG 2019

S. G. Gal, I. Sabadini, *Quaternionic Approximation*, Frontiers in Mathematics,

https://doi.org/10.1007/978-3-030-10666-9

[12] Aron, R. and Markose, D., On universal functions, *J. Korean Math. Soc.* **41** (2004), 65–76.

[13] Beck, B., Sur les équations polynomiales dans les quaternions, *Enseign. Math.* **25** (1979), 193–201.

[14] Bernstein, S.N., Sur les recherches récentes relatives à la meilleure approximation des fonctions continues par les polynômes, in: *Proc. of 5th Inter. Math. Congress*, Vol. **1**, pp. 256–266, 1912.

[15] Bernstein, S.N., *Leçons sur les Propriétés Extrémales et la Meilleure Approximation des Fonctions Analytiques d'une variable reélle*, Collection Borel, Paris, 1926.

[16] Birkhoff, G.D., Démonstration d'un théorème élémentaire sur les fonctions entières, *Comptes Rendus Acad. Sci., Paris* **189** (1929), 473–475.

[17] Bock, S. and Gürlebeck K., On a generalized Appell system and monogenic power series, *Math. Meth. Appl. Sci.* **33** (2010), 394–411.

[18] Bock, S., Gürlebeck K., Lavicka, R. and Soucek, V., Gelfand-Tsetlin bases for spherical monogenics in dimension 3, *Rev. Mat. Iberam.* **28** no. 4 (2012), 1165–1192.

[19] Blumenson, L.E., A derivation of n-dimensional spherical coordinates, *The American Mathematical Monthly* **67** no. 1 (1960), 63–66.

[20] Brackx, F., Delanghe, R. and Sommen, F., *Clifford Analysis*, Pitman Res. Notes in Math. **76**, 1982.

[21] Burckel, R.B., *An Introduction to Classical Complex Analysis. Volume 1*, Pure and Applied Mathematics **82**, Academic Press, Inc., New York–London, 1979.

[22] Butzer, P.L., On the singular integral of de la Vallée Poussin, *Arch. Math.* **7** (1956), 295–309.

[23] Cação, I., *Constructive approximation by monogenic polynomials*, PhD thesis, University of Aveiro, 2004.

[24] Cação, I., Complete orthonormal sets of polynomial solutions of the Riesz and Moisil–Teodorescu systems in $\mathbb{R}^3$, *Numer. Algorithms* **55** (2010), 191–203.

[25] Cação, I. and Gürlebeck K., A complete system of homogeneous monogenic polynomials and their derivatives, in: *Progress in Analysis*, vol. I, II (Berlin, 2001), pp. 317–324, World Sci. Publ., River Edge, NJ, 2003.

[26] Cação, I., Gürlebeck, K. and Bock, S., On derivatives of spherical monogenics, *Complex Var. Elliptic Equations* **51** (2006), 847–869.

[27] Cação, I., Gürlebeck K. and Malonek, H., Special monogenic polynomials and L^2-approximation, *Adv. Appl. Clifford Alg.* **11** (2001), 47–60.

[28] Cação, I. and Malonek, H.R., Remarks on some properties of monogenic polynomials, in: *Proceedings ICNAAM 2008*, AIP Conference Proceedings 1048, Amer. Inst. Phys., 647–650, 2009.

[29] Carathéodory, C., Über die gegenseitige Beziehung der Ränder bei der konformen Abbildung des Inneren einer Jordanschen Kurve auf einen Kreis, *Math. Ann.* **73** no. 2 (1913), 305–320.

[30] Carleman, T., Sur un théorème de Weierstrass, *Ark. Mat. Astr. Fys. 20B* **4** (1927), 1–5.

[31] Castillo Villalba, C.M.P., Colombo, F., Gantner, J. and González-Cervantes, J.O., Bloch, Besov and Dirichlet spaces of slice hyperholomorphic functions, *Complex Anal. Oper. Theory* **9** no. 2 (2015), 479–517.

[32] Cavaretta Jr, A.S., Sharma, A. and Varga, R.S., Interpolation on the roots of unity: An extension of a theorem of J.L. Walsh, *Resultate der Mathematik* **3** no. 2 (1980), 155–191.

[33] Cerejeiras, P., Ferreira. M., Kähler, U. and Vieira, N., Monogenic frames for an integral transform on the unit sphere, *Complex Variables and Elliptic Equations* **51** no. 1 (2006), 51–61.

[34] Chen, L., Definition of determinant and Cramer solution over quatemion field, *Acta Math Sinica, New Series* **7** no. 2 (1991), 171–180.

[35] Chen, Y. and Ren, G., Jackson's theorem in Q_p spaces, *Sci. China Math.*, **53** no. 2 (2010), 367–372.

[36] Cheney, E.W., *Introduction to Approximation Theory*, McGraw-Hill, New York, 1966.

[37] Chui, C. and Parnes, M.N., Approximation by overconvergence of a power series, *J. Math. Anal. Appl.* **36** (1971), 693–696.

[38] Colombo, F. and Gantner, J., *Fractional powers of quaternionic operators and Kato's formula using slice hyperholomorphicity*, Trans. Amer. Math. Soc. textbf370 no. 2 (2018), 1045–1100.

[39] Colombo, F. and Gantner, J., *Quaternionic Closed Operators, Fractional Powers and Fractional Diffusion Processes*, book submitted, 2018.

[40] Colombo, F., Gantner, J. and Kimsey D.P., *Spectral Theory on the S-Spectrum for Quaternionic Operators*, book submitted, 2017.

[41] Colombo, F., González-Cervantes, J.O., Luna-Elizarrarás, M.E., Sabadini, I., Shapiro, M., On two approaches to the Bergman theory for slice regular functions, Springer INdAM Series **1**, pp. 39–54, 2013.

[42] Colombo, F., Gonzáles-Cervantes, J.O. and Sabadini, I., Further properties of the Bergman spaces of slice regular functions, *Adv. Geom.*, **15** no. 4 (2015), 469–484.

[43] Colombo, F., Lavicka, R., Sabadini, I. and Soucek V., The Radon transform between monogenic and generalized slice monogenic functions, *Math. Annalen* **363** (2015), 733–752.

[44] Colombo, F. and Sabadini, I., Some remarks on the S-spectrum, *Complex Variables and Elliptic Equations* **58** (2013), 1–6.

[45] Colombo, F., Sabadini, I. and Sommen F., The Fueter mapping theorem in integral form and the $\mathcal{F}$-functional calculus, *Math. Meth. Appl. Sci.* **33** (2010), 2050–2066.

[46] Colombo, F., Sabadini, I. and Sommen F., The inverse Fueter mapping theorem, *Comm. Pure Appl. Anal.* **10** (2011), 1165–1181.

[47] Colombo, F., Sabadini, I. and Sommen F., The inverse Fueter mapping theorem using spherical monogenics, *Israel J. Math.* **194** (2013), 485–505.

[48] Colombo, F., Sabadini, I., Sommen F. and Struppa, D.C., *Analysis of Dirac Systems and Computational Algebra*, Progress in Mathematical Physics **39**, Birkhäuser Boston, Inc., Boston, MA, 2004.

[49] Colombo, F., Sabadini, I. and Struppa, D.C., Slice monogenic functions, *Israel J. Math.* **171** (2009), 385–403.

[50] Colombo, F., Sabadini, I. and Struppa, D.C., Duality theorems for slice monogenic functions, *J. Reine Angew. Math.* **645** (2010), 85–104.

[51] Colombo, F., Sabadini, I. and Struppa, D.C., *Noncommutative Functional Calculus. Theory and Applications of Slice Hhyperholomorphic Functions*, Progress in Mathematics **289**, Birkhäuser, Basel, 2011.

[52] Colombo, F., Sabadini, I. and Struppa, D.C., The Runge theorem for slice hyperholomorphic functions, *Proc. Amer. Math. Soc.* **139** (2011), 1787–1803.

[53] Colombo, F., Sabadini, I. and Struppa, D.C., Sheaves of slice regular functions, *Math. Nachr.* **285** (2012), 949–958.

[54] Colombo, F., Sabadini, I. and Struppa, D.C., Entire Slice Regular Functions, SpringerBriefs in Mathematics, Springer, 2016.

[55] Common, A.K., Axial monogenic Clifford-Padé approximants, *J. Approx. Theory* **68** (1992), 206–222.

[56] Coxeter, H.S.M., *Regular Polytopes*, Dover Publications, 3rd edition, New York, 1973.

[57] Davis, Ph.J., *Interpolation and Approximation*, Dover Publications, New York, 1975.

[58] De Bruijn, N.G., Inequalities concerning polynomials in the complex domain, *Indag. Math.* **9** (1947), 591–598.

[59] De Schepper, N., Qian, T., Sommen, F. and Wang, J., Holomorphic approximation of L_2-functions on the unit sphere in $\mathbb{R}^3$, *J. Math. Anal. Appl.* **416** (2014), 659–671.

[60] Delanghe, R., On regular-analytic functions with values in a Clifford algebra, *Math. Ann.* **185** (1970), 91–111.

[61] Delanghe, R. and Brackx, F., Runge's theorem in hypercomplex function theory, *J. Approx. Theory* **29** (1980), 200–211.

[62] Delanghe, R., Sommen, F. and Souček, V., *Clifford Algebra and Spinor-valued Functions*, Mathematics and Its Applications **53**, Kluwer Academic Publishers, 1992.

[63] DeVore, R.A. and Lorentz, G.G., *Constructive Approximation*, Springer, Berlin–Heidelberg–New York, 1993.

[64] Dong, B., Kou, K.I., Qian T. and Sabadini, I., On the inversion of Fueter's theorem, *J. Geom. Phys.* **108** (2016), 102–116.

[65] Duren, P.L., *Univalent Functions*, Grundlehren der Mathematischen Wissenschaften **259**, Springer-Verlag, New York, 1983.

[66] Duren, P. and Schuster, A., *Bergman Spaces*, American Mathematical Society, *Mathematical Surveys and Monographs* **100**, Providence, R.I., 2004.

[67] Faber, G., Über polynomische Entwickelungen, *Math. Ann.* **57** (1903), 389–408.

[68] de Fabritiis, C., Gentili, G. and Sarfatti, G., Quaternionic Hardy Spaces, *Ann. Sc. Norm. Super. Pisa Cl. Sci.* **18** (2018), 697–733.

[69] Falcão, M.I., Cruz, J.F. and Malonek H.R., Remarks on the generation of monogenic functions, in: *Proc. of the 17th International Conference on the Application of Computer Science and Mathematics in Architecture and Civil Engineering*, Weimar, Germany, 2006, 22 pages, 2007.

[70] Falcão, M.I. and Malonek H.R., Generalized exponential through Appell sets in $\mathbb{R}^{n+1}$ and Bessel functions, in: *Numerical Analysis and Applied Mathematics*, AIP Conference Proceedings 936, Amer. Inst. Phys., 750–753, 2007.

[71] Fuchs, W.H., *Théorie de l'Approximation des Fonctions d'une Variable Complexe*, Presse de l'Université de Montréal, 1968.

[72] Fueter, R., Die Funktionentheorie der Differentialgleichungen $\Delta u = 0$ und $\Delta\Delta u = 0$ mit vier reellen Variablen (German), *Comment. Math. Helv.* **7** no. 1 (1934), 307–330.

[73] Fueter, R., Zur Theorie der regulären Funktionen einer Quaternionenvariablen (German), *Monatsh. Math. Phys.* **43** no. 1 (1936), 69–74.

[74] Gaier, D., *Lectures on Complex Approximation*, Birkhäuser, Boston, 1987.

[75] Gal, S.G., Convolution-type integral operators in complex approximation, *Comput. Methods Funct. Theory* **1** no. 2 (2001), 417–432.

[76] Gal, S.G., Voronovskaja's theorem and iterations for complex Bernstein polynomials in compact disks, *Mediterr. J. Math.* **5** no. 3 (2008), 253–272.

[77] Gal, S.G., *Approximation by Complex Bernstein and Convolution-Type Operators*, World Scientific Publ. Co, Singapore–Hong Kong–London–New Jersey, 2009.

[78] Gal, S.G., Approximation by quaternion q-Bernstein polynomials, $q > 1$, *Adv. Appl. Clifford Alg.* **22** (2012), 313–319.

[79] Gal, S.G., Voronovskaja-type results in compact disks for quaternion q-Bernstein operators, $q \geq 1$, *Compl. Anal. Oper. Theory* **6** (2012), 512–527.

[80] Gal, S.G., *Overconvergence in Complex Approximation*, Springer, New York–Heidelberg–Dordrecht–London, 2013.

[81] Gal, S.G., Quantitative approximations by convolution polynomials in Bergman spaces, *Complex Anal. Oper. Theory* **12** no. 2 (2018), 355364.

[82] Gal, S.G., González-Cervantes, J.O. and Sabadini, I., Univalence results for slice regular functions of a quaternion variable, *Complex Var. Elliptic Equ.* **60** no. 10 (2015), 1346–1365.

[83] Gal, S.G., González-Cervantes, J.O. and Sabadini, I., On some geometric properties of slice regular functions of a quaternion variable, *Complex Var. Elliptic Equations* **60** no. 10 (2015), 1431–1455.

[84] Gal, S.G. and Sabadini, I., Approximation by polynomials in Bergman spaces of slice regular functions in the unit ball, *Math. Meth. Appl. Sci.* **41** no. 4 (2018), 1619–1630.

[85] Gal, S.G. and Sabadini, I., Approximation in compact balls by convolution operators of quaternion and paravector variable, *Bull. Belg. Math. Soc. Simon Stevin* **20** no. 3 (2013), 481–501.

[86] Gal, S.G. and Sabadini, I., Carleman type approximation theorem in the quaternionic setting and applications, *Bull. Belg. Math. Soc. Simon Stevin* **21** no. 2 (2014), 231–240.

[87] Gal, S.G. and Sabadini, I., Walsh equiconvergence theorems in the quaternionic setting, *Complex Var. Elliptic Equations* **59** no. 12 (2014), 1589–1607.

[88] Gal, S.G. and Sabadini, I., On Bernstein and Erdős–Lax's inequalities for quaternionic polynomials, *C.R. Acad. Sci. Paris* Ser. I **353** (2015), 5–9.

[89] Gal, S.G. and Sabadini, I., Arakelian's approximation theorem of Runge type in the hypercomplex setting, *Indag. Math. (N.S.)* **26** no. 2 (2015), 337–345.

[90] Gal, S.G. and Sabadini, I., Universality properties of the quaternionic power series and entire functions, *Math. Nachr.* **288** no. 8–9 (2015), 917–924.

[91] Gal, S.G. and Sabadini, I., Faber polynomials on quaternionic compact sets, *Complex Anal. Oper. Theory* **11** no. 5 (2017), 1205–1220,

[92] Gal, S.G. and Sabadini, I., Overconvergence of Chebyshev and Legendre expansions in quaternionic ellipsoids, *Adv. Appl. Clifford Algebras* **27** no. 1 (2017), 125–133.

[93] Gal, S.G. and Sabadini, I., Approximation by polynomials on quaternionic compact sets, *Math. Meth. Appl. Sci.* **38** (2015), 3063–3074.

[94] Gal, S.G. and Sabadini, I., On the Turán's inequality for quaternionic polynomials, submitted.

[95] Gal, S.G. and Sabadini, I., Approximation by polynomials in Bloch spaces and in Besov spaces of slice regular functions in the unit ball, in preparation.

[96] Gamelin, T., *Complex Analysis*, Undergraduate Texts in Mathematics, Springer Verlag, New York, 2001.

[97] Gelfand, I.M. and Tsetlin, M.L., *Finite-dimensional representations of groups of orthogonal matrices*, Dokl. Akad. Nauk. SSSR **71** (1950), 1017–1020.

[98] Gentili, G. and Stoppato, C., Power series and analyticity over the quaternions, *Math. Ann.* **352** (2012), 113-131.

[99] Gentili, G., Stoppato, C. and Struppa, D.C., *Regular Functions of a Quaternionic Variable*, Springer Monographs in Mathematics, Springer, Berlin–Heidelberg, 2013.

[100] Gentili, G. and Struppa, D.C., A new approach to Cullen-regular functions of a quaternionic variable, C.R. Math. Acad. Sci. Paris **342** (2006), 741–744.

[101] Gentili, G. and Struppa, D.C., A new theory of regular functions of a quaternionic variable, *Adv. Math.* **216** (2007), 279–301.

[102] Gentili, G. and Struppa, D.C., On the multiplicity of zeroes of polynomials with quaternionic coefficients, *Milan J. Math.* **76** (2008), 15–25.

[103] Ghiloni, R., Moretti, V. and Perotti, A., Continuous slice functional calculus in the quaternionic Hilbert spaces, *Rev. Math. Phys.* **25** (2013), 1350006, 83 pp.

[104] Ghiloni R. and Perotti, A., Slice regular functions on real alternative algebras, *Adv. Math.* **226** (2011), 1662–1691.

[105] Ghiloni, R. and Perotti, A., Power and spherical series over real alternative *-algebras, *Indiana Univ. Math. J.* **63** (2014), 495–532.

[106] Ghiloni, R. and Perotti, A., The quaternionic Gauss-Lucas theorem, *Ann. Mat. Pura Appl.*, 2018, online access, https://doi.org/10.1007/s1023

[107] Gilbert, J.E. and Murray, M.A.M., *Clifford algebras and Dirac operators in harmonic analysis*, Cambridge Studies in Advanced Mathematics **26**, Cambridge University Press, Cambridge, 1991.

[108] Giroux, A., Rahman, Q.I. and Schmeisser, G., On Bernstein's inequality, *Can. J. Math.* **31** (1979), 347–353.

[109] Golitschek, M.V., A short proof of Müntz's theorem, *J. Approx. Theory* **39** (1983), 394–395.

[110] Goluzin, G.M., *Geometric Theory of Functions of a Complex Variable*, Translation of Mathematical Monographs **26**, Amer. Math. Soc., Providence, R.I., 1969.

[111] Gordon, B. and Motzkin, T.S., On the zeros of polynomials over division rings, *Trans. Amer. Math. Soc.* **116** (1965), 218–226.

[112] Govil, N.K., On the derivative of a polynomial, *Proc. Amer. Math. Soc.* **41** (1973), 543–546.

[113] Gupta, V. and Agarwal, R.P., *Convergence Estimates in Approximation Theory*, Springer, New York, 2014.

[114] Gupta, V. and Tachev, G., *Approximation with Positive Linear Operators and Linear Combinations*, Springer, New York, 2017.

[115] Gürlebeck, K., *Über Interpolation und Approximation verallgemeinert analytischer Funktionen* (German), Technische Hochschule Karl-Marx-Stadt, Sektion Mathematik, Karl-Marx-Stadt, 21 pages, 1982.

[116] Gürlebeck, K., Interpolation and best approximation in spaces of monogenic functions, *Wiss. Z. Tech. Univ. Karl-Marx-Stadt* **30** no. 1 (1988), 38–40.

[117] Gürlebeck, K., Habetha K. and Sprössig, W., *Holomorphic Functions in the Plane and n-Dimensional Space*, Birkhäuser, Basel, 2008.

[118] Gürlebeck, K. and Malonek H., A hypercomplex derivative of monogenic functions in $\mathbb{R}^{n+1}$ and its applications, Compl. Var. Theory Appl. **39** (1999), 199-228.

[119] Hausner, A., A generalized Stone-Weierstrass theorem, *Arch. Math.* **10** (1959), 85–87.

[120] Hedenmalm, H., Korenblum, B. and Zhu, K., *Theory of Bergman Spaces*, Springer-Verlag, New York–Berlin–Heidelberg (2000).

[121] Hoischen, L., A note on the approximation theorem of continuous functions by integral functions, *J. London Math. Soc.* **42** (1967), 351–354.

[122] Holladay, J.C., A note on the Stone-Weierstrass theorem for quaternions, *Proc. Amer. Math. Soc.* **8** (1957), 656–657.

[123] Hou R.M., Zhao X.Q., Wang L.T., The double determinant of Vandermonde's type over quaternion field, *Appl. Math. Mech. (English Ed.)* **20** (1999), 1046–1053.

[124] Iftimie, V., Fonctions hypercomplexes, *Bull. Math. Soc. Sci. Math. R.S. Roumanie* **57** (1966), 279–332.

[125] Jakimovski, A., Sharma, A. and Szabados, J., *Walsh Equiconvergence of Complex Interpolation Polynomials*, Springer, New York, 2006.

[126] Kaplan, W., Approximation by entire functions, *Michigan Math. J.*, **3** (1955–56), 43–52.

[127] Kou, K.I., Qian, T. and Sommen, F., *Generalizations of Fueter's theorem*, Meth. Appl. Anal. **9** (2002), 273–290.

[128] Lam, T.Y., A general theory of Vandermonde matrices, *Exposition. Math.* **4** (1986), 193–215.

[129] Lam, T.Y., *A First Course in Noncommutative Rings*, Springer-Verlag, New York, 1991.

[130] Lam, T.Y. and Leroy, A., Vandermonde and Wronskian matrices over division rings, *J. Algebra* **119** (1988), 308–336.

[131] Lavicka, R., Complete orthogonal Appell systems for spherical monogenics, *Complex Anal. Oper. Theory* bf 6 (2012), 477–489.

[132] Laville, G. and Ramadanoff, I.P., Stone–Weierstrass theorem, in : *Generalizations of Complex Analysis and their Applications in Physics* (Warsaw/Rynia, 1994), 189–194, Banach Center Publ., **37**, Polish Acad. Sci., Warsaw, 1996.

[133] Lax, P.D., Proof of a conjecture of P. Erdős on the derivative of a polynomial, *Bull. Amer. Math. Soc.* **50** (1944), 509–513.

[134] Lorentz, G.G., *Bernstein Polynomials*, 2nd editon, Chelsea Publ. Comp., New York, 1987.

[135] Lorentz, G.G., *Approximation of Functions*, Chelsea Publ. Comp., New York, 1987.

[136] Loskot, P. and Beaulieu, N.C., On monotonicity of the hypersphere volume and area, J. Geom. **87** (2007), 96–98.

[137] Luh, W., Approximation analytischer Functionen durch überkonvergente Potenzreihen und deren Matrix-Transformierten, *Mitt. Math. Sem. Giessen* **88** (1970), 1–56.

[138] MacLane, G., Sequences of derivatives and normal families, *J. Anal. Math.* **2** (1952), 72–87.

[139] Malik, M.A., On the derivative of a polynomial, *J. London Math. Soc.* **1** no. 2 (1969), 57–60.

[140] Malonek H.R., A new hypercomplex structure of the Euclidean space $\mathbb{R}^{n+1}$ and the concept of hypercomplex derivative, *Complex Var. Theory Appl.* **14** (1990), 25–33.

[141] Malonek H.R., Power series representation for monogenic functions in $\mathbb{R}^{m+1}$ based on a permutational product, *Complex Var. Theory Appl.* **15** (1990), 181–191.

[142] Malonek, H.R. and Falcão, M.I., Special monogenic polynomials properties and application, in: *Numerical Analysis and Applied Mathematics*, AIP Conference Proceedings **936**, Amer. Inst. Phys., 764–767, 2007.

[143] Malonek, H.R. and Falcão, M.I., 3D-mappings by means of monogenic functions and their approximation, *Math. Meth. Appl. Sci.* **33** (2010), 423–430.

[144] Mejlihzon, A.Z., On the notion of monogenic quaternions (Russian), *Dokl. Akad. Nauk SSSR* **59** (1948), 431–434.

[145] Mergelyan, S.N., Uniform approximation to functions of a complex variable (Russian), *Uspehi Mat. Nauk (N.S.)* **7** no. 2(48) (1952), 31–122, translated in *A.M.S. Translations* **101** (1954).

[146] Moisil, Gr.C., Sur les quaternions monogènes, *Bull. Sci. Math. (Paris)* **LV** (1931), 168–174.

[147] Moisil, Gr.C. and Teodorescu, N., Functions holomorphes dans l'espace, *Mathematica* (Cluj) **5** (1931), 142–159.

[148] Morais, J., A complete orthogonal system of spheroidal monogenics, *JNA-IAM. J. Numer. Anal. Ind. Appl. Math.* **6** no. 3–4 (2011), 105–119.

[149] Morais, J., An orthogonal system of monogenic polynomials over prolate spheroids in $\mathbb{R}^3$, *Mathematical and Computer Modelling* **57** (2013), 425–434.

[150] Morais, J., Nguyen, H.M. and Kou, K.I., On 3D orthogonal prolate spheroidal monogenics, *Mathematical Methods in the Applied Sciences* **39** (2016), 635–648.

[151] Moskowitz, M. and Paliogiannis, F., *Functions of Several Real Variables*, World Scientific Publ. Co., Singapore–Hackensack NJ–London, 2011.

[152] Müntz, Ch.H., Über den Approximationssatz von Weierstrass, in: *H.A. Schwarz Festschrift*, Berlin, pp. 303–312, 2014.

[153] Narasimhan, R., *Complex Analysis in One Variable*, Birkhäuser, Boston, 1985.

[154] Niven, I., Equations in quaternions, *Amer. Math. Monthly* **48** (1941), 654–661.

[155] Opfer, G., Polynomials and Vandermonde matrices over the field of quaternions, *Elect. Trans. Num. Anal.* **36** (2009), 9–16.

[156] Ostrovska, S., q-Bernstein polynomials and their iterates, *J. Approx. Theory* **123** (2003), 232–255.

[157] Pál, J., Zwei kleine Bemerkungen, *Tohoku Math. J.* **6** (2014–2015), 42–43.

[158] Penã Penã, D. and Sommen, F., Some power series expansions for monogenic functions, *Comp. Meth. Funct. Theory* **7** (2007), 265–275.

[159] Phillips, G.M., Bernstein polynomials based on the q-integers, *Annals Numer. Math.* **4** (1997), 511–518.

[160] Pólya, G. and Schoenberg, I.J., Remarks on the de la Vallée Poussin means and convex conformal maps of the circle, *Pacific J. Math.* **8** (1958), 295–333.

[161] Pompeiu, M., Sur une classe de fonctions d'une variable complexe, *Rend. Circolo Mat. Palermo* **33** (1912), 108–113.

[162] Qian, T., Generalization of Fueter's result to $\mathbb{R}^{n+1}$, *Rend. Mat. Acc. Lincei* **8** (1997), 111–117.

[163] Qian, T., Singular integrals on star-shaped Lipschitz surfaces in the quaternionic space, *Math. Ann.* **310** (1998), 601–630.

[164] Qian, T., Fourier Analysis on Starlike Lipschitz Surfaces, *J. Funct. Anal.* **183** (2001), 370–412.

[165] Rahman, Q.I. and Schmeisser, G., *Analytic Theory of Polynomials*, Clarendon Press, London Mathematical Society Monographs, Oxford University Press, 2002.

[166] Ren, G. and Chen, Y., Gradient estimates and Jackson's theorem in Q_μ spaces related to measures, *J. Approx. Theory* **155** (2008), 97–110.

[167] Ren, G. and Wang, M., Holomorphic Jackson's theorems in polydiscs, *J. Approx. Theory*, **134** no. 2 (2005), 175–198.

[168] Ren, G. and Wang, X., Growth and distortion theorems for slice monogenic functions, *Pacific J. Math.* **290** (2017), 169–198.

[169] Riesz, M., Eine trigonometrische interpolationsformel und einige ungleichungen für polynome, *Jahresber. Dtsch. Math.-Ver.* **23** (2014), 354–368.

[170] Rinehart, R.F., Elements of a theory of intrinsic functions on algebras, *Duke Math. J.* **27** (1960), 1–19.

[171] Rosay, J.-P. and Rudin, W., Arakelian's approximation theorem, *The Amer. Math. Monthly* **96** no. 5 (1989), 432–434.

[172] Runge, C., Zur Theorie eindeutigen analytischen Funktionen, *Acta Math.* **6** (1885), 229–244.

[173] Ryan, J., Runge approximation theorems in complex Clifford analysis together with some of their applications, *J. Funct. Analysis* **70** (1987), 221–253.

[174] Saff, E.B. and Varga, R.S., A note on the sharpness of J.L. Walsh's theorem and its extensions for interpolation in the roots of unity, *Acta Math. Hung.* **41** no. 3–4 (1983), 371–377.

[175] Schafer, R.D., On the algebras formed by the Cayley-Dickson process, *Amer. J. Math.* **76** (1954), 435–446.

[176] Sce, M., Osservazioni sulle serie di potenze nei moduli quadratici, *Atti Acc. Lincei Rend. Fisica* **23** (1957), 220–225.

[177] Seidel, W. and Walsh, J.L., On approximation by euclidean and non-euclidean translations of an analytic function, *Bull. Amer. Math. Soc.* **47** (1941), 916–920.

[178] Seleznev, A.I., On universal power series (Russian), *Mat. Sb.(N.S.)* **28** (1951), 453–460.

[179] Sommen, F., On a generalization of Fueter's theorem, *Zeit. Anal. Anwen.* **19** (2000), 899–902.

[180] Stoppato, C., A new series expansion for slice regular functions, *Adv. Math.* **231** no. 3–4 (2012), 1401–1416.

[181] Stone, M.H., Applications of the theory of Boolean rings to general topology, *Trans. Amer. Math. Soc.* **41** (1937), 375–481.

[182] Sudbery, A., Quaternionic Analysis, *Math. Proc. Cambridge Philos. Soc.* **85** (1971), 199–225.

[183] Suetin, P.K., *Series of Faber Polynomials*, Gordon and Breach, Amsterdam, 1998.

[184] Szabados, J., Converse results in the theory of overconvergence of complex interpolating polynomials, *Analysis* **2** (1982), 267–280.

[185] Szász, O., Über die Approximation stetiger Funktionen durch lineare Aggregate von Potenzen, *Math. Ann.* **77** (1916), 482–496.

[186] Szegö, G., I. Bemerkungen zu einem Satz von J.H. Grace, *Math. Zeit.* **13** (1922), 28–55.

[187] Trent, T.T., A Müntz-Szász theorem for $C(\overline{D})$, *Proc. Amer. Math. Soc.* **83** no. 2 (1982), 296–298.

[188] Turán, P., Ueber die Ableitung von Polynomen, *Compositio Math.* **7** (1939), 89–95.

[189] de la Vallée-Poussin, Ch., Sur l'approximation des fonctions d'une variable réelle et de leurs dérivées par les polynômes et des suites limitées de Fourier, *Bull. Acad. Roy. de Belgique* **3** (1908), 193–254.

[190] Vlacci, P., The Gauss-Lucas theorem for regular quaternionic polynomials, in: *Hypercomplex Analysis and Applications*, Trends in Mathematics, Birkhäuser, pp. 275–282, 2011.

[191] Van Lancker, P., Approximation theorems for spherical monogenics of complex degree, *Bull. Belg. Math. Soc.* **6** (1999), 279–293.

[192] Walsh, J.L., *Interpolation and Approximation by Rational Functions in the Complex Domain*, American Matematical Society Colloquium Publications, Vol. **XX**, Providence, Rhode Island, 4th ed., 1965.

[193] Walsh, J.L., *Interpolation and Approximation by Rational Functions in the Complex Domain*, American Matematical Society Colloquium Publications, Vol. **XX**, Providence, Rhode Island, 5th ed., 1969.

[194] Wang, H. and Wu, X.Z., Saturation of convergence for q-Bernstein polynomials in the case $q \geq 1$, *J. Math. Anal. Appl.* **337** (2008), 744–750.

[195] Xu, Z., The Bernstein inequality for slice regular polynomials, arXiv: 1602.08545v1.

[196] Xu, Z. and Ren, G., Slice starlike functions over quaternions, *J. Geom. Anal.* **28** (2018), 3775–3806.

Index

© Springer Nature Switzerland AG 2019
S. G. Gal, I. Sabadini, *Quaternionic Approximation*, Frontiers in Mathematics,
https://doi.org/10.1007/978-3-030-10666-9

Printed in the United States
By Bookmasters